FUNDAMENTALS
OF PLANT VIROLOGY

FUNDAMENTALS
OF PLANT VIROLOGY

R. E. F. Matthews
Department of Cellular
and Molecular Biology
University of Auckland
Auckland, New Zealand

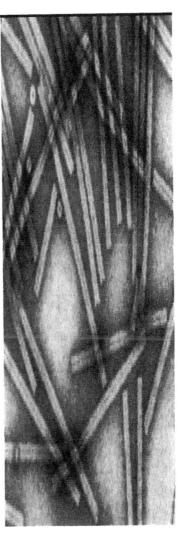

ACADEMIC PRESS, INC.
Harcourt Brace Jovanovich, Publishers

San Diego New York Boston London Sydney Tokyo Toronto

(**Cover photograph**) Resistance against virus infection in transgenic plants expressing a viral coat protein gene. Plants of *Nicotiana tobacum* cv xanthi expressing the gene encoding the coat protein of tobacco mosaic *Tobamovirus* (U_1 strain) and control plants not expressing the gene were photographed eight days after inoculation with a strain of the virus that causes yellow systemic symptoms. (Photograph courtesy of R. N. Beachy, Washington University, St. Louis, Missouri.) (**Top inset**) Structure of an icosahedral virus. Computer graphics representation of the arrangement of the small and large protein subunits in the shell of bean pod mottle *Comovirus*. The clusters of five small protein subunits are shown in blue. The large protrusions at the five-fold axes can be seeen at the circumference of the model. The two domains of the large protein subunit are shown in red and green. Three molecules of this subunit are located around each three-fold symmetry axis. (Photograph, courtesy of J. E. Johnson, Purdue University, West Lafayette, Indiana, is a version of a figure from Chen *et al.*,1989.) (**Bottom inset**) Structure of a helical virus. Computer graphics representation of about one twentieth of a particle of tobacco mosaic *Tobamovirus*. Protein subunits are colored yellow; RNA is brown. The RNA is shown extending from the protein helix to indicate its spatial arrangement within the helix. Protrusions on the RNA chain (one pointing out, two pointing up) represent the three nucleotides bound to each protein subunit. (Photograph, courtesy of G. Stubbs, Vanderbilt University, Nashville, Tennessee, is a color version of a figure from Namba *et al.*, 1989.)

Academic Press, Inc.
1250 Sixth Avenue, San Diego, California 92101-4311

United Kingdom Edition published by
Academic Press Limited
24–28 Oval Road, London NW1 7DX

Library of Congress Cataloging-in-Publication Data

Matthews, R. E. F. (Richard Ellis Ford). date
    Fundamentals of plant virology / R.E.F. Matthews.
       p. cm.
    Includes index.
    ISBN 0-12-480558-2 (hardcover)
    1. Plant viruses. I. Title.
QR351.M35 1992
581.2'34--dc20

                      91-42901
                        CIP

Printed and bound in the United Kingdom
Transferred to Digital Printing, 2011

CONTENTS

CHAPTER 7
Replication of Viruses with ss-Positive Sense RNA Genomes

CHAPTER 8
Replication of Other Virus Groups and Families

CHAPTER 9
Small Nucleic Acid Molecules That Cause or Modify Diseases

CHAPTER 10
Transmission, Movement, and Host Range

PREFACE

This book summarizes knowledge of all aspects of plant virology. It has been written for students of plant virology, plant pathology, or microbiology who have no previous knowledge of plant viruses, or virology in general. An elementary knowledge of molecular biology is assumed in covering the basic structures of DNAs, RNAs, and proteins; the genetic code; and the processes involved in protein synthesis. It is also expected that students will have a basic knowledge of structure and function in plants, including the main subcellular structures found in typical plant cells. The conventions used in naming viruses, virus groups, and virus families are given in the Appendix.

In Chapter 1, the basic properties of cells and viruses are compared to give students an appreciation of the place of viruses in the living world. Over the past 10 years, gene manipulation technology has played a key role in developing our understanding of the structure, functions, and replication of plant viral genomes. In Chapter 2, I have briefly outlined the principles of the more important techniques involved, to assist those who have no previous knowledge in this field. In recent years, serological techniques have become increasingly important in many aspects of plant virology. For this reason, Chapter 3 is devoted to a discussion of the basis for the various techniques currently in use, including monoclonal antibodies.

Chapters 4–16 follow fairly closely the content of the third edition of *Plant Virology* (Matthews, 1991), but have been substantially condensed. References to the literature are given only where they are essential, mainly in acknowledging sources of illustrations, for the additional reading lists given at the end of each chapter, or for very recent publications. Original references to the material discussed in Chapters 4–16 can be found in the corresponding sections of the third edition of *Plant Virology.*

In general, the subject matter is not dealt with on an historical basis. Chapter 17, however, gives a very brief historical summary. This final chapter concludes with some views on the prospects for research in plant virology in the coming decade.

I wish to thank my colleague R. C. Gardner for his expert advice on the content of Chapter 2. I thank my many colleagues who have provided illustrations, especially those for the cover. Figures are acknowledged individually in the text. I especially thank Jean Parrott for preparing the manuscript.

R. E. F. Matthews

WHAT ARE VIRUSES?

About 50 years ago the first viruses were being isolated, and character-ized chemically. They were viruses infecting plants. At the time there were many arguments about whether these agents belonged in the living or the nonliving world. This was because, on the one hand, they could be crystallized in the test tube, and on the other, they could orga-nize their own increase in number, when they were present in a suit-able host organism. With the application of the techniques and ideas of molecular biology over the past 30 years, it has become quite clear that viruses are part of the living world. In essence, they are very small obli-gate parasites that contain between one and several hundred genes of their own, which can mutate and evolve as do cellular genes.

1 VIRUSES AND CELLS COMPARED

In this section, it will be useful to consider some properties of viruses infecting other host groups, as well as those of viruses infecting plants.

1.1 SIZE

Most viruses are very much smaller than most cells, and require electron microscopy for their particles to be visualized. Only the pox viruses, which infect animals, can be seen as tiny dots in the light microscope. Relative sizes can be compared using various properties.

1.1.1 Dimensions

Figure 5.1 provides outline drawings of the different kinds of plant viruses. They vary widely in size and shape, but all are very small in relation to the size of the cells they infect. Thin sections of infected cells (e.g., Fig. 8.3) give some indication of relative sizes, but they are rather misleading. Such illustrations are two dimensional, whereas cells and viruses exist in three dimensions; thus, a more realistic procedure is to compare volumes. A typical tobacco mesophyll cell has a volume of about 2×10^{13} nm^3; a large plant virus, about 6×10^5 nm^3; and a small plant virus, about 2×10^4 nm^3. Thus in a typical plant cell containing 10 million particles of a small virus, only about 1% of the volume of the cell would be occupied by virus.

1.1.2 Size and Complexity of Genome

In making comparisons between different kinds of organisms, the most fundamental single character is the size of the genome. Figure 1.1 shows that the genomes of viruses infecting all kinds of hosts span a range of almost three orders of magnitude. The smallest are about the size of a cellular messenger RNA (mRNA). The largest viral genomes are about the same size as those of the simplest cells; thus, there is a very wide range of size and complexity.

A biologically more meaningful way of comparing genome sizes is

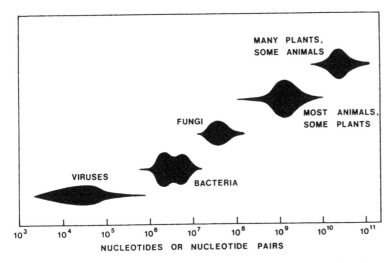

Figure 1.1 The range of genome sizes for viruses, prokaryotes, and eukaryotes. The vertical axis gives an approximate indication of the relative number of species within the size range of each group. From Matthews (1981); modified from Hinegardner (1976).

to consider their information content, that is to say, the number of genes they contain or the number of proteins they code for. Some examples are given in Table 1.1.

From this section we can conclude that (1) by several criteria, viruses in general span a very wide range of sizes; (2) viruses infecting angiosperms are confined to the small end of the size spectrum; and (3) plant viruses are indeed very small and genetically simple compared to the cells they infect.

1.2 COMPOSITION

1.2.1 Macromolecules and Structural Components

In cells, the most important molecules are those containing the genetic information—DNA—and the functional molecules synthesized using that information—RNA and proteins. A lipoprotein plasma membrane defines the boundary of the cells. In addition, in eukaryotes, many cell organelles are bounded by, or composed of, lipoprotein membranes, for example, the nucleus, mitochondria, chloroplasts, endoplasmic reticulum, Golgi apparatus, etc. Most plant cells have an outer wall made up largely of cellulose.

Table 1.1 Relative Sizes of Genomes Measured by the Number of Proteins Coded for

Type of organism	Example	Number of proteins
Higher plant	Average (approximate estimate)	50–100,000
Eubacteria	*Escherichia coli*	≃4100
Mycoplasma	—	≃820
Chlamydiae	Psittacosis	≃660
Large virus infecting vertebrates	Vaccinia	≃260
Virus infecting a chlorella-like alga	PBCV1	≃200–250
Large virus infecting angio-sperms	Wound-tumor virus	12
Small virus infecting angio-sperms	Tobacco mosaic virus	4
Smallest known virus	Satellite virus of tobacco necrosis virus	1

All viruses have their genetic information in the form of nucleic acid. Some use double-stranded DNA (dsDNA), as do cells. Others use single-stranded DNA (ssDNA), dsRNA, or ssRNA. Most known plant viruses uses ssRNA. As well as genomic DNA, all cells competent to divide contain RNAs of various kinds and functions, mainly ribosomal (rRNA), mRNA, and transfer (tRNA). Most virus particles contain only their genomic nucleic acid, which is either DNA or RNA. However, a few DNA viruses contain small amounts of RNA, either host or viral coded, with specific functions. During the replication of most viruses, mRNAs are transcribed from the viral genomic nucleic acid.

The genetic material of viruses is almost always protected by a covering or coat of protein molecules coded for by the virus. Some viruses, especially the larger ones, have an outer lipoprotein membrane as well. Larger viruses may also contain viral-coded enzymes involved in nucleic acid synthesis.

The DNAs and RNAs found in viruses are made up of the same four nucleotides that are found in the corresponding cellular nucleic acids. Viral proteins contain a selection of the same 20 amino acids that are found in cells. The proportions of different amino acids in a typical viral protein are usually quite similar to those found in typical soluble cellular proteins. In viruses with a lipoprotein envelope, the proteins in the membrane are coded for by the virus and may be glycoproteins containing carbohydrate. The lipid is usually borrowed from some cellular source. No viruses contain cellulose or other components found in plant cell walls.

1.2.2 Low-Molecular-Weight Materials

Besides water, cells contain a very large number and variety of low-molecular-weight substances. These are involved in such processes as intermediary metabolism, the production of the amino acids, nucleic acid bases, sugars, fatty acids, etc., used in the synthesis of macromolecules, and in the building of cellular structures. Many cells also contain low-molecular-weight substances involved in specialized processes. By contrast, viruses contain a very limited range of low-molecular-weight materials. Viruses with a lipoprotein envelope contain the same proportion of water as do cells (about $3-4g/g$ dry mass). Viruses without such an envelope usually have associated with them about the same proportion of water as soluble cellular proteins ($0.7-1.5g/g$ dry mass). Several plant viruses contain divalent metal ions, usually Ca^{2+}, at specific sites in their structure; these are important for the stability of the viral protein coat or shell surrounding the nucleic acids. Some plant viruses have a specific domain (a *zinc finger*) in a protein involved in nucleic acid binding, which specifically binds an atom of zinc. A few viruses contain polyamines, which help to stabilize the RNA within the virus particle.

In summary, viruses have the following characteristics compared with cells: (1) apart from water, viruses consist primarily of nucleic acid and protein, built from the same nucleotides and amino acids as are found in cells; (2) some viruses have a lipoprotein envelope with the lipid usually derived from a host membrane; (3) the amount of water associated with viruses having a lipoprotein envelope is like that of cells; for others, it is like that of soluble cellular proteins; (4) compared with cells, viruses contain hardly any low-molecular-weight materials, other than water. The small amounts of metal ions and polyamines that may be present have a structural rather than a metabolic function.

1.3 STRUCTURE

1.3.1 Cell Walls

Most prokaryotic cells except mycoplasmas have cell walls. All viruses lack such a wall.

1.3.2 Variability in Structure

Cells, even those from an apparently uniform tissue such as the palisade mesophyll cells of a leaf, vary in structure. For example, they change shape as they expand in the growing leaf. Mature cells contain differing numbers of organelles, such as chloroplasts and mitochondria. By contrast, viruses, especially those without a lipoprotein envelope, have a very uniform structure dictated by the geometrical arrangement of the major components of the particle.

1.4 INCREASE IN POPULATION SIZE

There are many single-celled organisms that, like the viruses, are obligate intracellular parasites. The smallest and simplest of these belong to a group called the Chlamydiae. They have cell walls but lack an energy-generating system. They have two phases in their life cycle. Outside the host cell, they exist as infectious *elementary bodies* about 300 nm in diameter. These have dense contents and are specialized for survival outside the host cell. They have two wall layers and a plasma membrane. They contain the genomic DNA, RNA, and lipid. An infecting elementary body enters the host cell, and within 8 hr is converted into a much larger, noninfectious *reticulate body*. This is bounded by a bilayer membrane provided by the host cell. The reticulate body divides by binary fission within this membrane, giving thousands of progeny within 40 to 60 hr. The reticulate bodies are converted to elementary bodies, which are released when the host cell lyses (Fig. 1.2).

   Thus the Chlamydiae, and other single-celled parasitic organisms, increase in number by a process of binary fission in which single cells

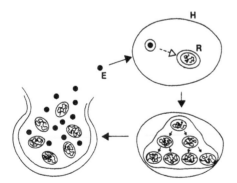

Figure 1.2 Schematic life cycle of the Chlamydiae. E, elementary body; R, reticulate body; H, host cell.

enlarge and then divide to give two daughter cells. During this process, the cytoplasm of the parasite is always separated from that of the host by at least the bilayer membrane of the parasitic cell itself.

By contrast, virus increase involves the synthesis in the host cell of pools of the required macromolecular components—proteins and copies of the viral genome. Virus particles are assembled from these pools of components in the absence of a membrane separating the site of synthesis and assembly from the rest of the host cell's cytoplasm or nucleoplasm.

1.5 BIOCHEMISTRY

Most cells have their own energy-producing systems to fuel their metabolic processes. Such a system is lacking in viruses and also in the Chlamydiae. Most prokaryotic cells can synthesize the amino acids and nucleotides they require. Viruses have no such synthetic abilities. Certain bacteria also require all of these compounds in the medium.

1.6 SUMMARY

From this discussion, we can conclude that three characteristics distinguish all viruses from the Chlamydiae and all other prokaryotic cells.

1. *Increase in population size*   Even the simplest cells increase in number by binary fission. All viruses increase in number by a process involving the synthesis of a pool of the required components, and the assembly of virus particles from this pool. For this reason, it is appropriate to apply two different terms for the process of increase: I shall use *multiplication* for cells, in which the population increases step-wise (by a factor of two at each step under ideal conditions); and *replication* for viruses, in which the process involves the more or less simultaneous production of many replicas of the original infecting particle from a pool of viral components. (Note: Some authors use *binary fission* for cells, *multiplication* for viruses, and reserve *replication* for nucleic acids.)

2. *Protein-synthesis machinery*   All cells contain their own protein-synthesizing system—ribosomes, mRNAs, tRNAs, and accessory enzymes and factors. No viruses contain such a system, although a few may contain some tRNAs or host ribosomes.

3. Presence of a membrane during multiplication or replication   All cellular parasites multiplying inside a host cell maintain a lipoprotein bilayer membrane that separates the cytoplasm of parasite and host. No such membrane exists during virus replication. The absence of such a membrane makes binary fission impossible, and allows a virus to use the protein-synthesizing machinery in the host cell cytoplasm.

2 DEFINITION OF A VIRUS

The original Roman meaning of *virus* was a poison, and the word was used in this sense for many centuries. The first property to be discovered about the agents we now know as viruses was their small size. The infectious agent could pass through filters whose pore size was so small that they held back all known cellular parasites from the filtrate. Thus these agents were called the *filterable viruses* for some decades after the first examples were discovered at the end of last century.

About 25 years ago, it came to be realized that some kinds of plant disease that were caused by infectious filterable agents were not attributable to viruses. The cellular parasites known as *mycoplasmas* and *spiroplasmas* are small enough to pass through bacteria-proof filters, and to be invisible under the light microscope. Electron microscopy of infected cells, together with other evidence, revealed that many plant diseases, particularly of the *yellows* and *witches-broom type,* are in fact caused by these small cellular parasites. At the other end of the size scale for filterable agents are infectious entities consisting only of a very small, circular RNA. These cause several important virus-like diseases, and are known as *viroids.* They are widely considered to belong to a class of infectious agents distinct from the viruses.

With modern knowledge we can now give a fairly precise definition of a virus:

A virus is a set of one or more genomic nucleic acid molecules, normally encased in a protective coat or coats of protein or lipoprotein, which is able to mediate its own replication only within suitable host cells. Within such cells, virus replication is (1) dependent on the host's protein-synthesizing machinery; (2) derived from pools of the required materials rather than from binary fission; and (3) located at sites not separated from the host cell contents by a lipoprotein bilayer membrane.

The properties of viruses can be clarified further by considering some features in more detail. These features are important but are not applicable to all plant viruses.

2.1.1 Transmission and Disease Induction

To be identified positively as a virus, an agent must normally be shown to be transmissible, and to cause disease in at least one host species. The *Cryptovirus* group of plant viruses, however, is an exception. These viruses rarely cause detectable disease, and are not transmissible by any mechanism except through the seed or pollen.

2.1.2 Properties of the Genomic Nucleic Acid

Chemical nature The nucleic acid of a viral genome may be DNA or RNA, and single- or double-stranded. If the nucleic acid is single-stranded, it may be of positive or negative sense. For viruses, *positive sense* is defined as the sequence that actually codes for a protein, that is, acts as an mRNA. *Negative sense* is the complementary sequence. The single-stranded nucleic acid of a few viruses is *ambisense*. In other words, it is positive sense in coding for one protein and negative sense, for another.

Information content In relation to viruses infecting other host groups, plant viral genomes are at the small end of the virus spectrum, coding for 1 to 12 proteins.

Segmented genomes Some plant viruses have segmented genomes. Each nucleic acid segment may be housed in a separate virus particle or all may be housed in one.

Mixed nucleic acids A few viruses with DNA genomes also contain small amounts of RNA.

Genetic change Point mutations occur, especially in ssRNA viruses, with high frequency, owing to nucleotide changes brought about by errors in the copying process during genome replication. Other kinds of genetic change may be the result of recombination, reassortment of pieces in viruses with segmented genomes, loss of genetic material, or acquisition of nucleotide sequences from unrelated viruses or the host genome.

Integration A few viruses share with certain nonviral nucleic acids the ability to be integrated into the host cell genome. No plant viruses of this sort are known so far.

2.1.3 Functions of Viral Proteins

Almost all viruses code for a protein or proteins necessary for replication of the viral genome. In some viruses, these enzymes are present in the virus particle. One or more viral-coded proteins provide a protective coat for the genomic nucleic acid. Viruses may also code for proteins with other functions, for example: (1) *movement,* facilitating virus movement from cell to cell; (2) *transmission,* enabling transmission by invertebrate or fungal vectors; or (3) *protein processing,* a protease activity specifically cleaving a large viral polyprotein into functional products.

2.1.4 Viroplasms

Replication of many kinds of viruses takes place in distinctive virus-induced regions of the cell cytoplasm known as *viroplasms.*

2.1.5 Dependence on Another Virus

A few viruses require the presence of another virus for their replication. These are known as *satellite viruses.* Some other viruses require a helper virus to be present in the tissue to allow transmission by invertebrates to take place.

FURTHER READING

Matthews, R. E. F. (1991). "Plant Virology," 3rd ed. Academic Press, New York.

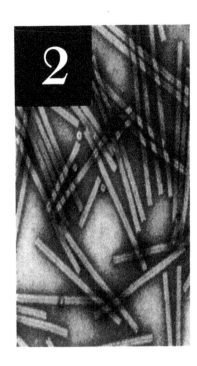

PRINCIPAL TECHNIQUES FOR THE STUDY OF VIRUS PARTICLE AND GENOME STRUCTURE

The study of a plant virus usually begins with investigation of two aspects, the structure of the virus particle and the organization of the viral genome. The most important methods used in such studies are outlined briefly in this chapter.

---

1 STRUCTURE OF VIRUS PARTICLES

The two major techniques for studying virus structure are X-ray crystallography and electron microscopy using various kinds of specimen preparation.

1.1 X-RAY CRYSTALLOGRAPHIC ANALYSIS

When X-rays pass through a crystal, the rays are scattered in a regular manner. The scattered radiation can be recorded photographically. What is recorded is, however, not a picture of the virus particle, but a very abstract-appearing collection of dots from which the structure of the virus particle is deduced by complex mathematical analysis. Inducing virus particles to form crystals suitable for X-ray crystallography is more of an art than a science, and generally requires many trials of salt and alcohol solutions and other precipitating conditions to obtain crystals of sufficient size and stability. Isometric particles will form true crystals. Rod-shaped, rigid virus particles often will form liquid crystals in which the rods are regularly arrayed in two dimensions. X-ray analysis can be applied to such crystals, but not to rod-shaped viruses with flexuous particles or to large virus particles with lipoprotein envelopes.

Where they can be applied, X-ray techniques provide the most powerful means of obtaining information about virus structure. Over the past 10 years, significant advances have allowed the application of X-ray crystallographic analysis to more viruses and at higher resolutions. With the definition of structures at atomic level, it has been possible to define interactions between the viral genome and the protecting protein coat, and to establish the positions of water molecules and divalent cations in the structure.

In summary, the major technical advances responsible for this progress have been (1) high-intensity, coherent X-ray sources that allow data to be recovered in a short time from delicate crystals; (2) an increase in the speed and capacity of computers, together with a reduced cost of computing; (3) noncrystallographic symmetry averaging, a process involving successive approximations that remove noise and enhance detail in the density map; and (4) the development of computer graphics, replacing the laborious manual model building that was required previously to refine structures.

There is a significant limitation for the study of small isometric viruses that can be crystallized. Such viruses crystallize because of regular symmetries in the protein shell. However, most of the nucleic acid inside the virus is not arranged in a regular manner with respect to these symmetries. Thus, very little information can usually be obtained about the conformation of the genome within the virus. Such information can be obtained for the rigid rod-shaped viruses, in which the RNA is arranged in a regular helix within a cylinder of protein.

1.2 ELECTRON MICROSCOPY

Development of images using electron microscopy depends on differences in electron scattering in different parts of the specimen. Virus particles themselves have very little contrast with respect to the scattering of electrons, compared with the carbon film on which they are usually mounted. For this reason, various specimen-preparation techniques have been used to enhance contrast. In early work, shadowing of the specimen at an angle with a vaporized heavy metal, such as gold, was employed. This procedure, however, obscured much detail. It was subsequently shown that various osmium, lead and uranyl compounds, and phosphotungstic acid (under certain conditions) react chemically with, and are bound to the virus. This procedure was called *positive staining*, but it was found to cause alteration in or disintegration of the virus structure. Today negative staining is the most widely used procedure for visualization of viruses in the electron microscope.

1.2.1 Negative Staining

Negative staining uses potassium phosphotungstate at pH 7.0, or uranyl acetate or formate near pH 5.0. The electron-dense material does not react chemically with the virus, but penetrates available spaces on the surface and within the virus particle. The virus structure stands out against the electron-dense background (Fig. 5.2). However, even in the best electron micrographs, fine details of structure tend to be obscured, first, by noise due to minor irregularities in the virus particle image and in the stain, and second, by the fact that contrast due to the stain is often developed on both sides of the virus particle to a varying extent. Thus electron micrographs of very high quality are essential in order to distinguish particles of small isometric viruses belonging to different virus groups. More detailed structural information may be obtained from a number of images of single negatively stained particles, by processing the image in one of several ways (e.g., Fig. 5.9).

1.2.2 Thin Sections

The diameter of small isometric viruses (20–30 nm) is much less than the thickness of a typical thin section ($\approx 40-100$ nm) used to study tissues by electron microscopy. Thus, no detailed structural information is revealed. However, some aspects of the structure of the larger viruses with lipoprotein envelopes can be studied using thin sections of infected cells (Fig. 5.16) or of a pellet containing the virus.

1.2.3 Cryoelectron Microscopy

Cryoelectron microscopy is a recently developed technique. The specimen is frozen extremely rapidly in an aqueous medium. The virus is suspended in a very thin film of liquid stretching across holes in a carbon grid, which is plunged into liquid ethane that contains some ethane ice (183°C). The freezing is so rapid that water molecules do not have time to form micro ice crystals; thus, the specimen is frozen in vitreous ice with no damage caused by crystallization of water. The method is being usefully applied to the study of virus structures or substructures that may be altered by other specimen-preparation techniques.

2 THE STRUCTURE OF VIRAL GENOMES

2.1 CLASSICAL PROCEDURES

The nature of a viral nucleic acid, whether it is DNA or RNA, and whether it is single-stranded or double-stranded, circular or linear, can be established by various standard physical, chemical and enzymatic methods. Chemical and enzymatic procedures allow known special structures at the 5'- or 3'-end of linear genomic nucleic acid to be identified. Electrophoresis of purified nucleic acid from virus particles usually will give good estimates of the RNA or DNA molecular weight and the number of different size classes of genomic nucleic acid for those viruses with split genomes. The application of gene-manipulating technology is also rapidly increasing our understanding of the structure of viral genomes and how they replicate.

2.2 GENE-MANIPULATION TECHNOLOGY

2.2.1 Importance

There are two situations in which viral genome sequence information is of great use. Each of these situations has theoretical and practical aspects.

The Virus in the Plant

Theoretical A knowledge of the viral genes and the products they code for is beginning to lead to an understanding of how viruses cause disease.

Practical The ability to identify and isolate particular viral genes and integrate them into the host-plant genome is providing novel methods for understanding virus gene function and, in some instances, for the control of virus diseases.

The Virus in Relation to Other Viruses

Theoretical A knowledge of the nucleotide sequences of many viral genomes is of very great assistance in virus classification. The nucleotide sequences are revealing unexpected relationships between viruses, and this information is beginning to give us an understanding of how viruses might have originated and how they evolved. Computer-aided comparison of a viral nucleotide sequence with those of other viruses, and sequences encoding cellular proteins can sometimes indicate possible viral protein functions.

Practical It is essential to be able to identify the virus, and often the strain of virus, before effective control measures can be developed for a virus-causing disease in a particular crop and location. Identification requires both an effective framework of classification for the relevant viruses and strains, and methods for identifying the virus being studied. Nucleotide sequences are proving to be of great use in both these aspects of disease control.

2.2.2 Nucleic Acid Hybridization

In a double-stranded (ds) nucleic acid, the two strands are held together by hydrogen bonding between the complementary base pairs: $G:C$ and $A:T$ for DNA, and $G:C$ and $A:U$ for RNA. When a ds nucleic acid is heated in solution, the secondary bonds between the bases are broken, and the strands separate. This process is known as *melting* or *denaturation.* If the mixture of strands is then incubated at a lower temperature, the ds structure is reformed; that is, the nucleic acid *renatures.* The temperature at which 50% of the sequences are denatured is called the *melting temperature* or T_m. The denaturation of nucleic acid secondary structure leads to a rise in the absorbance of the solution at 260 nm. This can be used to follow the denaturation and establish the T_m.

The T_m is affected mainly by the following factors: (1) base composition, a high $G + C$ content giving a higher T_m; (2) salt concentration. Raising the salt concentration raises the T_m; and (3) the presence of a hydrogen bond-breaking agent such as formamide will lower the T_m. For most ds plant viral RNAs, the T_m in 0.3 M NaCl lies between 88 and 93° C. For hybridization tests, temperature conditions are usually chosen to maximize the rate of association between complementary strands. This is about 65° C for most plant viral RNAs, i.e., 23–28° C below the T_m.

Hybridization will take place only between nucleic acids that have sequence complementarity. The degree of complementarity required before two sequences can hybridize depends on the temperature of the reaction. A rule of thumb is that the T_m is reduced by 1° C for each 1% of sequence mismatch. At the commonly used temperature of 65° C, sequences with ≥72–77% similarity will reanneal. Thus, reassociation experiments can be used to obtain an estimate of the extent of sequence similarity between two single-stranded (ss) nucleic acids. Over a range of conditions, association between strands reaches a dynamic steady state with strands associating and dissociating. More or less stringent conditions in solution can be used, which allow less or more hybridization to occur. Hybridization can occur when both nucleic acids are in solution or when one of the strands is immobilized on a solid matrix such as nitrocellulose paper.

2.2.3 Gel Electrophoresis

Electrophoresis in gels is an important technique for the separation and size determination of fragments of DNA. (See Fig. 2.3B) In this technique, an electrical field is applied along the gel, and charged molecules will migrate in the fluid in the pores of the gel. The rate of migration of DNA molecules will depend very much on their size (length). Shorter molecules are retarded less by the gel matrix and, hence, will move faster than longer ones. Marker molecules of known length (base pairs) can be run in one channel of the gel. The separated and purified fragments of DNA can be recovered undamaged from the gel.

This technique has many applications in the study of viral genomes and nucleic acids generally. Different types of gels containing agarose or polyacrylamide resolve nucleic acid fragments of different sizes. For example, gel conditions have been devised whereby 300–500 ssDNA fragments, each differing in length by only a single nucleotide, can be separated. This step is vital to the DNA-sequencing process.

2.2.4 Restriction Maps

Restriction endonucleases are enzymes found in a number of bacterial species. Their function in these cells is to cleave and thus inactivate foreign DNA, such as that from infecting viruses. Their important biochemical feature is that they cleave the DNA only at very specific nucleotide sequences, usually four to eight nucleotides long, known as *recognition* sequences (Fig. 2.1). The bacterium protects its own DNA by adding methyl groups to bases in potential cleavage sites. Over 100 restriction enzymes have been characterized, allowing for the possibility of cutting dsDNA specifically at many different sites. Using different restriction enzymes to cut a viral genome in DNA form, several distinct size classes are produced, which can be separated by gel electrophoresis. It is then possible to construct a restriction map of the genome showing the location of each cutting site in relation to the others. Such a map will be characteristic for a particular virus and can be used to give an approximate estimate of similarity to the genomes of related viruses. These enzymes also have other important uses, discussed in the following sections.

2.2.5 DNA Polymerase I

The availability of this enzyme (Pol I) and a knowledge of its mode of action has been crucial to the development of several molecular techniques, including the labeling of probes and DNA sequencing. Pol I was found in *Escherichia coli* and is involved in DNA replication. From the point of view of the study of viral genomes, its main properties are as follows: it can polymerize only deoxynucleoside triphosphates (dNTPs), and can do so only when copying a template DNA. To start replication, a *primer* is necessary. A primer is an oligonucleotide hydrogen-bonded by base-pairing to the template strand. The primer's terminal 3′ OH group must be available to react with an incoming dNTP. Nucleotides are not added to a free 5′ OH group. That is to say, growth of the new strand proceeds in the direction 5′ to 3′.

Besides its polymerase function, Pol I has two other functions: a 3′–5′ exonuclease activity involved in eliminating incorrect bases during strand growth, and a 5′–3′ exonuclease activity, which functions in the cell to remove RNA primers from the DNA. This 5′–3′ exonuclease activity at a single-stranded break (or nick) involves removal of a 5′ nucleotide. As a nucleotide is removed, it can be replaced by the polymerase activity of the enzyme. The nick moves 5′ to 3′ along the direction of strand synthesis. This process is known as *nick translation*. In experimental systems, nicks in the dsDNA substrate are produced using DNase in the incubation medium. Nick translation is not to be confused with *translation*, the process by which ribosomes and other factors synthesize proteins in a cell under the guidance of mRNA.

2.2.6 Labeled Nucleic Acid Probes

Sequences of radioactively labeled nucleic acids known as probes are used in a variety of ways in the study of viral genomes and viral replication. The radioactive nucleic acid may be DNA or RNA. The nucleic acid can be labeled in several ways. The following methods are commonly used.

End-Labeling

Polynucleotide kinase is an enzyme found in many kinds of cells, which catalyses the transfer of the γ phosphorus from adenosine triphosphate (ATP) to the 5'-OH at the 5' end of DNA and RNA molecules of all sizes. If the ATP is labeled with ^{32}P in the γ position, this will be transferred to the 5'-OH of the polynucleotide, thus labeling the 5' end of the molecule. If a 5' phosphate is already present, this is readily removed before the labeling step using the enzyme alkaline phosphatase, to expose a free 5'-OH.

Nick Translation

A common method of labeling probes is by a process known as *nick translation* (Section 5). Radioactive dNTPs added to the incubation mixture are incorporated into the growing strand by DNA Pol I to provide a radioactive probe.

Random Priming

This method involves the use of a mixture of random nucleotide primers on ssDNA or ssRNA. The DNA is then copied *in vitro* with a DNA polymerase or reverse transcriptase using radioactive dNTPs.

Strand-Specific Probes

This method also uses *in vitro* synthesis of nucleic acid from labeled dNTPs. For most purposes, a DNA-dependent RNA polymerase is involved (e.g., from T7 or SP6 bacteriophages). The enzyme is used to synthesize a labeled RNA copy of a DNA segment that has been cloned adjacent to a specific T7 or SP6 promoter. In this way it is possible to synthesize and label only one strand of nucleic acid.

2.2.7 Using Labeled Probes

Labeled probes are used in many different ways in the study of nucleic acids and in the manipulation of genes. Their usefulness depends on the ability of a probe to hybridize specifically with a complementary nucleotide sequence in another piece of DNA or RNA, as discussed in Section 2.2.2.

Southern Blotting

Southern blotting is named after E. M. Southern, who developed the procedure. DNA molecules in an agarose gel are separated into discrete bands by electrophoresis. The gel is then laid on a membrane of porous nitrocellulose or nylon. A flow of an appropriate buffer is then set up perpendicular to the direction of electrophoresis and toward the membrane. The flow or *blotting* procedure causes the DNA to be transferred by capillary action to the membrane where it is bound, forming a replica of the DNA bands in the gel. A labeled probe can then be exposed to the filter under renaturing conditions. DNA with complementary sequences will bind the labeled probe. Autoradiography of the membrane will then reveal which DNA bands, if any, have sequences complementary to those of the probe. With appropriate marker DNAs in the gel, the sizes of DNA in hybridizing bands can be estimated.

Northern Blotting

A technique analogous to Southern blotting has been developed for RNA and has been called *Northern blotting,* to confuse people who are not molecular biologists. A mixture of RNAs, viral or cellular or both, are separated on the basis of size by agarose gel electrophoresis usually under strongly denaturing conditions, which prevent the formation of intrastrand base pairing and thus the formation of loops in the RNA. As with DNA, the gel is then blotted onto nylon or nitrocellulose, which traps the RNA. The membrane is then exposed to a solution containing a labeled probe under renaturing conditions. Autoradiography will then reveal any bands of RNA that have sequences complementary to the probe, and their sizes can be estimated with respect to markers in the gel.

In Situ *Hybridization*

Viruses usually replicate and accumulate at particular sites within infected cells. Under appropriate circumstances, labeled cDNA or RNA probes can be used to hybridize with viral genomic nucleic acid or mRNAs present in thin sections of infected cells. If the probe is radioactively labeled, the sites of hybridization are located by radioautography. If they are labeled with fluorescent dyes, a fluorescence microscope is used.

2.2.8 Manipulation and Amplification of DNA Sequences

A knowledge of the nucleotide sequence of a viral genome is basic to our understanding of the structure and replication strategy of that genome, and of its relationships to other viruses.

Restriction Endonucleases

An important feature of the previously discussed restriction endonucleases is that some of them do not make a straight cut, but make staggered cuts in each strand at sites a few nucleotides apart (Fig. 2.1). These staggered cuts leave short single-stranded ends on both sides of the cut. These are called *cohesive ends* because they can hybridize to form complementary base pairs with the ends of any other DNA molecule that has been cut by the same restriction nuclease.

DNA Ligase

When two ends of DNA molecules that have been cut by the same restriction nuclease have joined together by complementary base pairing, the ends can be sealed by an enzyme called DNA ligase, which forms a covalent phosphodiester bond between the ends of the two DNAs.

DNA Cloning

In order to determine nucleotide sequences, or carry out other operations on the fragments of DNA produced by restriction nucleases, it is necessary to be able to amplify very substantially the amounts of these fragments. This can be done by inserting the DNA into a *plasmid*

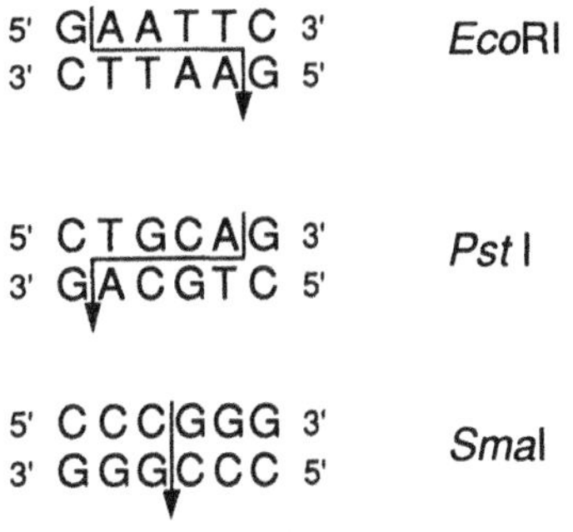

5′ G|A A T T C 3′   *Eco*RI
3′ C T T A A|G 5′

5′ C T G C A|G 3′   *Pst* I
3′ G|A C G T C 5′

5′ C C C|G G G 3′   *Sma*I
3′ G G G|C C C 5′

Figure 2.1 Recognition sequences for three restriction nucleases. In these examples, the sites are six base pairs long, and the nucleotide sequences 5′ → 3′ are the same in each strand. Three different kinds of termination can be generated: *Eco*RI gives a 5′ overhang; *Pst*I gives a 3′ overhang; *Sma*I gives a blunt-ended cut. *Eco*RI is from *E. coli*; *Pst*I, from *Providencia stuarti*; and *Sma*I, from *Serratia marcescens*. The names of the enzymes are abbreviated from the names of their host bacteria.

or a bacterial virus (*bacteriophage*) and then growing these in bacterial or yeast cells. Plasmids are small circular dsDNA molecules, found naturally in bacteria and yeast, that replicate in the host cell. Plasmid DNA is much smaller than the bacterial DNA and can therefore be easily separated from it. When plasmids are used as *cloning vectors*, fragments of viral DNA are inserted into the plasmid DNA.

The hybrid plasmid DNA molecules are then reintroduced into the host bacterium, and bacterial colonies allowed to grow on a culture plate. Only some of the cultures will contain the viral DNA of interest. Those containing inserted DNA are selected in the following manner. The plasmid used for cloning has had inserted into its DNA a gene for resistance to an antibiotic (often ampicillin). The plasmid also contains a gene allowing for the fermentation of lactose. The restriction enzyme site for insertion of the DNA of interest is in this gene. When the plasmid DNA molecules have been introduced into a bacterial culture, the bacteria are plated out on a medium containing the antibiotic and an analog of lactose that gives a blue color when it is fermented. Only cells that are resistant to the antibiotic and therefore contain the plasmid will grow to form colonies. Most of these will not contain the DNA of interest and will therefore give blue colonies. Any white colonies present will be those that have had the lactose enzyme gene inactivated by the insertion of the DNA of interest.

The well-known vector plasmid pBR322 contains an ampicillin and a tetracycline resistance gene. Several sites for different restriction nucleases lie within these genes. When a piece of foreign DNA is inserted into one of these genes by the process outlined in Fig. 2.2, that gene is inactivated. When the plasmid is introduced into the host bacterium, colonies resistant to one antibiotic and susceptible to the other are readily identified. Colonies containing the desired foreign DNA are multiplied up in bulk culture. The plasmid replicates along with the bacterial host cells. The plasmid DNA is then purified. Copies of the original inserted DNA may then be excised from the plasmid by a second treatment with the same restriction nuclease used to insert the molecule.

Certain bacteriophages can be used instead of a plasmid to carry a piece of foreign DNA into a bacterial cell. Phage M13 infecting *E. coli* is particularly useful for cloning DNA to be used for Sanger dideoxy sequencing (see Section 2.2.9). This is because the phage provides a single-stranded DNA template. The site at which the foreign DNA is inserted into M13 DNA is precisely known. A 10–20 nucleotide sequence complementary to this site can be synthesized chemically, and can be

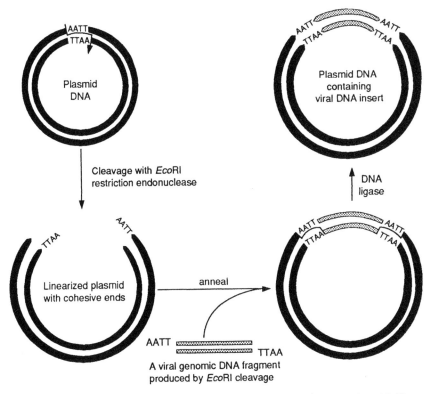

Figure 2.2   Incorporation of a fragment of viral DNA into a bacterial plasmid. The same restriction nuclease is used to cleave both DNAs. The fragments are annealed, and treated with DNA ligase to produce covalently closed hybrid plasmid DNA (a so-called *recombinant* DNA molecule).

used as a primer for the dideoxy sequencing of any foreign DNA sequence inserted adjacent to the primer site.

2.2.9 Determination of Nucleotide Sequences

Sequencing Viral DNA Genomes

Two major methods have been developed and used for sequencing DNA. The method of Maxam and Gilbert involves radioactively labeling a preparation of DNA at the 5′ end, and dividing the preparation into four samples. Each sample is then treated very gently with a chemical that specifically destroys one kind of base, for example G, or two kinds of bases, C and T, but only at one or a few locations of that base in any given molecule. The mixture of DNA molecules, each molecule with a specific kind of base destroyed at only one site, is chemically cleaved where the base has been removed. For the G cleavage reaction, for example, the resulting 5'-end-labeled fragments will form a set corresponding to all possible chain lengths from the labeled end to the site of a G nucleotide. The radioactive fragments are then separated by gel electrophoresis. From the pattern of bands, a nucleotide sequence can be deduced.

The second method was devised by Fred Sanger, and is known as the dideoxy chain termination procedure. Since it is now the preferred method for sequencing viral nucleic acids, it will be considered in a little more detail. Dideoxy nucleotide sequencing is based on the fact that a nucleoside triphosphate containing a pentose residue lacking an hydroxyl group at the both the 2' and 3' positions (a dideoxy nucleotide) cannot form a phosphodiester bond with a nucleotide adding on to a growing chain in the 3' direction. Thus, incorporation of a dideoxy nucleotide into a growing chain terminates extension of that chain. Sanger's procedure is illustrated in Fig. 2.3. Various modifi-

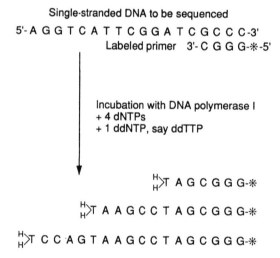

Single-stranded DNA to be sequenced

5'- A G G T C A T T C G G A T C G C C C -3'

　　　　Labeled primer　3'- C G G G -✳-5'

Incubation with DNA polymerase I
+ 4 dNTPs
+ 1 ddNTP, say ddTTP

$^H_H\!\rangle$T A G C G G G -✳

$^H_H\!\rangle$T A A G C C T A G C G G G -✳

$^H_H\!\rangle$T C C A G T A A G C C T A G C G G G -✳

Figure 2.3A Principle of the Sanger dideoxy sequencing procedure. The DNA fragment to be sequenced is annealed with a short, radioactively labeled primer at the 3' terminus. The DNA is mixed with the enzyme DNA polymerase I from *E. coli.* This enzyme can incorporate both normal deoxynucleotide triphosphates (dNTPs) and dideoxynucleotide triphosphates (ddNTPs) into the copy of the DNA being synthesized. Wherever a ddNTP is incorporated, elongation of that strand will cease. Thus, in this example, wherever an A occurs in the original strand, some copies will terminate in the dideoxy form of T.

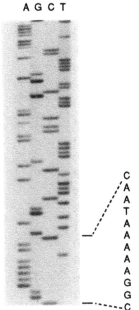

A G C T

C
A
A
T
A
A
A
A
G
G
C

Figure 2.3B The DNA strand to be sequenced, with its radioactively labeled primer, is split into four DNA polymerase reaction mixtures, each containing one of the four ddNTPs in an appropriate ratio with all four dNTPs. The labeled fragments formed after incubation are separated from the template DNA by boiling. They are then fractionated according to length on a polyacrylamide gel under conditions that do not allow rehybridization of the ssDNA (a denaturing gel). A radioautograph of the gel is made. The nucleotide sequence can be deduced directly from the pattern of fragments (the *sequencing ladder*). Photograph courtesy of G. Albertson.

cations have been made to Sanger's original procedure. For example, DNA-dependent DNA polymerases obtained from T7 bacteriophage and the bacterium *Thermus aquaticus* (Taq polymerase) have proved more efficient than the DNA Pol I enzyme used originally. In an important advance toward automated sequencing, in each of the four reactions, the products are labeled with a different-colored fluorescent tag, for example red, G; blue, C; etc. In this situation, the sequence can be read by an automated laser detector from a single lane on a gel (Fig. 2.3C).

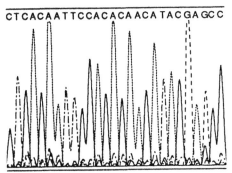

CTCACAATTCCACACAACATACGAGCC

Figure 2.3C Determination of a viral nucleotide sequence using a different-colored fluorescent tag for each of the four reaction mixes. The sequence is read from a single lane in the gel using automated laser fluorescence. In this diagram the four colors have been converted to black and white signals by computer. G, dashed line; A, dotted line; C, solid line; and T, dash-dot line. Courtesy of M. Gibbs.

Sequencing Viral RNA Genomes

In the early 1980s the first plant viral genomes were completely sequenced. These were viruses belonging to the *Caulimovirus* group, which have dsDNA genomes. Thus the sequencing methods developed for cellular DNAs could be applied to them. Most plant viruses have RNA genomes, however, to which the methods could not be directly applied. A very important development for plant virology was the isolation and characterization of an enzyme known as *reverse transcriptase* from the retroviruses that infect vertebrates. This enzyme catalyzes the synthesis of a complementary DNA chain on an RNA template. The RNA template can then be selectively destroyed by RNase H treatment. The single-stranded copy DNA (cDNA) molecules synthesized by the reverse transcriptase can then be converted into ds cDNA molecules using a DNA-dependent DNA polymerase. These can be inserted into plasmids, cloned, and sequenced as outlined above. The main steps in the process are summarized in Fig. 2.4. This procedure has been applied to other RNAs, such as cellular mRNAs that might be of interest in relation to plant viruses.

Strategies for Sequencing

The number of nucleotides that can be sequenced on a single gel is limited to about 300 to 500. Even small plant viral genomes are much longer than this; thus, a strategy must be developed for sequencing sections of the genome and then piecing these together in the correct order. Various strategies have been developed, only some of which can be briefly mentioned here:

1. The full-length genome can be sequenced from both ends, to give the 5′ and 3′ terminal sequences of the molecule. The genomic or cloned cDNA then can be cut into fragments with a restriction endonuclease and inserted into a plasmid vector to generate a set of random *subclones*. These are sequenced to obtain an overlapping set of sequences from which the full-length sequence can be deduced.

2. In another strategy, a restriction map of genomic DNA or cDNA from an RNA virus is prepared. Specific fragments at known positions can then be cloned or otherwise purified and sequenced.

3. The enzyme exonuclease III from *E. coli* can be used to remove nucleotides sequentially from the 3′ end of a DNA molecule at

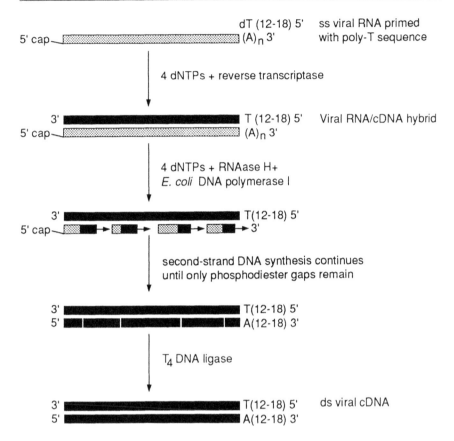

Figure 2.4 Steps in the preparation of a ds cDNA copy of a ss viral RNA. Reverse transcriptase synthesizes the first DNA strand. This strand is then used as the template in a nick translation reaction using RNase H. This produces gaps in the RNA strand, which are filled by DNA polymerase I. A DNA ligase then completes second-strand DNA synthesis.

the rate of about 200 nucleotides/min. By carefully timing the length of incubation, a nested set of DNA molecules with deletions from both ends of the molecule (which are shorter by steps of about 200 nucleotides) can be generated. These can be cloned and sequenced, thus building up the complete sequence in a systematic fashion.

4. Recently it has become possible to derive overlapping sequence data by synthesizing oligonucleotides that will prime at predetermined points. For example, it is possible to get 500 bp of sequence information at the end of a subclone, synthesize overnight an oligonucleotide that will prime the sequencing reaction

starting at the end of this sequence, and then derive the next 500 bp of sequence with this primer the next day. This method of *walking* along the clone has become possible with the advent of rapid automated oligonucleotide-synthesizing machines.

3 AMINO ACID SEQUENCES OF VIRAL PROTEINS

The amino acid sequences of some viral structural proteins were established by laborious direct methods many years before sequencing of the nucleotides in the DNA was possible. Automated machines using these chemical procedures are now available, and can sequence many amino acids in a day from very small amounts of protein. Today however, in most circumstances, it is much easier to sequence the viral genome and deduce the amino acid sequences of the protein products of the genome using the genetic code.

4 mRNAs

Plant viruses with DNA genomes give rise to mRNAs during replication in the cell; these are translated to produce viral-coded proteins. The viruses with ss positive sense RNA genomes can themselves act as mRNAs, but many also give rise to smaller subgenomic mRNAs during replication. In addition, cellular mRNAs stimulated or inhibited by virus infection may be of interest.

4.1 ISOLATION

There are two general methods by which mRNAs can be isolated from cells. In the first, the polyribosome fraction of the cell contents is isolated. This should contain all the mRNAs actually serving as messengers at the time of extraction. In the second method, advantage is taken of the fact that most mRNAs have a tract of A residues perhaps 30–200 long at their 3′ terminus. These can be selectively retained on a fractionating column that has oligo d(T) sequences attached. The oligo

d(T) residues anneal with the poly(A), and RNAs without such sequences are washed through the column.

4.2 *IN VITRO* TRANSLATION

Three systems have found wide application for the translation of eukaryotic mRNAs *in vitro,* or at least under conditions in which most or
all of the protein synthesis is directed by the mRNA of interest. The general procedures for applying these systems are as follows:

1. The RNA or RNAs of interest are purified in high degree, using
   density-gradient centrifugation of the virus, where appropriate,
   and/or polyacrylamide gel electrophoresis of mRNAs.
2. The RNAs are then added to the protein-synthesizing system in
   the presence of amino acids, one or more of which is radioactively labeled.
3. After the reaction is terminated, the polypeptide products are
   fractionated by electrophoresis on sodium dodecyl sulfate (SDS)
   polyacrylamide gels, together with markers of known size.
4. The products are located on the gels by means of the incorporated radioactivity.

   The three systems are

   1. *The rabbit reticulocyte system*   The cells from blood of rabbits
that have been made anemic are lysed in water and centrifuged. The
supernatant fluid is then used. This is a useful system because of the
virtual absence of RNase activity. Figure 2.5 illustrates the use of this
system, which is the most frequently used.
   2. *Toad oocytes,* which strictly do not constitute an *in vitro* system.
Intact live egg cells of *Xenopus* or *Bufo* are injected with the viral mRNA
and incubated in a labeled medium.
   3. *The wheat germ system*   In this system, the viral RNA is added,
in the presence of an appropriate label, to a supernatant fraction from
extracted wheat embryos from which the mitochondria have been removed.

Not infrequently, when purified viral genomic RNAs are translated in
cell-free systems, several viral-specific polypeptides may be produced
in minor amounts in addition to those expected from the open reading
frames (ORFs) in the genomic RNA. It is unlikely that such polypeptides
have any functional role *in vivo.*

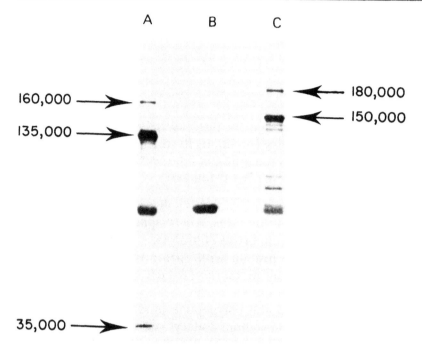

Figure 2.5 Translation of plant viral RNAs in the rabbit reticulocyte system. The poly-peptide products were fractionated by electrophoresis in a polyacrylamide gel and located by radioautography of the gel. (A) Products using TMV RNA as message. (B) Control with no added RNA. (C) Products using TYMV RNA. The unmarked smaller polypeptides may be incomplete transcripts of the viral message, or the result of endogenous mRNA. From Briand (1978).

4.3 HYBRID ARREST AND HYBRID SELECTION PROCEDURES

Hybrid arrest and hybrid selection procedures can be used to demonstrate that a particular cDNA clone contains the gene for a particular protein. The presence or absence of the protein after these procedures is usually established using some form of recognition by specific antibodies (see Chapter 3), or by its location in a gel after electrophoresis.

4.3.1 Hybrid Arrest

The cloned cDNA, complementary to a small region of the viral genome, is hybridized to mRNAs, and the mRNAs are translated in an *in vitro* system. The hybrid will not be translated because of its ds nature. Identification of the missing polypeptide defines the gene on the cDNA.

4.3.2 Hybrid Selection

The cDNA–mRNA hybrid is isolated from the unhybridized mRNAs, and is then dissociated. The mRNA is translated *in vitro,* and the protein coded for is identified.

5 INTRODUCTION OF A DNA STEP INTO THE LIFE CYCLE OF RNA PLANT VIRUSES

Until 1984, various workers had made cDNA from RNA genomes of plant viruses and introduced these into plasmids that replicated in *E. coli.* Then they transcribed the DNA back into RNA *in vitro,* but the products were not infectious when inoculated into plants. One reason was that the RNA strands did not start and terminate at the correct sites. Ahlquist *et al.* (1984) made a major technical advance in this field. They constructed a plasmid with special DNA sequences on either side of the restriction endonuclease site that was to be used for inserting the viral cDNA into the plasmid. These sequences allowed transcription of the viral cDNA sequence to begin precisely at the correct 5' nucleotide, and to terminate very close to the correct 3' nucleotides. *In vitro* RNA transcripts from their system were infectious. This introduction of an artificial DNA step into the life cycle of an RNA virus allowed manipulations that can be made only with DNA to be applied to RNA genomes, and for the effects of such specific manipulations on infectivity and other biological functions to be tested. Some examples of the kinds of experiments that can be carried out are given in the following sections.

5.1 RNA GENOMES WITH UNIFORM SEQUENCES

In nature, all cultures of RNA viruses consist of a mixture of particles containing nucleotide sequences that differ at one or more sites. *In vitro* transcription of infectious RNA from a single cDNA clone provides a source with a uniform sequence. This can be used in various ways, for example, to gain some estimate of the rate of mutation following replication in the natural host under various conditions.

5.2 REVERSE GENETICS

An important application of the technique introduced by Ahlquist *et al.* (1984) is the introduction of defined changes in individual RNA viral genes in order to study the biological effects of such changes, and thus to define gene functions. This approach is commonly known as *reverse genetics.*

The changes in sequence may consist of a change in a base at a defined site, an insertion or deletion of a base, or the insertion or deletion of short or long sequences of nucleotides. These changes can be brought about by a variety of *in vitro* procedures followed by replication of the altered cDNA in an *E. coli* plasmid. The cDNA is then isolated and transcribed *in vitro* to give a supply of the altered viral RNA genome.

One particularly useful procedure is known as *site-directed mutagenesis.* The steps in this procedure are outlined in Fig. 2.6. This procedure is highly versatile for producing point mutations in DNAs or viral cDNAs. These mutations can be used to establish gene function. For example, if a mutation in a particular viral gene gave rise to a virus unable to move from the initially infected cell, we could infer that the protein encoded by that gene was involved in virus movement from cell to cell. By studying the effects of sets of point mutations in a systematic way, it may be possible to define functional regions within the protein encoded by the gene. Similarly noncoding-controling elements within a viral genome may be delineated.

5.3 RECOMBINANT VIRUSES

Recombinant DNA technology can be used to construct a viable virus from segments of related virus strains that have different properties. By inoculating the hybrid virus into appropriate hosts, it is then possible to associate a particular part of the viral genome with the property being observed. It may even be possible to construct viable hybrids between unrelated viruses in the same virus family or group.

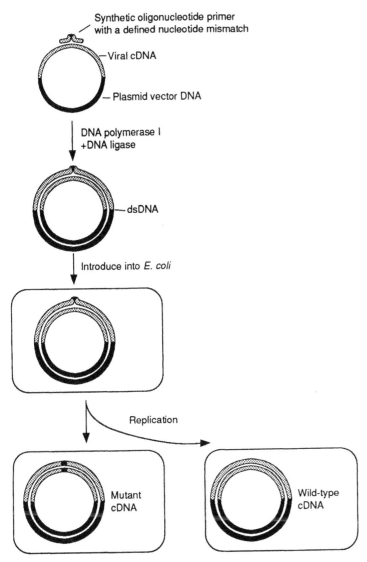

Figure 2.6 Site-directed mutagenesis. A short ($\simeq$ 15-nucleotide) synthetic oligonucleo-tide primer is made with a sequence complementary to a sequence in one viral cDNA strand, except for a base change in the desired position, which should be near the middle of the sequence. This sequence is incubated with viral cDNA under conditions in which annealing can occur in spite of the mismatched base. The oligonucleotide then acts as a primer for DNA polymerase I, which completes the complementary strand. When the re-sulting ds molecule is replicated in *E. coli*, both the wild-type and mutant sequences are replicated. The mutated DNA can be distinguished from wild-type, by hybridization with ^{32}P-labeled synthetic oligonucleotide, under stringent conditions, that is, close to the T_m, under which the probe hybridizes only with the mutant DNA.

6 TRANSGENIC PLANTS

Several methods have been developed in recent years whereby any new gene or piece of DNA can be introduced into the plant cell, and stably integrated into the cellular genome, where the gene can often be expressed, and also inherited in a Mendelian fashion. Plants that have received a new gene in this way are called *transgenic;* however, not all inserted sequences are compatible with a surviving plant.

The ability to transform plants stably with viral genes has opened up several important avenues of research on plant viruses. It provides another means for studying the roles of viral genes *in vivo*. More important, it has opened up new possibilities for the control of some virus diseases (Chapter 15).

6.1 THE *AGROBACTERIUM TUMEFACIENS* VECTOR SYSTEM

By far the most commonly used system for transferring genes into the host-plant genome is the one based on *A. tumefaciens*. This phytophathogenic bacterium contains a large plasmid, the Ti or tumor-inducing plasmid. When *A. tumefaciens* infects a plant, a segment of the Ti plasmid, the T DNA (for *T*ransferred DNA), is integrated into a host cell chromosome. Gene products of the T DNA stimulate the formation of tumors, and of compounds that can be used as a source of carbon and nitrogen by the bacterium. By appropriate manipulations of the plasmid DNA, it has been possible to (1) *disarm* the plasmid so that no tumor cells are produced; (2) create sites in the plasmid where any foreign DNA could be readily inserted; and (3) insert promoter sequences and other controlling elements that allow efficient expression of the introduced genes in plant cells. The resulting constructs are able to be integrated into the host cell DNA, usually with high frequency. The foreign DNA can be expressed in the transformed cell and its progeny.

In order to distinguish transformed cells from untransformed, it is usual to insert DNA for a selectable marker gene into the T DNA as well as the viral DNA of interest. These marker genes are normally antibiotic resistance genes derived from bacteria. A commonly used selectable dominant marker gene contains the coding sequence for neomycin phosphotransferase II. Expression of this gene confers resistance to kanamycin. Many plant species are susceptible to this antibiotic.

The following procedure is commonly used to introduce a disarmed Ti plasmid containing the viral DNA of interest together with a kanamycin resistance gene into plant cells: The *A. tumefaciens* cells are cocultivated with excised leaf disks. Cells near the cut edges are susceptible to infection. The disks are then grown on a selective medium containing kanamycin as well as hormones to induce shoot formation. Resistant shoots are then cultivated on a medium containing kanamycin, which allows root formation. The rooted transgenic plants are transferred to growth chambers and grown to maturity.

6.2 DIRECT GENE TRANSFER TO PLANT CELLS

Direct transfer of DNA into plant protoplasts can be brought about by incubation in a medium containing polyethylene glycol or by short electric pulses (electroporation). Alternatively the DNA is microinjected directly into the nucleus or cytoplasm of a cell. In a third procedure, the DNA is coated onto the surface of $1-2$ μm diameter gold or tungsten micro-projectiles that are then bombarded onto plant cells and some cells are penetrated by the DNA. In all three of these procedures, plants are then regenerated from the transformed cells.

7 THE POLYMERASE CHAIN REACTION

The polymerase chain reaction (PCR) is a major development in DNA technology. A limitation of cloning techniques using restriction endonucleases has been the difficulty in detecting DNA sequences occurring as a very low proportion of the DNA sample, for example, single copy genes, or nucleic acid associated with a low level of virus infection or integration. The PCR overcomes this difficulty by allowing a DNA sequence of interest to be amplified several million times *in vitro*, so that the target DNA can be visualized in a gel after electrophoresis.

To apply the technique, the nucleotide sequence of the DNA of interest, or sequences near it must be known. The PCR is a mechanism for amplifying the DNA between two oligonucleotide primers. Two primers are synthesized to be complementary to known sequences in the target DNA on opposite strands, usually about $150-500$ base pairs apart. Design of the primers is a very important aspect of the process. The dsDNA is first heated to separate the strands in the presence of the primers. When the temperature is reduced, the primers anneal. At a selected temperature, a thermophilic DNA polymerase extends the primers to copy the DNA. Successive cycles of denaturation and synthesis achieve logarithmic amplification of the DNA. In principle, each new cycle doubles the amounts of product DNA; however, the precise conditions in the incubation medium can have a marked effect on yield of product. A major technical problem is the possibility that nontarget DNA may be amplified.

FURTHER READING

Davies, J. W., Wilson, T. M. A., and Covey, S. N. (1991). "The Molecular Biology of Plant Viruses." Chapman and Hall, in press.

Matthews, R. E. F. (1991). "Plant Virology," 3rd ed. Academic Press, New York.

Sambrook, K. J., Fritsch, E. F., and Maniatis, T., (1989). "Molecular Cloning, A Laboratory Manual," 2nd ed. Cold Spring Harbor Laboratory Press, Cold Spring Harbor, New York.

Watson, J. D., Tooze, J., and Kurtz, D. T. (1983). "Recombinant DNA. A Short Course." Scientific American Books, New York.

Watson, J. D., Hopkins, N. H., Roberts, J. W., Steitz, J. A., and Weiner, A. M. (1987). "Molecular Biology of the Gene," 4th ed., Vol. 1. Benjamin/Cummings Publishing Co., Menlo Park, California.

SEROLOGICAL METHODS IN PLANT VIROLOGY

When plant viruses are injected into various species of vertebrates, they elicit the production of specific antibody proteins, which circulate in the blood stream. The reaction of such specific antibodies with intact viruses or dissociated viral components has been very important for various aspects of plant virology, including assay and detection of viruses, diagnosis of virus diseases, studies of virus structure, and the classification of viruses.

1 THE BASIS FOR SEROLOGICAL TESTS

1.1 ANTIGENS

Antigens are usually fairly large molecules or particles consisting of, or containing, proteins or polysaccharides that are foreign to the vertebrate species into which they are introduced. Most have a molecular weight greater than 10,000. There are two aspects to the activity of an antigen. First, the antigen can stimulate the animal to produce antibody proteins that will react specifically with the antigen. This aspect is known as the *immunogenicity* of the antigen. Second, the antigen must be able to combine with the specific antibody produced. This is generally referred to as the *antigenicity* of the molecule.

Large molecules are usually more effective immunogens than small ones; thus, plant viruses, being macromolecules containing protein, are often very effective in stimulating specific antibody production. The subunits of a viral protein coat are much less efficient. About 15 or fewer amino acids at the surface of a protein may be involved in an antigenic site. There are difficulties, however, in defining such sites precisely.

1.2 ANTIBODIES

One response of a vertebrate animal to the presence of a foreign antigen is the production of antibody proteins, an important type being immunoglobulin G (IgG). These circulate in the body and are capable of binding specifically to the antigen. The structure of an IgG molecule is shown schematically in Fig. 3.1. It consists of two light chains and two heavy chains joined together by disulphide bridges. The two identical antigen-binding sites are made from the variable regions of the light and heavy chains.

These variable regions consist of a framework of relatively conserved amino acid sequence interrupted by several regions of highly variable sequence. These hypervariable regions interact to form the two highly specific antigen-binding sites. These variable amino acid sequences arise in the first place through multiple combinational rearrangements of the DNA of the genes coding for the light and heavy

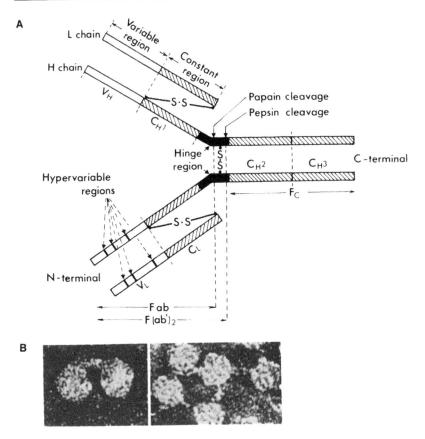

Figure 3.1 (A) IgG antibody molecule. Arrows indicate specific sites where the molecule is cleaved by the enzymes papain and pepsin to give the Fab, Fc, and (Fab')₂ fragments. L, light chains; H, heavy chains. The four polypeptide molecules are joined by disulphide bridges. The two antigen-combining sites are made up from the variable regions VH and VL of the H and L chains. From van Regenmortel (1982). (B) Antibody–virus complexes. Electron micrographs showing individual particles of the satellite virus of tobacco necrosis virus linked together by virus-specific IgG antibody molecules. Courtesy of S. Höglund.

chains within the cells of the B-lymphocyte lineage. These are the cells that secrete antibodies in response to the presence of an antigen. There are two general stages in the differentiation of B-cell lineages. The first stage, which is independent of the presence of antigen, begins in the fetus and is maintained continuously in the bone marrow of the adult. Each new B cell, as it is generated, expresses on its surface the IgG molecule of its particular structure and therefore antigen-binding potential. All its progeny express the same IgG. The B cells then migrate to other organs of the immune system such as spleen and lymph nodes, where they remain as resting cells in the absence of an appropriate antigen. The second stage of B-cell differentiation occurs when circulating antigen combines with the surface-receptor IgG molecule. This stimulates the B cells to proliferate and differentiate into plasma cells, which secrete into the blood stream large amounts of the IgG molecule of exactly the same specificity as that present on the surface of the progenitor B cell. During this proliferation, large numbers of point mutations occur in the variable region DNA. These mutants are under selection pressure from the antigen, and thus the fit between antibody and antigenic site is refined.

An antigenic protein molecule will have several, and perhaps many different structural sites, or *antigenic determinants,* on its surface that can be recognized by the surface-receptor IgG molecule of some B-cell lineage. Thus, for any given antigen, there may be many different B lymphocytes with differing binding sites that will be stimulated by the antigen. The resulting antiserum is *polyclonal,* because it contains many different antibodies combining with the antigen, each arising from a different clone of B lymphocytes.

Another important feature of IgG molecules of many species is the ability of the Fc region (Fig. 3.1) to bind protein A with very high affinity. Protein A is a molecule isolated from the cell wall of *Staphylococcus aureus.* This binding of protein A is used in several serological procedures.

B lymphocytes cannot be cultured *in vitro.* To overcome this problem, Köhler and Milstein (1975) took B lymphocytes from an immunized mouse and fused these *in vitro* with an *immortal* mouse myeloma cell line. Selection of appropriate single fused cells gave *hybridomas* producing only a single kind of antibody—a *monoclonal* antibody (MAb) derived from only a single B lymphocyte. The uses of monoclonal antisera in plant virology are discussed in Section III.

It has been assumed for a long time that plants could not produce antibodies. Although plants do not respond to immunogens in any way that is similar to the response of vertebrates, mouse light chain and heavy chain genes have been introduced into tobacco plants by transformation, and the resulting transgenic tobacco plants have produced the corresponding antibody in good yield.

1.3 PRODUCTION OF POLYCLONAL ANTISERA

Antisera have been produced against plant viruses in a variety of animals. Rabbits have been used most often, since they respond well to plant virus antigens, are easy to handle, and produce useful volumes of serum.

2 METHODS FOR DETECTING ANTIBODY–VIRUS COMBINATION

A wide variety of methods have been developed for demonstrating and estimating combination between antibodies and antigens. Most traditional methods for using antisera with plant viruses involved direct observation of specific precipitates of virus and antibody, either in liquid media or in agar gels. During about the past 10 years, most of these methods have been progressively superseded by the use of enzyme-linked immunoabsorbent assay (ELISA), immunoabsorbent electron microscopy, and *dot blots* employing either polyclonal or monoclonal antibodies. Immunodiffusion reactions in gels are still used for some purposes, especially to discriminate rapidly between different virus strains.

2.1 IMMUNODIFFUSION REACTIONS IN GELS

Plant viruses are *polyvalent,* i.e., each particle can combine with many antibody protein molecules. The actual number that can combine depends on the size of the virus antigen. Polyvalent antigen combines with divalent antibody to form a lattice-structured precipitate. Such precipitates can be observed in liquids or in gels.

The great advantages of gel diffusion tests, carried out on plates are (1) mixtures of antigenic molecules and their corresponding antibodies may be physically separated, either because of differing rates of diffusion in the gel, or because of differing rates of migration in an electric field (in *immunoelectrophoresis*), or by a combination of these factors; and (2) direct comparisons can be made of two antigens by placing them in neighboring wells on the same plate. It is usual to have the antiserum in a central well and the antigen solutions being tested in a series of wells surrounding the central well. Antigen and antibody diffuse toward each other in the agar, and after a time a zone will form where the two reagents are in suitable proportions to form a precipitating complex. Both reactants become insoluble at this point, and more antigen and antibody diffuse in to build up a visible line of precipitation, which traps related antigen and antibody.

2.2 ENZYME-LINKED IMMUNOSORBENT ASSAY (ELISA)

The microplate method of ELISA can be very effectively applied to the detection and assay of plant viruses. The method has come to be more and more widely used. Many variations of the basic procedure have been described, with the objective of optimizing the tests for particular purposes. The method is very economical in the use of reactants, and readily adapted to quantitative measurement. ELISA can be applied to viruses of various morphological types, both in purified preparations and crude extracts. The procedure is particularly convenient when large number of tests are needed, and it is very sensitive, detecting concentrations as low as 1 to 10 ng/ml.

Two general kinds of procedures have been developed: direct and indirect double-antibody sandwich methods. The steps in the direct procedure are summarized in Fig. 3.2. Figure 3.3 gives an example of the kinds of data obtained in an ELISA test. The direct procedure has been widely used, but suffers two limitations: (1) it may be very strain specific. For discrimination between virus strains, this can be a useful feature. For routine diagnostic tests, however, it means that different viral serotypes may escape detection. This high specificity is almost certainly owing to the fact that the coupling of the enzyme to the antibody interferes with weaker combining reactions with strains that are not closely related; (2) it requires a different antivirus enzyme–antibody complex to be prepared for each virus to be tested.

In the indirect procedure, the enzyme used in the final detection and assay step is conjugated to an antiglobulin antibody. For example,

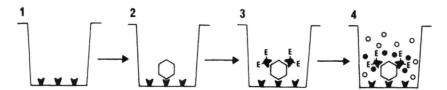

Figure 3.2 Principle of the ELISA technique for plant viruses. (1) The gamma globulin fraction of an antiserum is allowed to coat the surface of wells in a polystyrene microtiter plate. The plates are then washed. (2) The test sample containing virus is added and combination with the fixed antibody is allowed to occur. (3) After washing again, enzyme-labeled specific antibody is allowed to combine with any virus attached to the fixed antibody. (Alkaline phosphatase is linked to the antibody with gluteraldehyde.) (4) The plate is again washed and enzyme substrate, added. The colorless substrate, *p*-nitrophenyl phosphate (○) gives rise to a yellow product (●), which can be observed visually or measured spectrophotometrically at 405 nm. Modified from Clark and Adams (1977), with permission from Cambridge University Press.

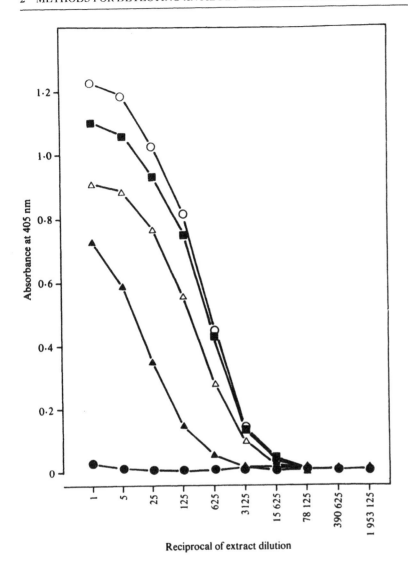

Figure 3.3 Example of data obtained in an ELISA test. Detection of lettuce necrotic yellow rhabdovirus (LNYV) in *Nicotiana glutinosa* plants, systemically infected for various periods. Uninoculated plants ●—●; and plants infected for 7 days (no systemic symptoms) ▲—▲; 15 days (prominent systemic symptoms) ■—■; 19 days (prominent systemic symptoms) ○—○; and 30 days (severe chlorosis and stunting) △—△. From Chu and Francki (1982).

if the virus antibodies were raised in a rabbit, a chicken antirabbit globulin might be used for conjugation to the enzyme; thus, one conjugated globulin preparation can be used to assay bound rabbit antibody for a range of viruses. Furthermore, indirect methods detect a broader range of related viruses with a single antiserum. Many variations of these procedures are possible.

2.3 DOT-BLOT IMMUNOBINDING ASSAYS

An important variation of the ELISA technique is known as dot-blot immunobinding. In these tests a nitrocellulose membrane or even plain paper is used as the solid substrate for ELISA tests. In one procedure, the virus in a plant extract is dot-blotted onto the membrane as the first step. For the final color development, a substrate is added, which the IgG-linked enzyme converts to an insoluble colored material. Intensity of the colored spot can be assessed by eye or by using a reflectance densitometer. An example of a dot-blot assay is shown in Fig. 3.4. The main advantages of the procedures are (1) speed, (2) low cost, and (3) small amounts of reagent that are needed. They are particularly useful for laboratories where an inexpensive and simple test is needed.

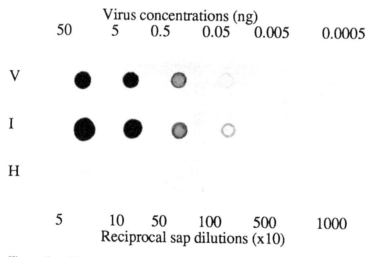

Figure 3.4 The sensitivity of a dot immunobinding assay. V, purified strawberry pseudo mild yellow edge *Carlavirus;* I, crude sap from infected leaves; and H, crude sap from healthy leaves. The conjugated enzyme was alkaline phosphatase, and the substrate for color development was Fast Red TR salt. From Yoshikawa *et al.* (1986).

2.4 ELECTROPHORESIS FOLLOWED BY ELECTROBLOT IMMUNOASSAY

The protein-fractionating power of electrophoresis in gels can be used together with the sensitivity and specificity of solid-phase immunoassay to identify and assay viral proteins. The main stages in the technique, which is often called *Western blotting,* are as follows: (1) fractionation of the proteins in infected plant sap by sodium dodecyl sulfate-polyacrylamide gel electrophoresis (SDS-PAGE); (2) electrophoretic transfer of the protein bands from the gel to activated paper (usually nitrocellulose); (3) blocking of remaining free binding sites on the paper with protein, usually serum albumin; and (5) detection of the antibody–antigen complex using an ELISA procedure or by some other detection system. A great advantage of this technique is that it identifies the virus by two independent properties of its coat protein—molecular weight and serological specificity.

2.5 SEROLOGICALLY SPECIFIC ELECTRON MICROSCOPY (SSEM)

This kind of procedure provides a diagnosis based on two poperties of the virus—serological reactivity with the antiserum used and particle morphology. The support film of an electron microscope (EM) specimen grid is coated with specific antibody for the virus being studied. Grids are then floated on a solution of the virus. Many virus particles that react with the antiserum will be trapped on the grid. The particles can then be observed in the electron microscope, after metal shadowing or negative staining.

In a modification of the general procedure known as the decoration technique, virus particles, after being adsorbed onto the EM grid, are coated with virus-specific antibody. This produces around the virus particles a halo of IgG molecules that can be readily visualized in negatively stained preparations. This procedure probably offers the most convincing demonstration by electron microscopy of specific combinations between virus and IgG. For a critical test, a second serologically unrelated virus of similar morphology should be added to the preparation (Fig. 3.5).

The main advantages of the methods are (1) the result is usually clear, in the form of virus particles of a particular morphology, and thus

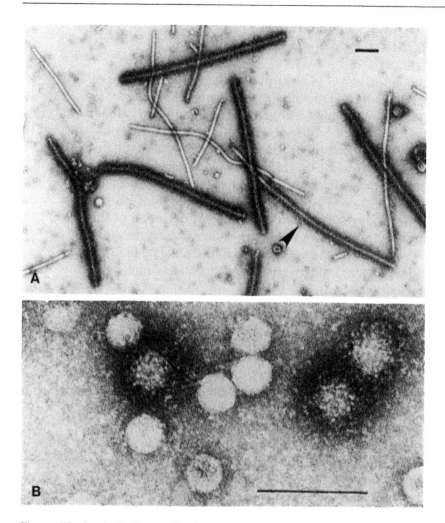

Figure 3.5 Serologically specific electron microscopy (SSEM). (A) A natural mixture of two potyviruses from a perennial cucurbit, *Bryonia cretica*. The antiserum used has decorated particles of only one of the viruses in the mixture. One particle near the center (arrow) is longer than normal and decorated for only part of its length. This particle probably arose by end-to-end aggregation between a particle of each of the two viruses. (B) Mixture of purified tobacco necrosis *Necrovirus* (TNV) and tomato bushy stunt *Tombusvirus* (TBSV) adsorbed to the grid, and then treated with saturating levels of antibodies to TBSV. TBSV particles are *decorated*. TNV particles remain clean, with sharp outlines. Antibody molecules appear in the background. From Milne (1991).

false-positive results are rare; (2) sensitivity may be of the same order as that of ELISA procedures and may be 1000 times more sensitive for the detection of some viruses than conventional electron microscopy; (3) when the support film is coated with antibody, much less host background material is bound to the grid; (4) antisera can be used without fractionation or conjugation, low-titer sera can be satisfactory, and only small volumes are required; (5) very small volumes ($\simeq 1 \mu l$) of virus extract may be sufficient; (6) antibodies against host components are not a problem inasmuch as they do not bind to virus; (7) one antiserum may detect a range of serological variants. On the other hand, the use of monoclonal antibodies may greatly increase the specificity of the test; (8) results may be obtained within 30 min; (9) when decoration is used, unrelated, undecorated virus particles on the grid are readily detected; and (10) prepared grids may be sent to a distant laboratory for application of virus extracts and returned to a base laboratory for further steps and EM examination.

Some disadvantages of the procedure are as follows: (1) it will not detect virus structures too small to be resolved in the EM (e.g., coat protein monomers); (2) sometimes the method works inconsistently or not at all for reasons that are not well understood; (3) it involves the use of expensive EM equipment, requires skilled technical work, and is labor intensive. For this reason it cannot compete with ELISA for large-scale, routine testing; and (4) when quantitative results are required, particle counting is laborious, variability of particle numbers per grid square may be high, and control grids are required.

In summary, SSEM cannot replace ELISA tests where large numbers of samples have to be tested. Its main uses are (1) in the identification of an unknown virus; (2) in situations where only a few diagnostic tests are needed; and (3) when ELISA tests are equivocal and need direct confirmation.

3 MONOCLONAL ANTIBODIES

During the 1980s there was an explosive growth of interest in the use of monoclonal antibodies (MAbs) for many aspects of plant virus research, and particularly for detection and diagnosis.

3.1 THE NATURE OF ANTIBODY SPECIFICITY
   IN RELATION TO MAbs

The binding site of an MAb, or any individual antibody in a polyclonal serum, is able to bind to many different antigenic determinants with varying degrees of affinity. That is, an individual binding site on an antibody is polyfunctional. Furthermore, antibodies against a single antigenic determinant (or epitope) may vary in their affinity from those that are scarcely measurable with standard techniques to those that are binding with an affinity that is 100,000-fold higher. Thus it is quite possible for an antibody to bind more strongly to an antigenic determinant differing from the one that stimulated its production. Such a phenomenon may well be obscured by the many different antibodies in a polyclonal antiserum, but with MAbs it will show up clearly. For example, some MAbs produced following immunization of a mouse with one strain of a virus may react more strongly with strains of the virus other than with the one used for immunization. This has important implications for the use of MAbs in the delineation of virus strains (Chapter 12).

Another aspect of the specificity of MAbs must be borne in mind. A particular protein antigen may have only one site for binding a particular MAb. If this single site is shared with another protein, a significant cross-reaction could occur even in the absence of general structural similarity between the two proteins. This is probably a very uncommon phenomenon. Nevertheless, cross-reactions detected with MAbs cannot be taken to indicate significant structural or functional similarity between two proteins without other supporting evidence.

3.2 THE PRODUCTION OF MAbs

An outline for the production of MAbs using mice is shown in Fig. 3.6.

3.3 TESTS USING MAbs

For detection and diagnosis, MAbs have been most commonly used in conjunction with ELISA tests. Again, exact conditions with respect to pH, etc., may be vitally important. For many applications it may be best to use the same ELISA protocol that was used to screen for MAbs during the isolation procedure, because quite different MAbs may be selected, depending on the ELISA procedure used. The reactivity of an MAb with

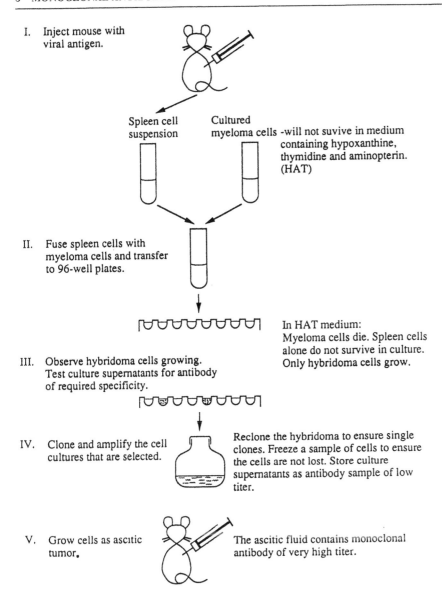

I.   Inject mouse with
     viral antigen.

     Spleen cell   Cultured
     suspension   myeloma cells -will not suvive in medium
                                         containing hypoxanthine,
                                         thymidine and aminopterin.
                                         (HAT)

II. Fuse spleen cells with
     myeloma cells and transfer
     to 96-well plates.

                                         In HAT medium:
                                         Myeloma cells die. Spleen cells
                                         alone do not survive in culture.
III. Observe hybridoma cells growing.   Only hybridoma cells grow.
     Test culture supernatants for antibody
     of required specificity.

IV. Clone and amplify the cell   Reclone the hybridoma to ensure single
     cultures that are selected.   clones. Freeze a sample of cells to ensure
                                         the cells are not lost. Store culture
                                         supernatants as antibody sample of low
                                         titer.

V.   Grow cells as ascitic   The ascitic fluid contains monoclonal
     tumor.   antibody of very high titer.

Figure 3.6 Steps in the production of monoclonal antibodies.

Note: The ability of fused cells to survive in hypoxanthine, aminopterin, thymidine (HAT) medium derives from the spleen cells, whereas the ability to multiply in culture derives from the myeloma cells. The kind of screening test used in step III to detect antibody is of critical importance. Most workers use some form of ELISA.

a given antigen may differ greatly depending on the kind of ELISA procedure used.

3.4 ADVANTAGES OF MAbs

1. *Requirements for immunization* Mice and rats can be immunized with small amounts of antigen ($\simeq$ 100 μg or less). If the virus preparation used is contaminated with host material or other viruses, it is still possible to select for MAbs that react only with the virus of interest.

2. *Standardization* MAbs provide a uniform reagent that can be distributed to different laboratories, eliminating much of the confusion that has arisen in the past from the use of variable polyclonal antisera. Furthermore, MAbs can be obtained in almost unlimited quantities in suitable circumstances.

3. *Specificity* MAbs combine with only one antigenic site on the antigen. Thus, they may have very high specificity, and can therefore provide a refined tool for distinguishing between virus strains (Chapter 12). MAbs can also be used to investigate aspects of virus architecture (Chapter 5) and transmission by vectors (Chapter 13).

4. *High affinity* The screening procedure for detecting MAbs permits selection of antibodies with very high affinity for the antigen. High-affinity antibodies may be used at very high dilutions, minimizing problems of background in the assays. They can also be used for virus purification by affinity chromatography.

5. *Storage of cells* Hybridomas can be stored in liquid nitrogen to provide a source of MAb-producing cells over a long period.

3.5 DISADVANTAGES OF MAbs

1. *Preparation* Polyclonal antisera are relatively easy to prepare. The isolation of new MAbs is labor intensive, time consuming, and relatively expensive. For any particular project, these realities must be weighed against the substantial advantages previously discussed.

2. *Specificity* MAbs may be too specific for some applications, especially in diagnosis; however, they may be mixed to give broader reactivity.

3. *Sensitivity to conformational changes* Because of their high specificity, MAbs may be very sensitive to conformational changes in

the antigen, brought about by binding to the solid phase or by other conditions in the assay. Thus, while MAbs provide an excellent tool for many aspects of plant-virus research, it is unlikely that they will replace the use of polyclonal antisera in all applications.

4 SEROLOGICAL METHODS IN THE STUDY OF VIRUS STRUCTURE

The reaction of specific antibodies with intact viruses or dissociated viral coat proteins has been used to obtain information relevant to virus structure. Monoclonal antibodies are proving to be particularly useful for this kind of investigation, although there are significant limitations.

Some workers distinguish two types of antigenic determinant in proteins. A *continuous determinant* is a continuous sequence of amino acids that is exposed at the surface of a protein, and that possesses a distinctive conformation. A *discontinuous determinant* is made up of amino residues that are close together on the protein surface, but are located at distant positions in the primary structure.

Three methods have been used to localize antigenic determinants: (1) peptides obtained from the protein by chemical or enzymatic cleavage are screened for their reactivity with antibodies; (2) short synthetic peptides representing known amino acid sequences in the protein can be similarly screened; and (3) immunological cross-reactivity is assayed between closely related proteins having one or a few amino acid substitutions at known sites.

There are substantial limitations for all these procedures. Peptides derived by cleavage or synthesis may not maintain the conformation in solution that the sequence possessed in the intact molecule. Furthermore, such peptides will only rarely represent all the amino acids in the original antigenic determinant. For these reasons, reactivity of peptides with antibodies to the intact protein is usually very low.

Structural interpretation of the results of cross-reactions between closely related proteins can be confused by the fact that the conformation of any antigenic determinant may be changed by an amino acid substitution occurring elsewhere in the protein. Furthermore, the ability of different MAbs to detect residue exchanges may be extremely variable. Nevertheless, serological techniques have produced some structural information, especially with viruses, for which the detailed protein structure has not been determined by X-ray crystallography. The location of the N-terminal and C-terminal regions of the coat proteins of several viruses at the surface of the virus was indicated by serological studies. Nevertheless, caution must be used in interpreting the results of ELISA tests in structural terms when the whole virus is the antigen.

FURTHER READING

Harlow, E., and Lane, D. (1988). "Antibodies: A Laboratory Manual." Cold Spring Harbor Laboratory Press, Cold Spring Harbor, New York.

Matthews, R. E. F. (1991). "Plant Virology," 3rd Ed. Academic Press, New York.

Milne, R. G. (1991). Immunoelectron-microscopy for virus identification. *In* "Electron Microscopy of Plant Pathogens" (K. Mendgen and D. E. Leseman, eds.), pp. 87–102. Springer-Verlag, New York.

Milstein, C. (1990). Antibodies; a paradigm for the biology of molecular recognition. *Proc. R. Soc. Lond. B* **239**, 1–16.

van Regenmortel, M. H. V. (1982). "Serology and Immunochemistry of Plant Viruses." Academic Press, New York.

van Regenmortel, M. H. V. (1986). The potential for using monoclonal antibodies in the detection of plant viruses. *In* "Developments and Applications of Virus Testing" (R. A. C. Jones and L. Torrance, eds.), pp. 89–101. Developments in Applied Biology I. Association of Applied Biologists, Wellsbourne, England.

ASSAY AND PURIFICATION OF VIRUS PARTICLES

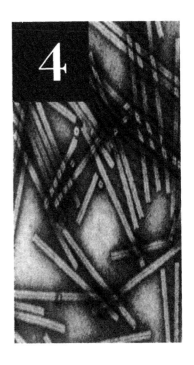

4

The ability to assay viruses is essential for many kinds of virological study. In particular, it would be very difficult or impossible to isolate a new virus of unknown properties unless we had some means of assaying for the virus. Assay methods are of four kinds depending on the properties of the virus being used.

1 ASSAY

1.1 METHODS USING INFECTIVITY

For many purposes, the infectivity of a virus preparation for a host plant is the most important property to use as an assay. On the other hand, infectivity assays, even under the best of conditions, are usually less precise than methods based on other properties of the virus, and they are more time consuming. Most infectivity assays involve mechanical inoculation to leaves, a process described in Chapter 10. A few involve the use of insect vectors.

1.1.1 Local-Lesion Assays

When leaves are rubbed with an infectious preparation of the virus, a process known as mechanical inoculation, some host plants will, in a few days' time, develop recognizable local lesions around points where successful virus infection has occurred. Wherever possible, such local-lesion hosts are used to assay the infectivity in samples containing a plant virus. Local lesions suitable for such assays are illustrated in Chapter 11 (Fig. 11.1). In setting up local-lesion assays, several important features must be considered.

Variation between Leaves

Many environmental and intrinsic factors affect the number of local lesions that will be produced on a set of leaves. There is much less variation between opposite halves of the same leaf than between different leaves. For this reason, two samples to be compared are inoculated to opposite half-leaves. To obtain reasonable accuracy in the assay, batches of leaves are inoculated with pairs of samples. For most experiments, 5–20 half-leaf comparisons would be sufficient. When several samples have to be compared, a standard virus dilution can be inoculated to half of every leaf, and different test samples, to sets of the opposite half-leaves.

Relation between Dilution of Inoculum and Lesion Number

When serial dilutions are made of a single concentrated preparation of virus, and inoculated to sets of leaves, the numbers of local lesions produced will fall with increasing dilution. A curve relating these two factors can be drawn (Fig. 4.1). The relationship, however, is not a simple one. For most host–virus combinations, the curve can be divided into three parts if a sufficient range of dilutions has been assayed: (1) a section at high concentrations where a change in concentration is accompanied by very little change in local-lesion number; (2) a section in the middle of the curve where a change in concentration is accompanied by a more or less proportional change in local-lesion number; and (3) a section at low virus concentrations where change in concentration has little effect on number of local lesions. Thus valid comparisons of infectivity can be made only in the region where lesion number is responding more or less proportionally to dilution.

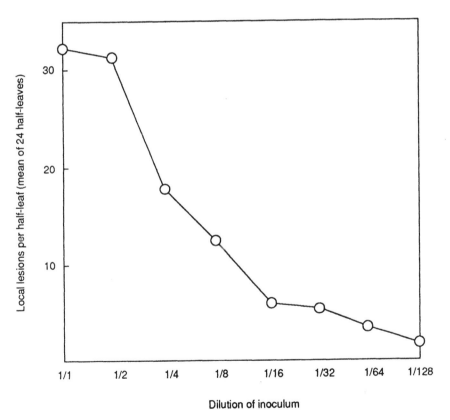

Figure 4.1 Effect of dilution of inoculum on number of local lesions produced by TBSV in *N. glutinosa*. Data from Kleczkowski (1950).

Variability in the Slope of the Dilution Curve

The slope of the dilution curve in its middle region is variable, being affected by many factors. Furthermore, the exact slope is unpredictable from experiment to experiment. This means that two samples should be compared at several dilutions (usually two-, five-, or tenfold).

Statistical Analysis of Results

For certain experiments, it may be desirable to apply a statistical analysis to comparisons of local-lesion numbers. However, neither the lesion numbers nor the logarithms of these numbers are normally distributed. The following transformation is satisfactory as a preliminary step in the analysis when mean local-lesion number is greater than about 10: $Y = \log_{10}(X + C)$, where X = number of local lesions, and C = a constant. Any value of C between about 5 and 20 is usually satisfactory.

Practical Consequences

The simplest way to minimize the problems noted above is as follows. The useful range where lesion number is most responsive to dilution should be determined in a preliminary experiment. The useful range often lies between 10 and 100 lesions per half-leaf for leaves with a size like those of *Nicotiana glutinosa*. The dilutions to be assayed are then arranged so that samples to be compared give nearly equal numbers of local lesions somewhere in the useful range.

It is important that the size and complexity of an assay should be appropriate to the needs of the experiment. It is a waste of labor to set up a very elaborate assay design when a very approximate estimate of relative infectivities will give the required answer. A more common failing is to draw conclusions from inadequately designed and analyzed experiments.

1.1.2 Assay in Insect-Vector Cell Monolayers

A few plant viruses that replicate in their insect vectors (Chapter 13) can be assayed by their infectivity on monolayers of insect-vector cells. The monolayers are usually grown on microscope coverslips, and the foci of infection located by light microscopy, using a fluorescent antibody technique. A focus is the equivalent of a local lesion on a plant leaf. The technique is sensitive, and the relationship between numbers of foci and dilution is linear over a wide range.

1.1.3 Quantal Assays Based on Number of Individuals Infected

When a virus has no known local lesion host or when it can be transmitted only by insect vectors, quantal assays may have to be used. This involves inoculating groups of plants with various dilutions of the virus preparation and observing the proportion of plants in each group that become systemically infected. With insect transmission, insects might feed on solutions of the virus through a membrane or be injected with such solutions. These kinds of procedures take longer, are usually less precise, and use many more plants than local-lesion assays.

1.2 METHODS DEPENDING ON PHYSICAL PROPERTIES OF THE VIRUS PARTICLE

Because of their nucleic acid content, viruses absorb quite strongly in ultraviolet (UV) light near 260 nm. UV absorption provides a useful assay for purified or semipurified virus preparations, provided that other criteria have eliminated the possibility of significant contamination with nonviral nucleic acids or proteins.

Density-gradient centrifugation can provide a sensitive assay for many viruses, even in crude plant tissue extracts. A centrifuge tube is partially filled with a solution having decreasing density from bottom to top of the tube. Sucrose solutions are often used to form the gradient. The liquid containing the virus is layered on top of the gradient. Following centrifugation under appropriate conditions, the contents are removed from the tube, usually by upward displacement with a more dense solution. The absorbance at 260 nm of the liquid column leaving the tube can be measured and recorded (Fig. 4.2).

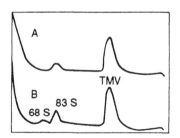

Figure 4.2 Density gradient centrifugation for the assay of viruses in crude extracts. An extract of 30 mg of epidermis (A) or the underlying tissue (B) from a tobacco leaf infected with TMV was sedimented in a 10–40% sucrose gradient at 35,000 rpm for 2 hr. Absorbancy at 254 nm through the gradient was measured with an automatic scanning device. Note absence of detectable 68 S ribosomes in the epidermal extracts. P. H. Atkinson and R. E. F. Matthews, unpublished.

1.3 METHODS DEPENDING ON PROPERTIES
OF VIRAL PROTEINS

1.3.1 Serological Procedures

Serological tests provide a rapid and convenient method for the estimation of plant viruses. The main advantages are (1) the specificity of the
reaction allows virus to be measured in the presence of host materials
or other impurities, although these may sometimes interfere with the
assay; (2) results can be obtained in a few minutes or hours compared
with days for infectivity assays; (3) the methods give an answer that is
directly proportional to virus concentration over a wide range of concentrations; (4) serological tests are particularly useful with viruses
that have no good local-lesion hosts or that are not sap transmissible;
and (5) antisera can be stored and comparable tests made over a period
of years.

Most serological procedures summarized in Chapter 3 can be
adapted for the assay of plant viruses. ELISA tests providing the kind of
data illustrated in Fig. 3.3 are currently one of the most popular forms
of serological assay. The dot immunobinding assay (Fig. 3.4) provides a
rapid, sensitive, and convenient assay that is less precise than ELISA.

1.3.2 Electrophoretic Procedures

During electrophoresis in a suitable substrate, such as a polyacrylamide
gel, proteins migrate according to their size and net charge at the pH
used. The position and relative amounts of protein can then be visualized and assayed approximately, either by a nonspecific procedure
such as staining with Coomassie brilliant blue, or with a specific procedure such as electroblot immunoassay as described in Chapter 3, Section 2.4.

1.4 METHODS DEPENDING ON PROPERTIES
OF THE VIRAL NUCLEIC ACID

Hybridization of a labeled nucleic acid probe (prepared as outlined in
Chapter 2, Section 2.2.6) to the virus nucleic acid provides a specific
and convenient method of assay. It can be used on crude extracts from

insects or plants. The main steps are as follows: (1) small amounts of extract are prepared; (2) the viral nucleic acid is denatured by heating if it is double stranded (ds); (3) a series of dilutions of the extracts to be compared may be made; (4) spots of the extracts and dilutions are applied to a nitrocellulose sheet or a nylon-based membrane; (5) the nitrocellulose is baked to bind the nucleic acid firmly to it; (6) non-specific binding sites on the membrane are blocked by incubation in a solution containing a protein, usually serum albumin together with small fragments of an unrelated RNA or DNA; (7) hybridization of a labeled probe nucleic acid to the test nucleic acid bound to the substrate; (8) washing off unhybridized probe; and (9) estimation of the amount of probe bound by a method appropriate to the kind of label in the probe. Electrophoretic procedures can be used to separate RNAs and DNAs on the basis of their molecular weights (MWs). Staining then allows the separated components to be identified.

1.5 SUMMARY

The choice of assay method will depend on the kind of experiment being carried out, the kind of property to be measured, and the sensitivity and accuracy required. Infectivity using local-lesion assays and serological tests, particularly some form of enzyme-linked immunosorbent assay (ELISA), will be appropriate in many circumstances. The facts that different kinds of assay may vary widely in sensitivity, and that different detailed protocols for the same kind of test may markedly affect sensitivity must always be taken into account when interpreting results.

2 PURIFICATION

To study the structure and other basic properties of a virus, it is essential to be able to obtain purified preparations that retain infectivity. Different plant viruses vary over a 10,000-fold range in the amount of virus that can be extracted from infected tissue (from about 0.4 to 4000 μg/gm fresh wt). They also vary widely in their stability with various physical and chemical agents and enzymes that may be encountered during isolation and storage. For these reasons there are no generally applicable rules for devising an isolation procedure. A procedure that works well for one virus may not be effective for an apparently similar virus. In this section we shall survey briefly the kinds of factors influencing virus isolation, and the main procedures that are used in various combinations. Wherever possible, a local-lesion assay host should be available to test the suitability of the infected tissue to be used as starting material, and then to monitor progress of the isolating procedure.

2.1 STARTING MATERIAL

The choice of host plant for propagating a virus is often of crucial importance for its successful isolation.

2.1.1 Interfering Substances

A very important property of the host species is that it should not contain high concentrations of substances that inactivate or irreversibly precipitate the virus. These substances include phenolic materials (as in many members of the Rosaceae), organic acids, mucilages and gums, and enzymes such as nucleases. A relatively small group of host plants have been used successfully to isolate a wide range of viruses. These include members of the Cucurbitaceae, the *Nicotiana* genus, *Petunia hybrida* (Vilm), cowpea (*Vigna sinensis* (Endl), *Chenopodium amaranticolor* (Coste and Reyn) and *C. quinoa* (Willd).

2.1.2 Starting Concentration

The other important factor is to maximize the concentration of virus in the starting material by (1) choice of host species; (2) time of harvesting after inoculation; (3) conditions of growth; and (4) choice of tissue. This is usually leaf, but it may be useful to discard large midribs, which often contain very low virus concentrations. With a few viruses that occur in special locations such as in tumors, dissection of starting material may be essential.

2.2 EXTRACTION MEDIUM

It is often necessary to design an extraction medium that will (1) preserve the virus particles in an infectious, intact, and unaggregated state during the various stages of isolation and storage; (2) disrupt or adsorb host organelles and proteins; and (3) release virus from particulate matter. Important factors in such a medium may be

1. *pH and choice of buffer system* Many viruses are stable only over a fairly narrow pH range;

2. *Divalent metal ions* Some viruses require the presence of divalent ions such as Ca^{2+}, whereas others are aggregated by them;

3. *Reducing agents,* such as sodium sulphite may protect some viruses against oxidation;

4. *Protection against phenolic compounds* has been achieved by the addition of a variety of substances such as various soluble proteins to the extraction medium;

5. *Removal of plant proteins and ribosomes* For some stable viruses, removal of host components can be achieved by heating the initial extract to 65°C or by extracting tissue in a low-pH buffer such as 0.1M sodium acetate, pH 5.0. For others, emulsification with an organic solvent such as chloroform can be very effective. Ethylenediaminetetracetic acid (EDTA) can dissociate ribosomes that otherwise might copurify with the virus;

6. *Release of virus from host components* Enzymes such as pectinase and cellulase may release virus bound in fibrous material. Detergents such as Triton X100 may also assist in the release of virus particles from insoluble cell components.

2.3 EXTRACTION PROCEDURE

The method used to crush the starting tissue and to ensure rapid contact of cell contents with the extraction medium will usually depend on the scale of the preparation: (1) for small-scale preparations, a pestle and mortar with some added sand may be useful; (2) on an intermediate scale, various types of domestic food blenders or extractors may be suitable; (3) where many kilograms of tissue are involved, commercial meat mincers or colloid mills may be needed. The liquid from the crushed tissue is usually expressed through muslin to remove most of the fiber.

2.4 ISOLATION OF THE VIRUS

2.4.1 Clarification of the Extract

The crude extract is usually subjected to low-speed centrifugation to remove host components that may have been coagulated or precipitated by one of the procedures noted previously.

2.4.2 Subsequent Purification

The steps in isolation of the virus from the clarified extract will be designed to concentrate the virus and to remove low-molecular-weight and any larger host components. The choice of procedures will depend primarily on (1) the inherent stability of the virus; (2) its initial concentration; and (3) the degree of *purity* required. The following are the most commonly employed procedures, which may be used in various combinations.

High-Speed Sedimentation

Provided the virus is not denatured or irreversibly aggregated by the process, sedimentation in a high-speed centrifuge is a very useful step. It both concentrates the virus and removes the host materials that are left behind in the supernatant fluid.

Precipitation with Polyethylene Glycol (PEG)

Viruses can be preferentially precipitated in a single-phase PEG system. Precipitation with PEG, usually at a final concentration of 2 to 8%, has become one of the commonest procedures used in the isolation of plant viruses. The method is widely applicable, even to rather fragile viruses. Its application to the isolation of any particular virus is empirical. Exact conditions for precipitation will depend on pH, ionic strength, and concentration of macromolecules.

Density-Gradient Centrifugation

This process features in the isolation procedure for many viruses. Gradients are usually made with sucrose solutions, with a wide choice of tube sizes available. The process can be used both to concentrate the virus and to remove host constituents without the pelleting and potential damage involved in high-speed sedimentation. A gradient made with cesium chloride may be useful as a late step in purification.

Salt Precipitation or Crystallization

This is a valuable step for viruses that are stable in strong salt solutions. Ammonium sulphate at a concentration of one third of saturation is most commonly used (Fig. 4.3), although many other salts will precipitate viruses or allow them to form crystals. After standing for some hours or days, the crystals are sedimented at low speed, redissolved in a small volume, and dialyzed to remove remaining salt.

2.5 STORAGE OF PURIFIED VIRUSES

Unless appropriate conditions are provided, many viruses are inactivated and denature quite rapidly on storage at 4°C. Many viruses can be stored effectively in a solution of 50% glycerol as a liquid at −20°C. Alternatively, with appropriate additions of protective substances such as ethylene glycol, a protein, sugar, or polysaccharide, it may be possible to store a virus for quite long periods in a frozen solution or as a freeze-dried powder.

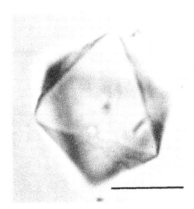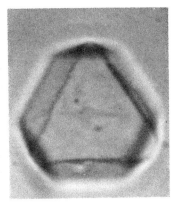

Figure 4.3 Crystals of TYMV formed in ammonium sulfate solutions. Bar = 25 μm. Courtesy of T. Hatta.

2.6 IDENTIFICATION OF THE INFECTIOUS VIRUS PARTICLE

With a newly discovered virus, the main objective of purification is to identify the virus particle causing the disease, and to characterize it at least in a preliminary way. The most suitable way positively to identify the infectious particle (or combination of particles in a multiparticle virus) is to subject a sample of the preparation to sucrose density-gradient fractionation. Fractions from the gradient can be assayed for infectivity singly, and in various combinations if a multiparticle virus is involved. UV-absorption spectra can be obtained on the fractions. This procedure is illustrated in Fig. 4.4. Samples can also be examined for characteristic particles by electron microscopy.

For the reasons noted earlier, viruses are frequently propagated for isolation purposes in hosts other than the one in which they originally occurred. The virus finally isolated should be checked for the following: (1) that it reproduces the disease in the original host, since a mixture of virus strains may have been present, and an atypical strain isolated, or the disease may have been caused by two viruses, only one of

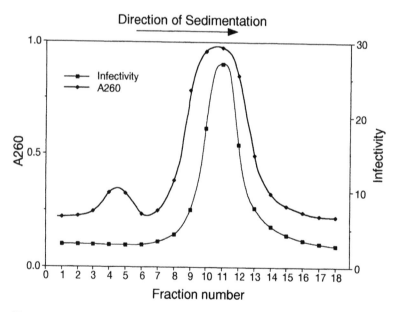

Figure 4.4 Use of sucrose density-gradient fractionation to identify the infectious virus particle in a partially purified virus preparation. Infectivity, number of local lesions/half-leaf. This preparation contained a slowly sedimenting, UV-absorbing, noninfectious component.

which was isolated; (2) that the virus of interest has not become contaminated with an extraneous virus, since contamination of greenhouse-grown plants with unwanted viruses is not uncommon. They are usually rather stable ones such as tobacco mosaic virus (TMV), potato virus X (PVX), and tobacco necrosis virus (TNV).

2.7 THE PURITY OF ISOLATED VIRUS PREPARATIONS

In a chemical sense, there is no such thing as a *pure* virus preparation, even though the virus may be present in a beautifully crystalline form. There are many reasons for this, for example: (1) many virus preparations contain particles that have an incomplete complement of nucleic acid; (2) almost all virus isolates consist of mixtures of closely related strains; and (3) the inorganic and small organic cations found associated with the virus will vary with the medium and method of extraction.

Thus purity and homogeneity are operational terms defined by the virus and the methods used. A virus preparation is pure for a particular purpose if the impurities or variations in the particles present do not affect the particular properties being studied, or can be taken account of in the experiment.

FURTHER READING

Matthews, R. E. F. (1991). "Plant Virology," 3rd Ed. Academic Press, New York.

5

VIRUS STRUCTURE

A knowledge of the detailed structure of viruses is essential for understanding many other aspects of plant virology, for example, how viruses infect and begin replication in cells, how they survive outside cells, and how different viruses are related to one another. The methods used to elucidate virus structures were outlined in Chapter 2.

    The term *capsid* has been used for the closed shell or tube of protein in a virus, and the term *capsomere*, for clusters of subunits as seen in electron micrographs. The mature virus has been termed the *virion*. In viruses with an outer membrane, the inner nucleoprotein core has been called the *nucleocapsid*. These names seem to be unnecessary and at times confusing. For example, what is the capsid or the virion in a virus whose genome is divided between several particles? Which is the capsomere in a virus in which different parts of the same protein subunit are clustered in different ways? The term *encapsidation*, however, is now widely used to refer to the process whereby a viral genome becomes encased in a coat of protein. This term serves a useful purpose. I shall use the term *protein subunit* or structural subunit to refer to the covalently linked polypeptide chain in a viral protein shell. The term *morphological subunit* will refer to the clusters of subunits revealed by electron microscopy or X-ray crystallography.

1 PHYSICAL PRINCIPLES IN THE ARCHITECTURE OF SMALL VIRUSES

1.1 REGULAR ARRAYS OF PROTEIN SUBUNITS IN ROD-SHAPED AND ISOMETRIC VIRUSES

Crick and Watson (1956) put forward a hypothesis concerning the structure of small viruses that has since been generally confirmed. Using the knowledge then available for turnip yellow mosaic virus (TYMV) and TMV, namely, that the viral RNA is enclosed in a coat of protein and (for TMV only, at that stage) that the naked RNA is infectious, they assumed that the basic structural requirement for a small virus is the provision of a shell of protein to protect its ribonucleic acid. They considered that the relatively large protein coat might be made most efficiently if the virus directed synthesis of a large number of identical small protein molecules, rather than one or a few very large ones. Crick and Watson pointed out that, if the same bonding arrangement is to be used repeatedly in the particle, the small protein molecules would then aggregate around the RNA in a regular manner. On this basis, there are only a limited number of ways in which the subunits can be arranged. All the viruses with coat protein subunits packed in regular arrays are either rods or isometric particles.

There is no theoretical restriction on the number of protein subunits that can pack into a helical array in rod-shaped viruses. From crystallographic considerations, Crick and Watson concluded that the protein shell of a small *isometric* virus could be constructed from identical protein subunits arranged with cubic symmetry, for which case, the number of subunits would be a multiple of 12. In such isometric viruses, the protein shell provides an internal space that contains the viral nucleic acid. In viruses with helical symmetry (rods), the nucleic acid is embedded in the helical array of subunits. Figure 5.1 illustrates the fact that most plant viruses have rod-shaped or small isometric particles. A few are larger, with lipoprotein envelopes.

1.2 HELICAL SYMMETRIES IN ROD-SHAPED VIRUSES

There are two kinds of rod-shaped plant viruses, those with rigid rods, such as those of the *Tobamovirus* group and those with flexuous rods, such as members of the *Closterovirus* and *Potexvirus* groups. Figure 5.2 illustrates an example of each of these types. Figure 5.3 illustrates, in outline, the structure of TMV, which is the most studied rod-shaped virus. The structure of TMV is discussed in more detail later in Section 2.1.1. All the rod-shaped viruses have a helical structure of the same general sort as that of TMV; that is, a helical array of protein subunits in which the single-strand RNA (ssRNA) is embedded, in the helix formed by the protein subunits. Thus the RNA is protected from the environment by the protein helix both on the outside of the particle and on the inner surface of the axial hole. The diameter of the rod and the parameters of the helix vary with different viruses and depend on the properties of the coat protein subunit.

It is not known for certain, on a molecular basis, why some helical rods are rigid and others are flexuous, or why some flexuous rods are more flexible than others. However, flexibility probably depends on the extent of bonding interactions between protein subunits in successive layers of the helix (that is, in an axial direction). In rods such as TMV, there are strong interactions for some distance in a radial direction (Fig. 5.8); these impart stiffness to the rod. In flexuous rods, there is probably a gradient in the strength of subunit interactions, with the highest density of interactions occurring at a low radius. At higher radius there is probably an increasingly *spongy* quaternary structure with sparse axial subunit contacts and water-filled cavities, allowing for overall flexibility.

Most of the polypeptide chain making up the TMV protein subunit is in the form of four α-helices, which lie more or less radially as the subunits are stacked in the rod. It is probable that the polypeptides in other

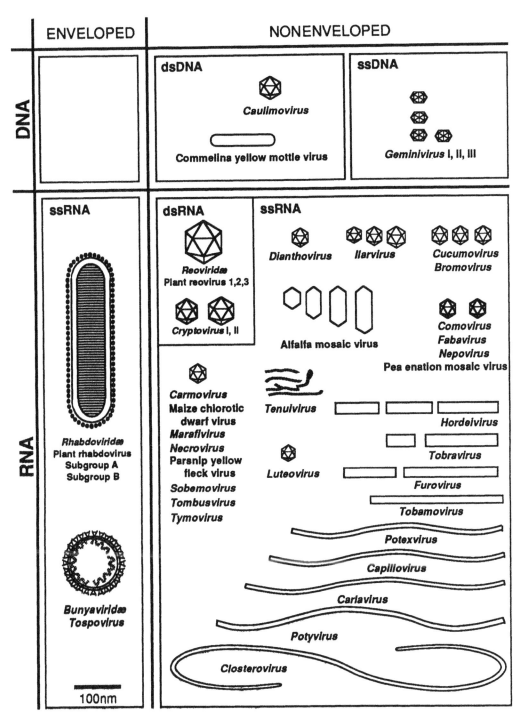

Figure 5.1 The families and groups of viruses infecting plants. Outline diagrams drawn approximately to scale. From Francki *et al.*, 1991.

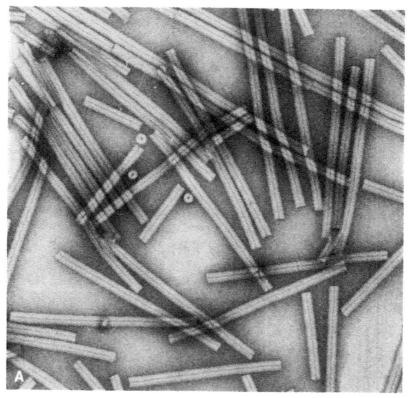

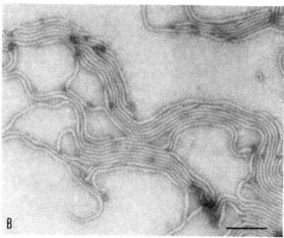

Figure 5.2 Electron micrographs of negatively stained rod-shaped plant viruses. (A) A rigid rod, tobacco mosaic *Tobamovirus*. The axial hole filled with stain can be seen clearly. A few protein disks showing the central hole are also present. Bar = 0.1 μm. (B) A flexuous rod. Lilac chlorotoc leafspot *Closterovirus*. Bar = 0.1 μm. Photos courtesy of R. W. Horne.

Figure 5.3 Helical arrangement of the subunits in relation to the RNA in a TMV rod. Each protein subunit is shaped roughly like a shoe. In successive turns of the helix, the subunits form a groove in which the RNA is situated. The helix has a pitch of 2.3 nm (pitch, the axial distance traveled for 1 turn of the helix). Bar = 10 nm. The repeat distance in the helix is three turns. There are three nucleotides per protein subunit or 49 per turn of the protein helix. Thus, there are $16\frac{1}{3}$ protein subunits per turn of the helix. There is an axial hole down the center of the rod. The nucleotide chain shown as a string of beads at the top of the diagram is to illustrate the conformation of the RNA inside the rod. Free RNA could not maintain such a defined configuration. From Klug and Caspar (1960), based mainly on the work of R. E. Franklin.

rod-shaped viruses have a similar overall conformation. In the rods that have been studied sufficiently, both the N- and C-termini of the polypeptide chain are exposed at the surface of the virus particle.

1.3 POSSIBLE ICOSAHEDRA

Crick and Watson (1956) postulated that cubic symmetry is the most likely form to lead to an isometric virus particle. There are three types of cubic symmetry: tetrahedral (2:3); octahedral (4:3:2); and icosahedral (5:3:2). Thus, an icosahedron has fivefold, threefold, and twofold rotational symmetry. For a virus particle, these three types of cubic symmetry would imply 12, 24, or 60 identical subunits arranged on the surface of a sphere. These subunits could be of any shape.

Figure 5.4 shows a regular icosahedron (20 faces). With three units in identical positions on each face, this icosahedron gives 60 identical subunits. This is the largest number of subunits that can be located in identical positions in an isometric shell. Some viruses have this structure, but many have much larger numbers of subunits, so that all subunits cannot be in identical environments.

Caspar and Klug (1962) enumerated all the possible icosahedral surface lattices, and the number of structural subunits involved. The basic icosahedron (Fig. 5.4) with $20 \times 3 = 60$ structural subunits, can be subtriangulated according to the formula:

$$T = P.f^2$$

where T is called the *triangulation number*.

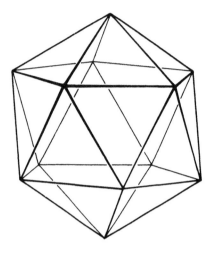

Figure 5.4 The regular icosahedron. This solid has 12 vertices with fivefold rotational symmetry; the center of each triangular face is on a threefold symmetry axis; and the midpoint of each edge is on a twofold symmetry axis. There are 20 identical triangular faces. Three structural units of any shape can be placed in identical positions on each face, giving 60 structural units. Some of the smallest viruses have 60 subunits arranged in this way.

1.3.1 The Meaning of Parameter f

The basic triangular face can be divided by lines joining equally spaced divisions on each side:

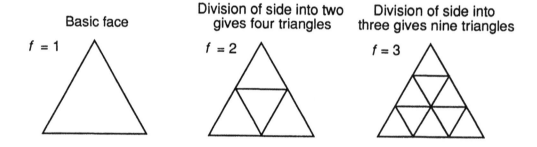

Thus f is the number of subdivisions of each side, and f^2 is the number of smaller triangles formed.

1.3.2 The Meaning of Parameter P

There is another way in which subtriangulation can be made, and this is represented by P. It is easier to consider a plane network of equilateral triangles:

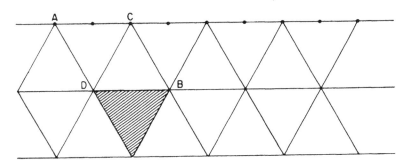

Such a sheet can be folded down to give the basic icosahedron by cutting out one triangle from a hexagon (e.g., cross-hatching) and then joining the cut edges to give a vertex with fivefold symmetry.

However, if each vertex is joined to another by a line not passing through the nearest vertex, other triangulations of the surface are obtained. In the simplest case, the *next but one* vertices are joined:

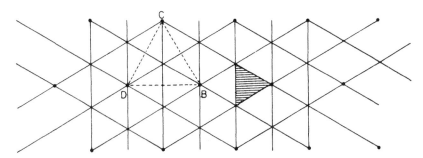

This gives a new array of equilateral triangles. This plane net can be folded to give the solid in Fig. 5.5C by removing one triangle from each of the original vertices (e.g., the shaded triangle), and then folding in to give a vertex with fivefold symmetry. It can be shown by simple trigonometry that each of the small triangles has one third the area of the original faces. This can be seen by inspection, by noting that there are six new half-triangles within one original face (dotted lines C.B.D.). In this example $P = 3$. In general

$$P = h^2 + hk + k^2$$

where h and k are any integers having no common factor.

For $h = 1$, $k = 0$, $P = 1$
For $h = 1$, $k = 1$, $P = 3$
For $h = 2$, $k = 1$, $P = 7$

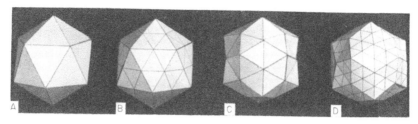

Figure 5.5 Ways of subtriangulation of the basic icosahedron shown in Fig. 5.2, to give a series of deltahedra with icosahedral symmetry (icosadeltahedra): (A) The basic icosahedron, with $T = 1$ (P = 1, f = 1). (B) With $T = 4$ (P = 1, f = 2). (C) With $T = 3$ (P = 3, f = 1). (D) With $T = 12$ (P = 3, f = 2). From Caspar and Klug (1962).

Since each of the triangles formed with the P parameter can be further subdivided in f^2 smaller triangles, T gives the total number of subdivisions of the original faces, and $20T$ the total number of triangles. Figure 5.5 gives some examples. Thus the number of structural subunits in an icosahedral shell = $20 \times 3 \times T = 60T$. Where $P \geq 7$, the icosahedra are skew, and right-handed and left-handed versions are possible. Most of the groups of plant viruses with small isometric particles have their subunits arranged in an icosahedral shell with $P = 3, f = 1$, and therefore $T = 3$ (Fig. 5.5C). Thus the number of protein subunits in these shells is $60 \times 3 = 180$.

1.4 SURFACE FEATURES IN SMALL PLANT VIRUSES WITH ICOSAHEDRAL SYMMETRY

The actual detailed structure of the surface of an icosahedral virus will depend on how the protein subunits are packed together.

1.4.1 Clustering of Subunits

In several groups of plant viruses, the protein subunits in the shell are clustered around the five- and sixfold symmetry axis. These clusters form morphological subunits that can be visualized in electron micrographs of negatively stained particles. Since there are always 12 vertices with fivefold symmetry in icosahedra, we can calculate the number of morphological subunits (M) (assuming clustering into pentamers and hexamers) as follows:

$$M = \frac{(60T - 60)}{6} \text{ hexamers } + \frac{60}{5} \text{ pentamers}$$

$$= 10 \, (T - 1) \text{ hexamers } + 12 \text{ pentamers}$$

The smallest known virus, the satellite virus of tobacco necrosis virus (STNV) has icosahedral symmetry with $T = 1$, giving 60 subunits. These are arranged in five clusters of 12 around the 12 fivefold symmetry axes (Fig. 5.6). In several virus groups, such as the tymoviruses with 180 subunits, these are clustered in 12 groups of five around the fivefold symmetry axes, and 20 groups of six around the threefold symmetry axes (Fig. 5.9).

1.4.2 Nonequivalence of Different Positions in Icosahedral Shells

In the basic icosahedron (Fig. 5.4), a feature located halfway along any edge of a triangular face is positioned on an axis of rotation for the whole solid. Thus this point is on a *true* or icosahedral symmetry axis. This is a true dyad. In more complex icosahedrons where $T > 1$ more than one kind of twofold symmetry exists. For example, in a $P = 3$, $T = 3$ shell (Fig. 5.5C), the center of one edge on each of the 60 triangular faces is on a true dyad axis relating to the solid as a whole. The center positions of the other two edges of a face have only local twofold symmetry. These are called *quasi* dyads.

The fact that all the positions in an icosahedral shell are not in equivalent symmetry positions could mean that the chemical bonding between subunits was not always the same. For a time it was believed that the bonds between subunits might be deformed in slightly different ways in different nonsymmetry-related environments. This led to the idea of *quasi equivalence* in the bonding between subunits, but later work has shown that the relatively undistortable nature of protein domains will not permit the angular shifts required. However, many viruses have evolved ways of overcoming this problem. For example, viruses such as tomato bushy stunt *Tombusvirus* (TBSV) have evolved multidomain subunits that, to a large extent, adjust to the different

Figure 5.6 STNV. Electron micrographs of two virus particles oriented close to the twofold axis, and a model of 12 morphological subunits shaped to approximate to the same pattern. Courtesy of J. T. Finch.

symmetry-related positions in the shell by means other than distortion of intersubunit bonds (Fig. 5.11). Other viruses have evolved different variations on the icosahedral theme. For example, members of the *Comovirus* group and Reoviridae have different polypeptides in different symmetry environments within the shell (Fig. 5.13). The surfaces of such viruses may appear relatively smooth in negatively stained preparations, or may reveal smaller or larger protuberances, depending on the detailed conformation of the proteins involved.

2 EXAMPLES OF PLANT VIRUSES WITH DIFFERENT KINDS OF ARCHITECTURE

2.1 SMALL REGULAR VIRUSES WITH ssRNA GENOMES

2.1.1 TMV—A Rod-Shaped Virus

TMV has an extremely stable structure, having been reported to retain infectivity in nonsterile extracts at room temperature for at least 50 years. The stability of naked TMV RNA is no greater than that of any other ssRNA. Thus, stability of the virus with respect to infectivity is a consequence of the interactions between neighboring protein subunits and between the protein and the RNA.

The structure of TMV was illustrated in outline in Fig. 5.3. The rod is 19 nm in diameter and 300 nm long. The RNA of the common strain has 6395 nucleotides. The coat protein subunit has 158 amino acids, which form a somewhat elongated molecule. There are about 2130 of these subunits in a virus particle. Studies using negatively stained particles have shown that the TMV rod at the 3' end of the RNA is slightly convex, and is concave at the 5' end. The central canal has a radius of about 2 nm. Although the length of infectious rods is 300 nm, nearly all TMV preparations contain a proportion of shorter rods. Most of these are of no special significance, except that rods of all lengths tend to aggregate end-to-end to give a wide distribution of rod lengths.

The Double Disk

The protein subunit monomer of TMV can aggregate in solution in various ways, depending on pH, ionic strength, and temperature. There are two important forms: (1) *A protein,* which exists in solution at pH values > 7.5. A protein is mainly a mixture of monomers, trimers, and pentamers; and (2) the *double disk.* This contains two rings of 17 protein subunits. These double disks have been very important for studying TMV structure, because they can form true three-dimensional crystals, which can be used for X-ray crystallographic analysis. Such studies using the double disk and also virus preparations have provided the structure of TMV to a resolution of 2.9 Å. Such a structure generates a great deal of information down to atomic level. Here only some of the main features can be mentioned.

Much of the polypeptide chain is in four α helices known as the right radial (RR), left radial (LR), right slewed (RS) and left slewed (LS). The arrangement of subunit polypeptides in a section through a double disk is illustrated in Fig. 5.7.

The Virus Rod

As is shown in Fig. 5.7, when TMV protein subunits are in the form of the double disk, the innermost part of the polypeptide chain is in a flexible state with no ordered structure. The RNA-binding site lies at a somewhat higher radius (4 nm), marked in Fig. 5.7. Subunits in the

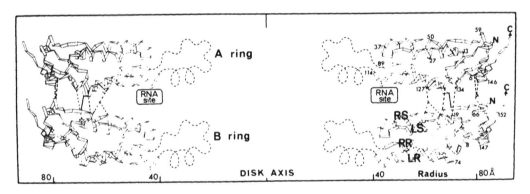

Figure 5.7 Structure of the double disk of TMV protein. Side view through the disk showing the relative disposition of subunits in the two rings and the regions of contact between them. Three regions of contact are indicated by a solid line for the hydrophobic contact of Pro 54 with Ala 74 and Val 75, by dashed lines for the hydrogen bonds between Thr 59 and Ser 147 and 148, and further dashed lines for the extended salt-bridge system. The low-radius region of the chain, which has no ordered structure in the absence of RNA, is shown schematically. At high radius, both the N- and C-termini of the polypeptide chain are exposed. Courtesy of A. Klug and A. C. Bloomer.

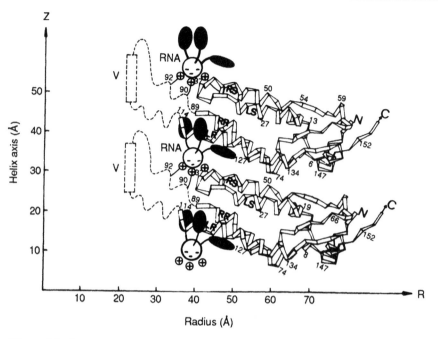

Figure 5.8 Structure of TMV. Schematic outline of the arrangement of the RNA and protein in the virus as seen in the side view of two subunits. V, the vertical helix not seen in the double disk (Fig. 5.7). The bases of the RNA are shown as black ovals. Three are associated with each protein subunit. The three negatively charged phosphate groups have been thought to form salt bridges with three arginine residues as indicated. However, more recent high-resolution studies suggest that the electrostatic interactions between the protein and RNA are best considered as complementarity between two electrostatic surfaces rather than as simple ion pairs. Photo courtesy K. C. Holmes.

double disk can be looked on as jaws ready to close on the RNA as the virus is being assembled (Chapter 7, Section 3.9). When this closure happens, several changes occur. The subunits, which lie more or less at right angles to the axis of the double disk, become significantly tilted with respect to the axis in the virus rod (Fig. 5.8). At the same time, much of the polypeptide, which is unordered at low radius, assumes a helical form with an orientation parallel to the axis of the rod (Fig. 5.8). These vertical helices form a *wall* lining the inside of the axial hole in the virus, which protects the RNA from the medium. High-resolution data show that water molecules are distributed throughout the surface of the protein subunit, on both the inner and outer surfaces of the virus, and in the subunit interfaces.

The Stability of TMV

The bonds of various sorts between all the surfaces of adjacent protein subunits, together with the interactions between the RNA bases and the protein subunits, give rise to a very stable macro-molecular structure. Nevertheless, it must be possible for the structure to disassemble *in vivo* to expose the viral RNA and allow virus replication to begin. Similarly, there must be molecular mechanisms for self-assembly of the virus from protein and RNA. It is probable that these processes are controlled by negative charges on neighboring groups, which can create an electrostatic potential that can trigger disassembly. These sites are probably carboxyl–carboxylate, or carboxyl–phosphate pairs that can bind a Ca^{2+} or H^+ ion.

The Specificity of Binding between RNA and Protein

TMV RNA has an origin-of-assembly sequence that interacts specifically with a double disk of protein subunits (Chapter 7, Section 3.9.1). However, there is no specificity as to which bases bind in the three sites associated with each subunit in the virus rod. This must be so, since every combination of bases must occur along the 6500 nucleotides of the viral RNA. Indeed, under appropriate conditions *in vitro,* TMV protein will form rods with almost any RNA, but it will not form rods with DNA.

2.1.2 Small Icosahedral Viruses

At present, we can distinguish four kinds of structures among the protein shells of small icosahedral plant viruses whose architecture has been studied in sufficient detail. However, high-resolution structures are available only for the second cluster of viruses. When more detailed information is available for the other groups, the arrangement suggested here may change:

1. *Banana-shaped subunits* Tymoviruses, bromoviruses, and cucumoviruses have 180 somewhat banana-shaped protein subunits clustered into pentamers and hexamers to form a $T = 3$ protein shell.

2. *β-barrels* Tombusviruses and sobemoviruses have 180 subunits in a $T = 3$ lattice, whereas STNV has 60 subunits in a basic $T = 1$ ico-

sahedron. Different domains can be distinguished in the protein subunits of these viruses.

3. *Bacilliform particles*   Alfalfa mosaic virus has several particles
of differing lengths.

4. *Two different coat proteins*   Comoviruses have two different proteins in the shell occupying different symmetry-related positions. The
following sections give an example of each of these four types.

Banana-Shaped Subunits

TYMV is a well-studied example of this type. The basic icosahedral
structure is shown in negatively stained particles and a model from X-
ray crystallographic data (Fig. 5.9). The polypeptide chain of the European strain of TYMV consists of 189 amino acid residues. The arrangement of this polypeptide within the coat protein is not known precisely,
but it appears to have ≈10% α–helix, ≈40% β-sheet, and about 50% in a
disordered conformation. However, X-ray studies at higher resolution
may change this picture. The diameter of the particle is ≈28 nm. The
space inside the protein shell has a diameter of ≈20 nm. The ssRNA of
the virus, which consists of 6318 nucleotides in the European strain,
lies within this interior space. There is very little interpenetration be-

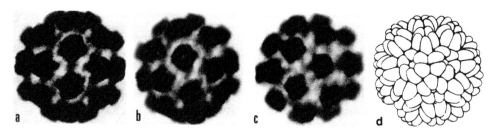

Figure 5.9 Surface features revealed by negative staining in an icosahedral virus (turnip
yellow mosaic *Tymovirus*) with 180 subunits clustered into 12 groups of 5 and 20 groups
of 6. The clarity of the images was enhanced by modeling the density distribution on
stacks of photographic plates, and by displaying only the top three quarters of the model.
The views illustrate the three symmetry axes of a $T = 3$ icosahedron: (a) view approximately down a twofold axis; (b) a threefold axis; and (c) a fivefold axis. The central morphological unit in (b) is clearly a hexamer, and in (c), it is a pentamer. At the periphery of
the particles, the clusters of subunits can be seen projecting from the surface with their
bases forming connections to neighboring clusters. From Mellema and Amos (1972).
(d) Drawing of the outside of the particle showing clustering of protein subunits into
groups of five and six. Orientation approximately as in (a). From Finch and Klug (1966).

tween protein and RNA. Studies using various physical techniques suggest that about 60 to 70% of the RNA could be in an internally based-paired configuration inside the virus.

About one third to one fifth of the particles found in a TYMV preparation are empty protein shells containing no nucleic acid. These are known as the T or top component because they sediment more slowly than the virus (B, or bottom component). The T-component protein shells are identical in structure to those of the virus. They are not artefacts of the isolation procedure, but the reason for their formation *in vivo* has not been established. Their existence as stable particles demonstrates that in TYMV, much of the virus stability is owing to strong interactions between the protein subunits.

Although the protein shells of members of the *Bromovirus* and *Cucumovirus* groups have a similar arrangement of 180 subunits and are about the same size as TMV, they are much less stable. This must reflect the strength of the bonding between protein subunits, and perhaps the arrangement of the RNA within the virus.

β-Barrels

The shell or S domains of tomato bushy stunt *Tombusvirus* (TBSV) Southern bean mosaic, Sobemovirus (SBMV) and the satellite virus of tobacco necrosis virus (STNV) have all been revealed at high resolution by X-ray crystallography. These protein subunits have in common a β-barrel structure. These are more or less wedge-shaped configurations of the polypeptide chain whose sides are made up of β-strands. A β-strand is a maximally stretched polypeptide chain. We can envisage side-by-side positioning of the β-strands to generate a β-sheet, stabilized by noncovalent forces. Bending the sheet into the wedge-like structure produces a β-barrel.

The S domain of TBSV can be described as an eight-stranded, antiparallel β-barrel consisting of two back-to-back four-stranded shells (Fig. 5.10). The domain is narrower at one end, making it wedge-shaped, thus allowing close approach of the subunits in the shell. In the virus shell, the wedge-shaped end of the S domain in all three viruses is close to the five- or sixfold vertices.

The structure of TBSV has been determined crystallographically to 2.9 Å resolution. It contains 180 protein subunits of $M_r = 43,000$ arranged to form a $T = 3$ icosahedral surface lattice, with prominent dimer clustering at the outside of the particle, the clusters extending to a radius of about 17 nm.

TBSV

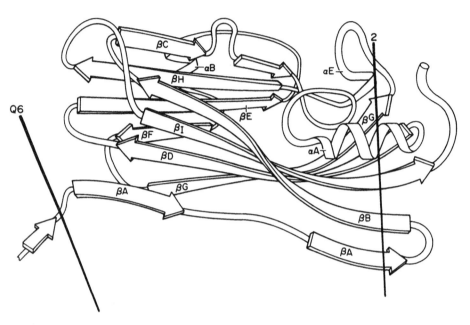

Figure 5.10   Structure of the β-barrel in TBSV coat protein. Diagrammatic representation of folding of the polypeptide backbone. From Rossmann *et al.* (1983).

The essential features of the structure are summarized in Fig. 5.11. The protein subunit is made up of two distinct globular parts (domains P and S) connected by a flexible hinge involving five amino acids (Fig. 11B). Each P domain forms one half of the dimer-clustered protrusions on the surface of the particle. The S domain forms part of the icosahedral shell. Each P domain occupies approximately one third of one icosahedral triangular face. The shell is about 3 nm thick, and from it protrude the 90 dimer clusters formed by the P domain pairs.

In addition to domains P and S, each protein subunit has a flexibly linked N-terminal arm containing 102 amino acid residues. This has two parts, the R domain and the connecting arm called a (Fig. 5.11B).

The coat protein assumes two different large-scale conformational states in the shell that differ in the angle between domains P and S by about 20° (Fig. 5.11D). The conformation taken up depends on whether the subunit is near a quasi dyad or a true dyad in the $T = 3$ surface lattice. The state of the flexible N-terminal arm also depends on the symmetry position.

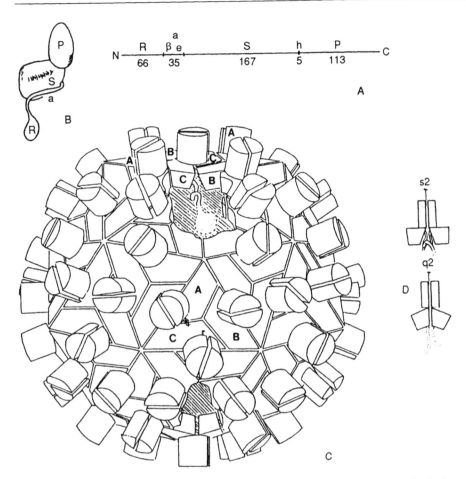

Figure 5.11 Architecture of TBSV particle. (A) Order of domains in polypeptide chain from N-terminus to C-terminus. The number of residues in each segment is indicated below the line. The letters indicate R domain (possible RNA-binding region), arm a (the connector that forms the β-annulus and extended arm structure on C subunits and that remains disordered on A and B), S domain, hinge, and P domain. (B) Schematic view of folded polypeptide chain, showing P, S, and R domains. (C) Arrangement of subunits in particle. A, B, and C denote distinct packing environments for the subunit. S domains of A subunits pack around fivefold axes; S domains of B and C alternate around threefold axes. The differences in local curvature can be seen at the two places where the shell has been cut away to reveal S domain packing near strict (top) and quasi (bottom) dyads. (D) The two states of the TBSV subunit found in this structure, viewed as dimers about the strict (s2) and local (q2) twofold axes. Subunits in C positions have the interdomain hinge *up* and a cleft between twofold-related S domains into which fold parts of the N-terminal arms. Subunits in the quasi-twofold-related A and B positions have hinge *down*, S domains abutting, and a disordered arm. (A), (B), and (C) from Olson *et al.* (1983); (D) from Harrison *et al.* (1978); (D) reprinted by permission from *Nature (London)* **276,** 370.

The N-terminal arms originating near a strict dyad follow along in the cleft between two adjacent S domains. On reaching a threefold axis, such an arm winds around the axis in an anticlockwise fashion (viewed from outside the particle). Two other N-terminal arms originating at neighboring strict dyads will be at each threefold axis. The three polypeptides overlap with each other to form a circular structure, called the β-annulus, around the threefold axis. Thus, 60 of the 180 N-terminal arms form an interlocking network that, in principle, could form an open $T = 1$ structure without the other 120 subunits. The annulus is made up of 19 amino acids from each arm. The R and a domains arising from the 120 subunits at quasi-dyad positions hang down into the interior of the particle in an irregular way, so that their detailed position cannot be derived by X-ray analysis. The RNA of TBSV is tightly packed within the particle, and these N-terminal polypeptide arms very probably interact with the RNA (see Section 3).

Bacilliform Particles

Purified preparations of alfalfa mosaic virus (AMV) contain four major classes of particle as illustrated in outline in Fig. 5.12. The structure of AMV coat protein is not yet known at atomic resolution. It behaves as a water-soluble dimer stabilized by hydrophobic interactions between the two molecules. This dimer is the morphological unit out of which the viral shells are constructed. The Ta particle is assumed to be made up of 120 protein subunits in which 60 dimers form a $T = 1$ icosahedron. In the cylindrical sections of the bacilliform particles, the protein dimers may be arranged with threefold symmetry, whereas the curved ends have $T = 1$ symmetry.

The genomic RNA is in three pieces, which are housed in the Tb, M, and B particles. The Ta particles contain a subgenomic mRNA derived from part of the Tb RNA sequence. A low-resolution model for the distribution of RNA and protein in these particles is given in Fig. 5.12.

The virus particles are unstable and are readily dissociated into protein and RNA at high salt concentrations. The viral components are held together in the virus particle mainly by protein-RNA interactions.

Two Different Coat Proteins

Cowpea mosaic *Comovirus* (CPMV) has a diameter of about 28 nm and an icosahedral structure made up of two different coat protein subunits located in different symmetry environments. There are 60 mole-

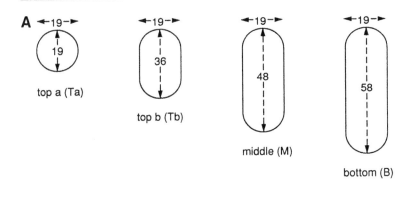

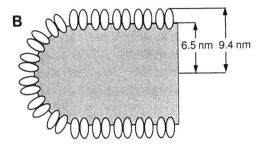

Figure 5.12 AMV particles. (A) Outlines for the four major classes of particle found in AMV preparations. Dimensions in nm. (B) A schematic representation of the distribution of protein and RNA in AMV B component. RNA is indicated by the hatched area; the protein molecules are represented by ellipsoids. The model is derived from an analysis by small-angle neutron scattering. (B) redrawn from Cusack *et al.* (1981).

cules of the larger protein (MW = 42×10^3) clustered at the 12 positions with fivefold symmetry. Sixty smaller proteins (MW = 22×10^3) form 20 clusters of three around the positions with threefold symmetry (see cover photograph insert).

Although the particle of CPMV is clearly a $T = 1$ icosahedral structure, the fact that it has two proteins in different symmetry environments gives it some overall resemblance to a $T = 3$ structure. The structure of CPMV at 3.5 Å shows that the two coat proteins produce three distinct β-barrel domains in the icosahedral asymmetric unit, although two of the β-barrels are covalently linked.

Purified preparations of CPMV contain three classes of particle, known as T, M, and B, with identical protein shells that can be separated by centrifugation. The T particle is an empty protein shell. The M and B particles contain the two pieces of genomic RNA found in viruses of this group.

2.2 VIRUSES WITH dsRNA GENOMES

Different members of the Reoviridae family, all of which have dsRNA genomes, infect vertebrates, invertebrates, and plants. They all have a similar general structure. Particles are isometric and about 70 nm in diameter, as observed in negatively stained preparations.

Two genera have been established for the plant reoviruses; *Phytoreovirus* for those with 12 pieces of dsRNA, no spikes on the outer coat, type member wound tumor virus (WTV); and *Fijivirus* for those with 10 pieces of RNA, spikes present on the outer coat, type member Fiji disease virus (FDV).

Almost all the information on structure of plant reovirus particles has been derived from electron microscope observations on individual negatively stained particles and on subviral structures produced by partial breakdown of the intact virus. Reoviruses contain an RNA-dependent RNA polymerase as part of the particle. This enzyme activity has not yet been identified with any of the polypeptides revealed by gel electrophoresis for the plant reoviruses. However, it is located in the subviral particle, not in the outer envelope.

2.2.1 Phytoreovirus

There are seven different proteins in the particle of WTV. The particles consist of an outer shell of protein and an inner core containing protein and the 12 pieces of dsRNA. However, there is no protein in close association with the RNA. The particles are readily disrupted during isolation, by various agents. Under suitable conditions, subviral particles can be produced that lack the outer envelope and reveal the presence of 12 projections at the five-fold vertices of an icosahedron.

2.2.2 Fijivirus

A model of the structure of FDV based on electron microscope observations of intact and partly degraded particles is shown in Fig. 5.13.

2.3 VIRUSES WITH DNA GENOMES

Two groups of plant DNA viruses have been studied structurally. They differ markedly in the architecture of their particles.

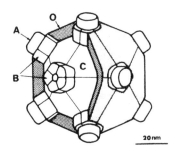

Figure 5.13 Structure of Fiji disease virus. A scale model of the particle. (A), A spike; (B), B spike. Part of the outer shell (O) of the particle and one of the A spikes have been removed to expose the core (C) and the structure and arrangement of the B and A spikes. The arrangement of distinct morphological subunits within the A spikes, outer shell, and core has not been established. From Hatta and Francki (1977).

2.3.1 The *Caulimovirus* Group

The type member of this group, cauliflower mosaic virus (CaMV), has a very stable isometric protein shell about 50 nm in diameter, containing a single molecule of dsDNA within it. Electron microscopy reveals a relatively smooth protein shell with no structural features. It probably has a $T = 7$ icosahedral structure.

2.3.2 The *Geminivirus* Group

These viruses contain ssDNA and one type of coat polypeptide. Particles in purified preparations consist of twinned or geminate icosahedra (Fig. 5.14). From a study of negatively stained preparations, together with models of possible structures, these particles probably consist of two $T = 1$ icosahedra joined together at a site where one morphological subunit is missing from each, giving a total of 22 morphological units in the geminate particles.

2.4 VIRUSES WITH LIPOPROTEIN ENVELOPES

2.4.1 Rhabdoviridae

This is another family of viruses whose members infect vertebrates, invertebrates, and plants. They have a complex structure. Rhabdoviruses from widely differing organisms are constructed on the basic plan shown in Fig. 5.15.

Some animal rhabdoviruses may be bullet-shaped, but most and perhaps all plant members are rounded at both ends to give a bacilliform shape. Electron microscopy on thin sections and negatively stained par-

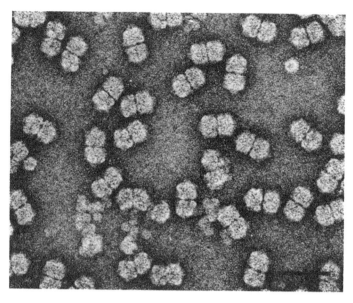

Figure 5.14 Purified *Geminivirus* from *Digitaria* negatively stained in 2% aqueous uranyl acetate. Bar marker represents 50 nm. From Dollet *et al.* (1986).

ticles has been used to determine size and details of morphology. The proteins present in plant rhabdoviruses are discussed Chapter 8, Section 4.2.

2.4.2 Bunyaviridae

Tomato spotted wilt *Tospovirus* (TSWV) is the only plant member of the Bunyaviridae family. This is a very large family of viruses, most of which infect vertebrates. TSWV is very unstable and difficult to purify. Its structure has been studied in thin sections of infected cells, and in partially purified preparations. Isolated preparations contain many deformed and damaged particles. As seen in thin sections, particles are about 100 nm in diameter (Fig. 5.16). By analogy with an enveloped animal virus of similar size, they are probably much larger in the living cell in the hydrated state. The central core of the particle contains the RNA. Outside the core is a layer of dense material, surrounded by a typical lipoprotein bilayer membrane. The RNAs and structural proteins of TSWV are discussed in Chapter 8, Section 5.

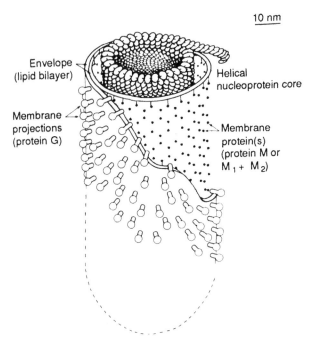

10 nm

Envelope
(lipid bilayer)

Helical
nucleoprotein core

Membrane
projections
(protein G)

Membrane
protein(s)
(protein M or
$M_1 + M_2$)

Figure 5.15 Generalized diagram for the structure of rhabdoviruses. From Francki and Randles (1980), reprinted with permission.

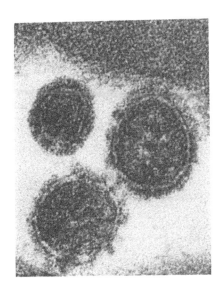

Figure 5.16 Tomato spotted wilt *Tospovirus* in a thin section of tomato. Courtesy of R. G. Milne.

3 INTERACTION BETWEEN RNA AND PROTEIN IN SMALL ISOMETRIC VIRUSES

Negatively charged phosphate groups on the RNA inside a small isometric virus would tend to destabilize the particle unless a substantial proportion of the charges were neutralized in some way. Current knowledge suggests that two types of interaction can overcome this problem.

3.1 VIRUSES IN WHICH BASIC AMINO ACIDS BIND RNA PHOSPHATES

In TBSV coat protein, 21 lysine and arginine residues are available to bind phosphates. Since there are about 26 nucleotides in the RNA for each protein subunit, most of the RNA charges can be neutralized by these amino acids. The flexible connection to the R arm allows it to conform to irregularities in RNA packing. Similar arrangements are known or proposed for representatives of several other groups of small isometric viruses, for example SBMV, brome mosaic virus (BMV), and AMV.

3.2 POLYAMINES

There is little interpenetration of RNA and protein in TYMV; however, there are sufficient amounts of the polyamines spermine and spermidine in the particles of this virus, together with divalent cations, to neutralize a significant proportion of the charged RNA. Comoviruses such as CPMV also contain polyamines that probably serve a similar function.

FURTHER READING

Harrison, S. C. (1983). Virus Structure: High-resolution perspectives. *Adv. Virus Res.* **28**, 175–240.
Matthews, R. E. F. (1991). "Plant Virology," 3rd Ed. Academic Press, New York.
Rossman, M. G., and Johnson, J. E. (1989). Icosahedral RNA virus structure. *Annu. Rev. Biochem.* **58**, 533–573.

INTRODUCTION TO THE STUDY OF VIRUS REPLICATION

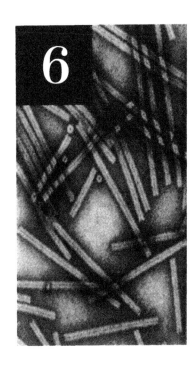

Our understanding of the ways in which plant viruses replicate has increased remarkably over the past few years. This is mainly because new techniques have allowed the complete nucleotide sequences of many plant viral genomes to be established. This in turn has allowed the number, size, and amino acid sequences of putative gene products to be determined. We now have this information for representatives of many plant virus groups and families. Developments in gene-manipulation technology have permitted an artificial DNA step to be introduced into the life cycle of RNA viruses. Thus, infectious genomic RNA transcripts with a uniform nucleotide sequence can be produced *in vitro*. In turn, this has allowed the application of site-directed mutagenesis in experiments to determine functional regions in the noncoding regions of genomic nucleic acid, and the functions of gene products and putative gene products. Determination of the functions and properties of gene products has also been greatly assisted by the use of well-established *in vitro* translation systems.

Studies on virus replication *in vivo* using isolated protoplast preparations have provided new information on many aspects of virus replication. The ability to produce transgenic plants in which every cell contains a functional DNA copy of a single viral gene or genome segment is opening up new possibilities for the study of virus replication *in vivo*. In this chapter I shall discuss, in general outline, the methods used to establish viral genome structure, expression, strategy, and the details of *in vivo* replication. Most attention will be given to viruses with single-stranded (ss)-positive sense RNA genomes, as 29 of the 35 recognized groups and families of plant viruses have genomes of this type. In Chapters 7 and 8, details concerning the replication of representative virus groups and families are discussed.

1 GENERAL PROPERTIES OF PLANT VIRAL GENOMES

1.1 INFECTIVITY OF VIRAL NUCLEIC ACIDS

A viral genome contains all the information necessary for the replication *in vivo* of complete virus particles. Whether or not a viral nucleic acid preparation will be infectious depends on two factors, physical integrity and the kind of nucleic acid.

1.1.1 Physical Integrity

The genomic nucleic acid must be intact, with no phosphodiester bonds broken by physical, chemical, or enzymatic means.

1.2.2 The Kind of Nucleic Acid

ss-Positive Sense RNAs

This type of genome is ready to function as a messenger RNA (mRNA) on entry into a suitable host cell, and is therefore usually infectious. However, infectivity is usually much lower than with intact virus, probably because nucleases can rapidly destroy much of the RNA. A few viruses have a small protein covalently linked to the 5' end of the RNA. For some of these, the protein is essential for infectivity. A few other ss-positive sense viruses such as alfalfa mosaic virus (AMV) require the presence of coat protein or a coat protein mRNA to initiate infection.

Negative Sense ssRNA Viruses

A negative sense genome has to be copied into positive sense mRNAs before it can function. This copying is carried out by a viral-coded enzyme found in the virus particle. If this is absent, the genomic RNA is noninfectious.

dsRNA Viruses

Viruses with double-stranded RNA (dsRNA) genomes also contain a viral-coded enzyme, which copies the genomic RNAs into mRNAs.

When the enzyme is removed during RNA isolation, the RNA becomes unable to initiate infection.

DNA Viruses

Plant viruses with ds- or ssDNA genomes use host enzymes in the initial events that give rise to functional mRNAs. These DNAs are therefore able to initiate infection in the absence of virus-coded proteins.

1.2 INFORMATION CONTENT

In theory, a single nucleotide sequence in a viral genome could code for up to 12 polypeptides. There could be an open reading frame (ORF) in each of the three reading frames of both the positive and negative sense strands, giving six polypeptides. If each of these had a leaky termination signal, they could give rise to a second read-through polypeptide. In principle, many more permutations exist, if we consider the possibility of translational frameshift mechanisms for generating additional proteins. In nature, however, there must be severe evolutionary constraints on such high multiple use of a nucleotide sequence, because even a single base change could have consequences for several gene products. Two overlapping genes in different reading frames do occasionally occur, as do genes on both positive and negative sense strands. Read-through proteins are quite common, and some examples of translational frameshift exist. Thus viruses do have considerable potential for flexibility in expression of their genomes.

The number of genes found in plant viruses ranges from one for the satellite virus, satellite tobacco necrosis virus (STNV), to 12 for some reoviruses. Most of the ss positive sense RNA genomes encode four to seven proteins. In addition to coding regions for proteins, genomic nucleic acids contain *cis*-acting nucleotide sequence elements with recognition and control functions important for virus replication. These control and recognition functions are mainly found in the 5′ and 3′ noncoding sequences of the ssRNA viruses, but they may also occur internally, even in coding sequences.

1.3 THE STRATEGIES OF PLANT VIRAL GENOMES

Genome strategy is a useful but ill-defined term. Central to the concept is the overall pattern whereby information in the viral genome is transcribed and translated into viral proteins, and the general aspects of the mechanisms whereby the viral genome is replicated. The nature of these processes depends very much on the physical structure of the genome. In this section, I shall discuss the strategies found in viruses with ss-positive sense RNA genomes. The strategies for five families and groups of plant viruses that have other kinds of genomic nucleic acid are discussed in Chapter 8. Several factors have probably been involved in the evolution of genome strategies among the ss-positive sense RNA viruses.

1.3.1 The Eukaryotic Protein-Synthesising System

In eukaryote protein-synthesizing systems, in most circumstances, mRNAs are monocistronic. Thus, 80 S ribosomes are adapted to translate primarily an ORF immediately downstream from the 5′ region of an mRNA. ORFs beyond (3′ to) the first in a viral mRNA normally remain untranslated. Much of the variation in the way gene products are translated from RNA genomes of viruses infecting eukaryotes appears to have evolved to meet this constraint.

On current knowledge, there are five main strategies by which RNA viral genomes ensure that all their genes are accessible to the eukaryotic protein synthesis system:

1. *Subgenomic RNAs* The synthesis of one or more subgenomic mRNAs enables the 5′ ORF on each such RNA to be translated.

2. *Polyproteins* Here the coding capacity of the RNA for more than one protein, and sometimes for the whole genome is translated from a single ORF. The polyprotein is then cleaved at specific sites by a viral-coded protease, or proteases, to give the final gene products.

3. *Multipartite genomes*, in which the 5′ gene on each RNA segment can be translated.

4. *Read-through proteins* The termination codon of the 5′ ORF may be *leaky*, allowing a proportion of ribosomes to carry on translation to another stop codon downstream from the first, giving rise to a second longer functional polypeptide.

5. *Transframe proteins* Another mechanism by which two proteins may commence at the same 5′ AUG is by a switch of reading frame near the termination codon of the 5′ ORF to give a second longer *transframe* protein. Another possible mechanism of this sort is premature termination of translation in the absence of a termination codon. In two

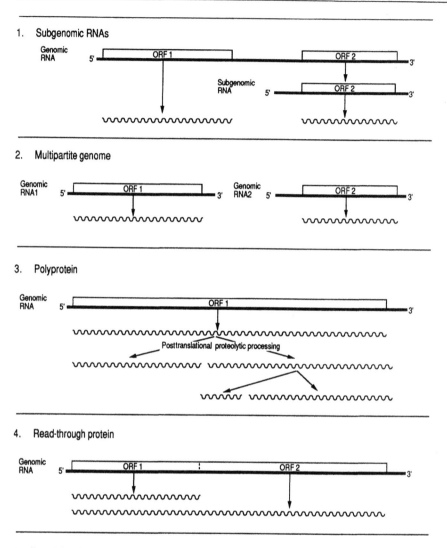

5. Translational frameshift

The ribosome bypasses a stop codon in Frame 0 by switching back one nucleotide to Frame -1 at a
UUUAG sequence before continuing to read triplets in Frame -1 to give a fusion or transframe protein.

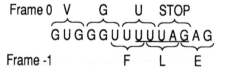

Figure 6.1 Five strategies used by plant viruses to allow protein synthesis in a eukaryotic system from a positive sense RNA genome containing more than one gene.

plant virus groups, it has been suggested that an internal ORF may be translated by internal initiation even though the system is a eukaryotic one. These are the *Luteovirus* and *Comovirus* groups, but the idea has not been substantiated. In TYMV, two polypeptides initiate at two closely spaced 5′ AUGs in different reading frames.

The five main strategies are illustrated in Fig. 6.1. A few plant virus groups and families have made use of only one of these devices to develop a successful genome strategy, but most viruses use two or three different strategies in combination. The strategies used by the virus groups with well-studied representatives are summarized in Table 7.1.

All these various combinations of translation strategies must have survival value. It is easy to recognize the value of some features. Thus all these small viruses require much more of the coat protein than any other gene product. Most of them use a coat-protein mRNA originating from the 3′ region of the genomic RNA. These small monocistronic RNAs are highly efficient messengers for coat protein. Furthermore, transcription of the mRNA from a minus-strand template is an additional mechanism for increasing the production of coat protein relative to other gene products.

On the other hand, it is difficult to visualize how the polyprotein strategy of the potyviruses can be efficient. The coat-protein gene is at the 3′ end of the genome (Fig. 7.1). Thus, for every molecule of the 20 K coat protein produced by tobacco etch virus (TEV), a molecule of all the other gene products has to be made, totaling about 320 K. Since about 2000 molecules of coat protein are needed for each virus particle, this appears to be a very inefficient procedure. Indeed, large quantities of several gene products, apparently in a nonfunctional state, accumulate in infected cells (Chapter 7, Section 1). Nevertheless, the potyviruses are very successful, as judged by the large number of viruses in the group.

1.3.2 Other Selection Pressures

Besides the eukaryotic protein-synthesizing system, there must have been other factors at work in the evolution of genome strategies. Multipartite genomes have other selective advantages, such as: (1) genetic flexibility. In viruses with several pieces of RNA in one particle, genetic reassortment can take place during virus replication. In addition to reassortment at this stage, the existence of the RNA in several separate particles allows for the possibility of selection and reassortment at other stages as well, for example, during transmission, entry into the cell, and movement through the plant; (2) separation of early and late gene functions; (3) separation of functions that are required to be carried out in different parts of the host cell; (4) separation of genes functioning only in the plant or insect hosts of a virus; (5) facilitation of encapsidation of the genome by keeping particles, especially icosahedral ones, relatively small; and (6) multiplicity reactivation. Genome subdivision might be an adaptive response to the high error rates in RNA virus replication, resulting mainly from the lack of error-correcting mechanisms.

Subgenomic RNAs might facilitate production of gene products at different sites or stages during virus replication. Read-through and transframe proteins give more gene products from a small genome.

1.4 ECONOMY IN THE USE OF GENOMIC NUCLEIC ACIDS

Plant viruses share with viruses of other host groups a number of features that indicate very efficient use of their genomic nucleic acids: (1) coding sequences are usually very closely packed, with a rather small number of noncoding nucleotides between genes; (2) coding regions for two different genes may overlap in different reading frames, or one gene may be contained entirely within another in a different reading frame; (3) read-through of a leaky termination codon may give rise to a second, longer read-through polypeptide that is coterminal at the amino end with the shorter protein. This is quite common among the virus groups with ss positive sense RNA genomes. In a few viruses there may be a second read-through protein giving rise to a set of three proteins coterminal at the amino end. *Transframe* proteins, in which the ribosome avoids a stop codon by switching to another reading frame, have a result that is similar to a leaky termination signal; (5) A functional viral enzyme may use a host-coded protein in combination with a viral-coded polypeptide; (6) regulatory functions in the nucleotide sequence may overlap with coding sequences; (7) in the 5′ and 3′ noncoding sequences of the ssRNA viruses, a given sequence of nucleotides may be involved in more than one function. Examples of these various features will be found in Chapters 7 and 8.

1.5 THE FUNCTIONS OF VIRAL GENE PRODUCTS

The known functions of plant viral gene products may be classified as follows.

1.5.1 Structural Proteins

These are the coat proteins of the small viruses; the matrix or core proteins of the reoviruses and those viruses with a lipoprotein membrane; and proteins found within such membranes.

1.5.2 Enzymes

Proteases

These are present in those virus groups in which the whole genome or a segment of the genome is first translated into a single polyprotein, which then has to be cleaved at specific sites.

Enzymes Involved in Nucleic Acid Synthesis

It is now highly probable that all plant viruses, except satellite viruses, code for one or more proteins that have an enzymatic function in nucleic acid synthesis, either genomic nucleic acid or mRNAs or both. The general term for these enzymes is *polymerase.* There is some inconsistency in the literature in relation to the terms used for different polymerases. I shall use the various terms with the following meanings. Polymerases that transcribe RNA from an RNA template have the general name *RNA-dependent RNA polymerase.* If such an enzyme makes copies of an entire RNA genome, it is called a *replicase.* Replicase enzymes may also be involved in the synthesis of subgenomic mRNAs. If an RNA-dependent RNA polymerase is found as a functional part of the virus particle as in the Rhabdoviridae and Reoviridae, it is called a *transcriptase.* The enzyme coded by caulimoviruses that copies a full-length viral RNA into genomic DNA, is called an *RNA-dependent DNA polymerase* or *reverse transcriptase.*

Two features have made the viral-coded RNA-dependent RNA polymerase enzymes difficult to study. First, they are usually associated with membrane structures in the cell, and on isolation, the enzymes usually become unstable. They are, therefore, difficult to purify sufficiently for positive identification of any viral-coded polypeptides. Second, tissues of healthy plants may contain low amounts of enzyme with similar activities in the soluble fraction of the cell. The amounts of such enzyme activity may be stimulated by virus infection.

Conserved amino acid sequence motifs have now been identified in the RNA polymerase genes of many ssRNA plant viruses. Indeed the presence of such a motif is used to identify putative polymerase genes in newly sequenced viral genomes. Recently, the discovery of another series of amino acid motifs from a wide range of organisms has led to the definition of a large group of proteins, the nucleic acid helicases. Their functions are believed to include nucleic acid unwinding, and acting in the replication, recombination, repair, and expression of DNA and RNA genomes. These helicase sequence motifs have been identified in a wide range of ssRNA plant viral genomes and can be assumed to function in plant viral RNA replication. In some viruses, the polymerase and helicase motifs appear to be in the same protein, but in others they are separate. These variations are discussed in relation to virus classification and evolution in Chapter 16, Section 5.

Two kinds of RNA structures have been isolated from viral RNA-synthesizing systems. One, known as replicative form (RF), is a fully base-paired ds structure, whose role is not certain. For example, RF structures may represent RNA molecules that have ceased replicating. The other, called replicative intermediate (RI) is only partly ds, and contains several ss tails (nascent product strands) (Fig. 6.2). This structure is

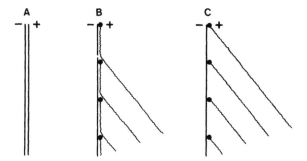

Figure 6.2 Forms of association between positive and negative sense strand viral RNA. (A) Replicative form (RF). A base-paired structure with full-length positive and negative sense strands. (B) Replicative intermediate (RI). A partially base-paired structure with polymerase molecules (●) and ss tails of nascent progeny positive sense strands. (C) Probable true state of the RI *in vivo*. The progeny positive sense strands and the template negative sense strand are almost entirely ss.

closely related to the one actually replicating the viral RNA. It is thought that the RI as isolated may be derived from a structure like that in Figure 6.2C by annealing of parts of the progeny strands to the template.

1.5.3 Virus Movement and Transmission

For several groups of viruses, a specific viral-coded protein has been identified as an essential requirement for cell-to-cell movement and for systemic movement within the host plant. Other gene products called *acquisition factors* are essential for successful transmission by invertebrate vectors. Viral gene products may also be involved in transmission by fungi.

1.5.4 Nonenzymatic Role in RNA Synthesis

The 5′ VPg (virus protein, genome linked) found in some virus groups is thought to act as a primer in RNA synthesis.

1.5.5 Coat Protein of AMV

The coat protein of AMV and the corresponding protein in ilarviruses have an essential role in the initiation of infection by the viral RNA.

1.5.6 Protein-Recognizing Host Cells

To identify markers on the surface of appropriate host cells, most viruses infecting bacteria and animals have recognition proteins on their surface. Such recognition proteins are lacking for most plant viruses. This is probably related to two properties of the host organism. A recognition protein would be no use to a plant virus for recognizing the surface of a suitable host plant because the virus cannot penetrate the surface layers unaided. Furthermore, a plant virus, once it infects a host cell, say in a leaf, can move from cell to cell via plasmodesmata and vascular tissue throughout almost the entire plant. Plant viruses therefore have no need for a special recognition protein on the surface of the host cell, but it is possible that specific recognition proteins exist inside cells. Possible exceptions among the plant viruses are those with a lipoprotein envelope, which can replicate in their insect vectors, and which may have surface proteins that recognize suitable insect host cells.

1.5.7 Summary

On present knowledge, only two functions appear to be common to all plant viruses except satellite viruses. These are the coat proteins and the enzymes involved in genome replication. Increasing numbers of viruses are being shown to require a protein with a cell-to-cell movement function. This may turn out to be another universal requirement for viruses that can move freely through various plant tissues, replacing, in a sense, the surface recognition proteins of bacterial and animal viruses noted above.

1.6 END-GROUP STRUCTURES IN ss-POSITIVE SENSE RNA VIRUSES

Many plant viral ssRNA genomes contain specialized structures at their 5′ and 3′ termini. This section summarizes the nature of these structures, whereas their biological functions are discussed in Section 5.1. Other viruses simply have free sugar hydroxyl or phosphate groups at their 5′ and 3′ termini.

1.6.1 The 5′ Cap

Many mammalian cellular messenger RNAs and animal virus messenger RNAs have a methylated blocked 5′-terminal group of the form:

$$\text{m7G}^{5'}\text{ppp}^5\text{X}^{(m)}\text{pY}^{(m)}\text{p}\dots.$$

where $X^{(m)}$ and $Y^{(m)}$ are two methylated bases.

Some plant viral RNAs have this type of 5′ end, known as a *cap*, but in the known plant viral RNAs bases, X and Y are not methylated. Virus groups with 5′ capped RNAs are the tobamoviruses, tobraviruses, tymoviruses, bromoviruses (also cucumoviruses and AMV), carmoviruses, furoviruses, potexviruses, and hordeviruses.

1.6.2 5' Linked Protein

Members of several plant virus groups have a viral-coded protein of relatively small size ($\approx$3500–24,000 MW) covalently linked to the 5' end of the genome RNA. These are known as VPgs (short for virus protein, genome linked). If a multipartite RNA genomes possesses a VPg, all the RNAs will have the same protein attached. VPgs have been found in viruses belonging to the following groups: potyviruses, comoviruses, nepoviruses, sobemoviruses, and luteoviruses.

1.6.3 3' Poly(A) Tracts

Polyadenylate sequences have been identified at the 3' terminus of the mRNAs of a variety of eukaryotes. Such sequences have been found at the 3' terminus of plant plant viral RNAs that can act as messengers. Members of the following groups have 3' terminal poly(A) sequences varying in length between about 25 and 400 residues: potyviruses, potexviruses, capilloviruses, comoviruses, nepoviruses, and furoviruses. The length of the poly(A) tract may vary for different RNA molecules in the same preparation.

1.6.4 3' tRNA-like Structures

The 3' terminal nucleotide sequences in the genomes of tobamoviruses, tymoviruses, hordeiviruses, bromoviruses, and cucumoviruses can fold into transfer RNA (tRNA)-like structures, which can accept specific amino acids. For example TYMV RNA accepts valine, whereas TMV RNA accepts histidine. Various tRNA-like secondary and tertiary structures have been proposed for these 3' terminal sequences, but all are to some extent speculative.

2 HOST FUNCTIONS USED BY PLANT VIRUSES

Like all other viruses, plant viruses are intimately dependent on the activities of the host cell for many aspects of replication.

2.1 COMPONENTS FOR VIRAL SYNTHESIS

Viruses use amino acids and nucleotides synthesized by host cell metabolism to build viral proteins and nucleic acids. Certain other more specialized components found in some viruses, for example polyamines, are also synthesized by the host.

2.2 ENERGY

The energy required for the polymerization involved in viral protein and RNA synthesis is provided by the host cell, mainly in the form of nucleoside triphosphates.

2.3 PROTEIN SYNTHESIS

Viruses use the ribosomes, tRNAs, and associated enzymes and factors of the host cell's protein-synthesizing system for the synthesis of viral proteins using viral mRNAs. All plant viruses appear to use the 80 S cytoplasmic ribosome system. Most viruses also depend on host enzymes for any posttranslational modification of their proteins—for example, glycosylation.

2.4 NUCLEIC ACID SYNTHESIS

Almost all viruses code for an enzyme or enzymes involved in the synthesis of their nucleic acids. However, they may not contribute all of the polypeptides involved in genome replication. For example, in the first phase of the replication of caulimoviruses, the viral DNA enters the host cell nucleus and is transcribed into RNA form by the host's DNA-dependent RNA polymerase II.

2.5 STRUCTURAL COMPONENTS OF THE CELL

Structural components of the cell, particularly membranes, are involved in virus replication.

2.6 MOVEMENT WITHIN THE PLANT

There is no doubt that many viruses encode a specific protein that is necessary for cell-to-cell movement. Nevertheless, since viruses lack any form of independent motility, they must depend on host mechanisms for movement within infected cells and for long-distance transport in the phloem. Simple diffusion would be quite inadequate to account for observed rates of movement.

3 GENERALIZED OUTLINE FOR THE REPLICATION OF A SMALL ss-POSITIVE SENSE RNA VIRUS

Figure 6.3 illustrates schematically the main steps in the replication of a small ss-positive sense RNA virus. Only the two most essential gene products—polymerase and coat protein—are considered.

4 METHODS FOR DETERMINING GENOME STRUCTURE AND STRATEGY

4.1 STRUCTURE OF THE GENOME

The first step is to establish the physical and chemical nature of the genome (DNA or RNA; ss or ds; linear or circular; in one piece or more than one piece) and the nature of any special terminal structures. The final step in determining structure is knowledge of the full nucleotide sequence, which is essential for understanding genome structure and strategy. The methods used were outlined in Chapter 2. With the help of an appropriate computer program, the nucleotide sequence is searched for ORFs in each of the three reading frames of both positive and negative sense strands. All ORFs are tabulated, as is illustrated in Fig. 6.4 for a *Tymovirus*.

As shown in Fig. 6.4, a large number of ORFs may be revealed. Those ORFs, which could code for polypeptides of MW less than 7 to 10

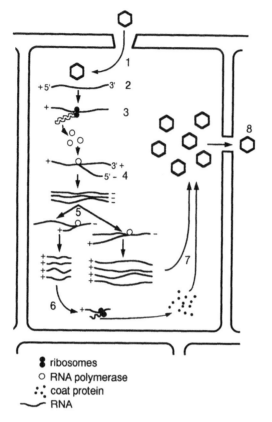

Figure 6.3   Generalized outline for the replication of a small ss positive sense RNA virus.

1. The virus particle enters a cell through a wound made in the cell wall.
2. The RNA escapes from the protein coat.
3. The infecting RNA becomes associated with host ribosomes and a viral-specific, RNA-dependent RNA polymerase is synthesized.
4. The polymerase synthesizes negative sense genomic-length strands of RNA.
5. The polymerase then synthesizes genomic-size RNA copies and coat protein mRNAs using an internal initiation site on the genomic negative sense RNA.
6. Host cell ribosomes synthesize large numbers of coat protein molecules using the coat mRNAs.
7. Coat protein is assembled around genomic RNA molecules to produce progeny virus particles, which accumulate in the cell.
8. A few particles migrate to neighboring cells through plasmodesmata.

Note: (a) The features in the diagram are not drawn to scale. (b) There are many variations in detail in the stages illustrated. (c) Not shown in this diagram is the fact that the processes illustrated often take place in specialized parts of the cytoplasm.

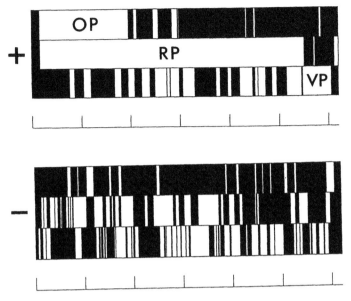

Figure 6.4 Diagram of the three triplet codon phases of the plus- and minus-strand RNAs of ononis yellow mosaic *Tymovirus* genomic RNA, showing as white boxes all ORFs that begin with an AUG and terminate with UGA, UAG, or UAA. There are three significant ORFs labeled OP, overlapping protein; RP, replicase protein; and VP, coat protein. From Ding *et al.* (1989).

K, are usually not given consideration. The amino acid sequence and MW of the potential polypeptide for each ORF of interest can then be determined from the nucleotide sequence and the genetic code. ORFs that would give rise to proteins of highly improbable amino acid composition are usually ignored. Various parts of the genome, and particularly the 5′ and 3′ noncoding sequences, are also searched for relevant regulatory and recognition signals, as discussed in Section 5.

4.2 DEFINING FUNCTIONAL ORFs

Some of the ORFs revealed by the nucleotide sequence will code for proteins *in vivo*. Others may not. The functional ORFs can be unequivocally identified only by *in vitro* translation studies using viral mRNAs, and by finding the relevant protein in infected cells. Functional ORFs have not been found in the negative sense strand of ss-positive sense RNA viruses. However, such ORFs have been found in some of the viruses discussed in Chapter 8.

4.2.1 Identification of Functional mRNAs

The genomes of DNA viruses must be transcribed into one or more mRNAs. These must be identified in nucleic acids isolated from infected tissue and matched for sequence with the genomic DNA (e.g., cauliflower mosaic virus (CaMV), Chapter 8, Section 1). Many plant viruses with ss-positive sense RNA genomes have some ORFs that are translated only from a subgenomic RNA. These RNAs must also be identified in order to establish the strategy of the genome.

Isolation from Virus Preparations

Not infrequently, genuine viral subgenomic mRNAs are encapsidated. These can then be isolated from purified virus preparations and characterized. When the sequence of the genomic nucleic acid is known, there are two techniques available to locate precisely the 5′ terminus of a presumed subgenomic RNA. In the S1 nuclease protection procedure, the mRNA is hybridized with a complementary DNA sequence that covers the 5′ region of the subgenomic RNA. The ss regions of the hybridized molecule are removed with S1 nuclease. The DNA that has been protected by the mRNA is then sequenced. In the second method, primer extension, a suitable ss primer molecule is annealed to the mRNA. Reverse transcriptase is then used to extend the primer as far as the 5′ terminus of the mRNA. The DNA produced is then sequenced.

Isolation from Infected Host Material

Possible viral-specific mRNAs can be isolated from infected host material and characterized by the methods described in Chapter 2, Section 4.

4.2.2 Identification of Proteins Found *in Vivo* with Particular Viral ORFs

Viral-coded proteins, other than those found in virus particles, may be difficult to detect *in vivo*, especially if they occur in very low amounts, and only transiently during a particular phase of the virus-replication cycle. However, a battery of methods is now available for detecting viral-coded proteins *in vivo* and matching these with the ORFs in a sequenced viral genome. In particular, the nucleotide sequence information gives a precise estimate of the size and amino acid composition of the expected protein; knowledge of the expected amino acid sequence can be used to identify the *in vivo* product either from a partial amino acid sequence of that product, or by reaction with antibodies raised against a synthetic polypeptide that matches part of the expected amino acid sequence.

Proteins Found in the Virus

Coat proteins are readily allocated to a particular ORF by several criteria: (1) Amino acid composition compared with that calculated for the ORF; (2) amino acid sequence of part or all of the coat protein; (3) serological reaction of an *in vitro* translation product with an antiserum raised against the virus; (4) for a few viruses such as TMV, assembly of an *in vitro* translation product into virus particles when mixed with authentic coat protein. The rhabdoviruses may be exceptional in that five of the six gene products, corresponding to five of the ORFs in the genome, are found in purified virus preparations (Chapter 8, Section 4).

Direct Isolation from Infected Tissue or Protoplasts

Healthy and virus-infected cell extracts are fractionated by appropriate procedures, and proteins are separated by gel electrophoresis. Protein bands appearing in the samples from infected cells and not from healthy cells may be identified with the expected product of a particular ORF by comparing their mobility and pattern of tryptic peptides with those of an *in vitro* translation product of the ORF. With appropriate *in vivo* labeling, partial amino acid sequencing of the isolated protein may allow the precise location of its coding sequence in the genome to be established.

Serological Reactions

Antisera provide a powerful set of methods for recognizing viral-coded proteins produced *in vivo,* and identifying them with the appropriate ORF in the genome. Experimental use of antisera often involves blotting procedures, such as those described in Chapter 3, Section 2.

Antisera against Synthetic Peptides A synthetic peptide can be prepared corresponding to part of the amino acid sequence predicted from an ORF. An antiserum is raised against the synthetic peptide and used to search for the expected protein in extracts of healthy and infected tissue or protoplasts.

Antisera against in Vitro ***Translation Products*** If an mRNA is available that is translated *in vitro* to give a polypeptide product clearly identified with a particular ORF or genome segment, antisera raised against the *in vitro* product can be used to search for the same protein in extracts of infected cells or tissue.

Antisera against Proteins Derived from Cloned Genes Antibodies can be raised against protein derived *in vitro* from a cloned gene, and used to search for the corresponding protein in extracts of infected plants.

Immunogold Labeling Antibodies produced against a synthetic peptide, corresponding to part of a particular ORF in the genomic nucleic acid and labeled with gold, can be used to probe infected cells for the presence of the putative gene product.

4.2.3 Comparison with Genes Known
To Be Functional in Other Viruses

Size, location in the genome, and nucleotide-sequence similarities with known functional genes may give a strong indication that a particular ORF codes for a functional protein *in vivo*.

4.2.4 Presence of Appropriate Regulatory Signals in the RNA

AUG triplets that are used to initiate protein synthesis may have a characteristic sequence of nucleotides nearby (Section 5.1.2). Upstream of the AUG triplet, there may be identifiable ribosome recognition signals. Presence of these sequences would indicate that the ORF is functional.

4.2.5 Reoviruses

The reoviruses are a special case with respect to establishing functional ORFs. Each ds genome segment is transcribed *in vitro* to give an mRNA that codes for a single protein product.

4.3 RECOGNIZING ACTIVITIES OF VIRAL GENES

4.3.1 Gene Products Found in the Virus

Fraenkel-Conrat and Singer (1957) reconstituted the RNA of one strain of TMV in the protein of another strain that was recognizably different. The progeny virus produced *in vivo* by this *in vitro* "hybrid" had the coat protein corresponding to the strain that provided the RNA. Since this classical experiment, it has been universally assumed that viral genomes code for coat proteins. Likewise, it has usually been assumed that other proteins found as part of the virus particle are also virus-coded—for example, those found in reoviruses and rhabdoviruses.

4.3.2 Classical Viral Genetics

Two kinds of classical genetics studies have identified many biological activities of viral genomes, and both these procedures are still useful in appropriate circumstances.

Allocation of Functions in Multiparticle Viruses

With ss positive sense RNA viruses that have been shown to have their genome in two or three separate pieces of nucleic acid, it is possible to carry out reassortment experiments. The nucleic acid components of two strains of a virus with a difference in some observable property are physically separated, and then mixed *in vitro* in different combinations. These mixtures are then inoculated into a suitable host plant. Appearance of the property, following inoculation with a particular mixture or mixtures, can identify the genome segment that contains the relevant gene. This kind of procedure has been useful in allocating the coat-protein gene and activities responsible for symptoms in a given host to a particular genome segment.

Virus Mutants

The study of naturally occurring or artificially induced mutants of a virus has allowed various virus activities to be identified. Again, many of the activities involve biological properties of the virus. Mutants that grow at a normal (permissive) temperature but that replicate abnormally or not at all at the nonpermissive (usually higher) temperature are particularly useful. Such temperature-sensitive (*ts*) mutants are easy to score and manipulate, and most genes seem to be potentially susceptible to such mutations. They arise when a base change (or changes) in the viral nucleic acid gives rise to an amino acid substitution (or substitutions) in a protein, which results in defective function at the nonpermissive temperature. Alternatively, the base change might affect the function of a nontranslated part of the genome—a control element, for example. The experimental objective is to collect and study a series of *ts* mutants of a particular virus.

To be useful for studies on replication, *ts* mutants must possess certain characteristics: (1) they must not be significantly leaky at the nonpermissive temperature, and (2) the rate of reversion to wild type must be low enough to allow extended culture of the mutant at both the permissive and nonpermissive temperatures.

4.3.3 Use of Recombinant DNA Technology

Many of the techniques that use some aspect of recombinant DNA technology to match a gene activity with a functional ORF may at the same time lead to recognition of a new function for that gene (Section 4.4.2).

4.4 MATCHING GENE ACTIVITIES WITH FUNCTIONAL ORFs

4.4.1 Direct Testing of Protein Function

Some viral-coded proteins besides coat proteins, such as viral proteases, can be identified directly in *in vitro* tests. The tests can be made on a protein isolated from infected cells or on the *in vitro* translation product of a particular segment of viral RNA.

4.4.2 Approaches Depending on Recombinant DNA Technology

Reverse Genetics

As discussed in Chapter 2, Section 5, the introduction of a DNA stage in an RNA virus life cycle allows the application of recombinant DNA technology to study the replication and biological properties of RNA viruses. Furthermore, it allows the production of genetically well-defined virus isolates. One important application of this technique is the introduction of defined changes in particular RNA viral genes in order to study their biological effects, and thus to define gene functions. The changes that can be introduced at defined sites include point mutations and small or large deletions, or insertion of known nucleotide sequences.

Location of Spontaneous Point Mutations

Knowledge of nucleotide sequences in natural virus variants allows a point mutation to be located in a particular gene even if the protein product has not been isolated. In this way the changed or defective function can be allocated to a particular gene.

Expression of the Gene in a Transgenic Plant

By introducing a single viral gene into the host-plant genome, it is possible to study the gene's function in the absence of the expression of other viral genes and of other elements of the viral genomic RNA. Such transformed plants provide a new method for physical dissection of the viral genome and assignment of functions to the ORFs identified by computer.

Recombinant Viruses

Recombinant DNA technology can be used to construct viable viruses from segments of related virus strains that have differing properties, and thus to associate that property with a particular viral gene. In some circumstances, it is also possible to construct viable recombinant hybrids between different viruses.

Hybrid Arrest and Hybrid Selection Procedures

Hybrid arrest and hybrid selection procedures (Chapter 2, Section 4.3) can be used to demonstrate that a particular cDNA clone contains the gene for a particular protein. In appropriate circumstances, these procedures can be used to identify gene function.

Sequence Comparison with Genes of Known Function

As noted (Section 4.2), sequence comparisons can be used to obtain evidence that a particular ORF may be functional. The same information may also give strong indications as to the actual function. For example, the gene for an RNA-dependent RNA polymerase has been positively identified in poliovirus. There are amino acid sequence similarities between this poliovirus protein and proteins coded for by several plant viruses. This similarity implies quite strongly that these plant viral-coded proteins also have a polymerase function. The conserved amino acid sequence is $GXXXTXXXN(X)_{20-40}GDD$, where X represents any amino acid. The sequence is found in the proposed RNA-dependent RNA polymerase of all plant and most animal ss positive sense RNA viruses.

Functional Regions within a Gene

Spontaneous mutations and deletions can be used to identify important functional regions within a gene. Mutants obtained by site-directed mutagenesis and deletions constructed *in vitro* can give similar information in a more systematic and controlled manner.

5 THE REGULATION OF VIRUS PRODUCTION

There is ample evidence for many plant viruses that overall virus production is usually under some form of control. Host cell factors must interact with viral regulatory processes to modulate virus replication.

However, we have almost no knowledge of the host factors involved. This section summarizes knowledge about viral regulatory processes.

5.1 REGULATORY AND RECOGNITION SIGNALS IN RNA VIRAL GENOMES

5.1.1 The Role of Terminal Structures

The 5' and 3' terminal structures described in Section 1.6 probably have regulatory or recognition functions, but the evidence is not conclusive. A major question is why small RNA viruses that have many otherwise similar properties differ in their 5' or 3' terminal features.

The 5' Cap

The role of the cap structure can be considered in relation to three properties:

1. *Infectivity* For some viruses, the cap is essential for infectivity. With others its removal merely lowers infectivity.

2. *Stability* The cap protects RNAs *in vitro* from exonuclease activity, and may have this role *in vivo*.

3. *Efficiency of translation* The cap has little effect on *in vitro* translation of most viral mRNAs; however, it may have a role in differential translation of the RNAs of some viruses with multipartite genomes.

The VPg

Removal of the VPg abolishes the infectivity of some viruses but not others. Although proof is lacking, the VPg may play a role in the initiation of RNA synthesis.

The 3' Poly(A) Sequence

For one plant virus, the 3' poly(A) sequence appears to be coded for in the genome by a poly(U) sequence at the 5' terminus of the negative sense strand. For most, however, it is added after RNA synthesis, probably by a host poly(A) polymerase. Some viruses of this sort have a nucleotide sequence that serves as a signal for polyadenylation near the 3' terminus. This is (AUAAA), which precedes the 3′ terminus by about 10 to 15 nucleotides. Poly(A) sequences have been shown to promote translational stability in *in vitro* protein-synthesizing systems, and they may have this role *in vivo*.

3′ tRNA-like Structures

A variety of functions have been proposed for the tRNA-like structures at the 3′ termini of some viral RNAs. The following four possible functions and origins are not necessarily mutually exclusive: (1) accepting and donating an amino acid in some aspect of protein synthesis; (2) facilitating translation; (3) acting as the replicase recognition site; and (4) representing molecular fossils from a very early stage of evolution.

Some viruses have been shown to have their 3′ tRNA-like termini charged with an amino acid *in vivo*. Other experiments involving site-directed mutagenesis in the relevant nucleotide sequence showed that aminoacylation was not essential for infectivity. The 3′ sequence in all the ssRNA viruses must be involved in the recognition site used by the viral replicases to initiate minus-strand RNA synthesis. This aspect is discussed in Section 5.1.2.

5.1.2 The Role of Nucleotide Sequences

Specific nucleotide sequences play regulatory and recognition roles in both protein and nucleic acid synthesis.

In Protein Synthesis

Both coding and noncoding nucleotide sequences may have recognition and regulatory roles in the translation of viral RNAs.

Ribosome Recognition Sequences   As discussed in Section 1.3, plant virus RNAs, with rare exceptions, behave like eukaryotic mRNAs in that they are translated only from an AUG near the 5′ terminus and not from internal initiation sites, as happens in prokaryotes. Therefore we expect to find one or more ribosome recognition sites between the 5′ terminus and the first functional AUG. The 5′ untranslated region of viral RNAs usually has a low G + C content, suggesting a low degree of secondary structure. A number of plant viruses have been shown to bind two or three ribosomes along this sequence, forming disome or trisome initiation complexes.

Selection of AUG Initiation Codons   Since there are usually AUG triplets in an mRNA other than the one used for initiation of translation, there must be a mechanism whereby the correct AUG is recognized by the ribosome and/or associated factors. The recognition lies in the nucleotides immediately surrounding the correct AUG triplet, with an A in position −3 and a G in position +4 perhaps being particularly important.

Sequences Enhancing Translation Rate   Sequences from the 5′ leaders of several plant viruses will act in cis to enhance the translation of foreign RNAs both *in vitro* and *in vivo,* and in prokaryote or eukaryote systems. Mutational analysis has suggested that the enhancement of translation by such sequences may be owing to reduced secondary structure, either within the leader sequence or between this sequence and downstream coding regions. Such secondary structure could inhibit or delay translation initiation by interfering with the binding of ribosome subunits or other protein factors.

Read-Through Termination Signals   In several ss positive sense RNA plant viruses, the first ORF from the 5′ terminus of the genome, or genome segment, may terminate in a stop codon that is sometimes suppressed. This suppression allows the ribosome to read to the next stop codon, giving rise to a read-through protein of greater length (Fig. 6.1). The synthesis of a read-through protein depends primarily on the presence of appropriate suppressor tRNAs. The proportion of read-through protein produced may be modulated by nucleotide sequences around the suppressed termination codon. A sequence containing two or three A residues immediately 5′ to a UAG stop codon has been found in several viruses, for example, 5′—CAAA*UAG*—3′.

Sequences Controlling Transframe Proteins   It has been proposed that the nucleotide sequence involved in the translational frameshift of a plant viral RNA is UU immediately 5′ to the *UAG* stop codon (Fig. 6.1).

In RNA Synthesis

Promoter Sequences   The term *promoter* was first used to designate nucleotide sequences in cellular DNA, just upstream of a gene, that were recognized by a DNA-dependent RNA polymerase as the initial event in transcription of mRNA for the gene. With RNA viruses, the term promoter is used to indicate nucleotide sequences recognized by the viral RNA-dependent RNA polymerase for the synthesis of both genomic RNAs and subgenomic mRNAs.

Experiments in which expressible cDNA clones were used to construct partial deletions throughout a viral RNA showed that the coding region could be changed or deleted without affecting replication. However, the 5′ and 3′ untranslated regions are essential for RNA replication.

Minus Strand Promoters   These are most likely to be near the 3′ end of the plus-strand genomic RNA. In viruses with segmented genomes, the 3′ terminal region usually shows a high degree of sequence homology. In addition, the 3′ sequences of related viruses usually show more sequence homology than the genome as a whole. This marked conservation of sequence undoubtedly reflects an important function or functions. One function must be the viral replicase recognition sequence. Mutational analysis of the tRNA-like structures at the 3′ termini of some viruses indicates that much of the three-dimensional structure may be important in promoter activity.

Promoters Near 5′ Termini   If the same viral replicase initiates minus-strand synthesis and plus-strand synthesis using the same recognition sequence, then we would expect the replicase recognition sequences at the 5′ and 3′ ends of the genomic RNA to be complementary. For most RNAs, there does not appear to be any such significant complementary relationship.

Since far more plus-strand than minus-strand RNA is synthesized during virus replication, the replicase must use the minus-strand 3′ promoter much more frequently than the plus-strand 3′ promoter. It is not known how this imbalance is maintained, but there are several possibilities: (1) two different viral-coded replicases might contain different recognition sites. However, only one region with amino acid sequence similarity to RNA-dependent RNA polymerases is present, at least in some ssRNA virus polymerases; (2) one viral RNA polymerase may recognize both 3′ minus-strand and 3′ plus-strand promoter sequences either with the same or with different active sites; (3) some host or viral-coded protein factor may affect the specificity of a single recognition site in the replicase; and (4) the promoter specificity of the polymerase may be altered by different subunit associations.

For viruses with multipartite genomes, roughly equal amounts of genomic RNA segments are usually synthesized. ssRNA viruses with such genomes show some sequence homology near the 5′ terminus, but it is usually less extensive than at the 3′ end.

Internal Promoters   The subgenomic mRNAs used by many RNA plant viruses for coat-protein synthesis are not necessary for infectivity. Therefore, the subgenomic RNA must be transcribed from a genomic RNA. Those that have been studied arise by internal initiation on a negative sense genomic strand. Figure 6.5 illustrates one internal promoter that has been precisely defined.

In Virus Assembly

There are two general mechanisms whereby viral coat proteins could package only viral nucleic acids into particles: (1) if synthesis and assembly of virus particles takes place in a cell compartment that does not contain free host nucleic acids, or (2) if virus assembly is initiated

5′...UCUGAGUUAUUAUUA<u>AAAAAAAAAAAAAAAAAAAAA</u>GAUCUAUGUCCUAAUUCAGC GUAUUAAUA<u>AUG</u>UCGAC...3′

RNA4:   m7GpppGUAUUAAUA<u>AUG</u>UCGAC...3′

Figure 6.5   Nucleotide sequence of an internal promoter. The genome of brome mosaic *Bromovirus* (BMV) consists of RNAs 1, 2, and 3. RNA3 contains two genes. The gene for the coat protein is located to the 3′ end of the RNA3. Synthesis of plus-strand coat protein mRNA (RNA4) is initiated by an internal promoter sequence in the genomic RNA3 minus strand. The sequences above illustrate the relationship of the promoter region to the 5′ sequence of RNA4. The positive sense sequence packaged into virus particles is shown. The actual promoter sequence for RNA4 plus-strand synthesis is complementary to the sequence shown. The upper sequence is the promoter that lies mainly 5′ to the beginning of the RNA4 sequence indicated by the arrow, but extends some way into the untranslated sequence of RNA4. The lower sequence is the 5′ region of RNA4 with its AUG initiation codon underlined. From French and Ahlquist (1988).

by a specific recognition process between coat protein and viral RNA. The second process is known to operate for some viruses and is probably widespread. For example, the initiation of TMV rod assembly involves a specific internal nucleotide sequence in the RNA (Chapter 7, Section 3.1).

5.2 REGULATORY AND RECOGNITION ROLES FOR VIRAL PROTEINS

It seems certain that viral-coded proteins must have roles concerning regulation and recognition in relation to the virus life cycle and to the effects of infection on the host cell. These functions will involve interactions between viral-coded proteins, between such proteins and viral nucleic acids, and between viral proteins or nucleic acids and host components. At present this is an important area where little meaningful information is available.

6 EXPERIMENTAL SYSTEMS FOR STUDYING VIRAL REPLICATION *IN VIVO*

6.1 INTACT LEAVES

Inoculated leaves or systemically infected leaves have both advantages and disadvantages. The main advantages are ease of preparation and the fact that virus is replicating in a natural system. The major disadvantage is the marked asynchrony in the times at which cells become infected in the tissue sampled. The synchrony of infection in the young, systemically infected leaf can be greatly improved by manipulating the temperature. The lower inoculated leaves of an intact plant are maintained at normal temperatures ($\simeq 25-30°$ C), while the upper leaves are kept at 5 to 12° C. Under these conditions, systemic infection of the young leaves occurs, but replication does not. When the upper leaves are shifted to a higher temperature, replication begins in a fairly synchronous fashion. The main requirement is for a systemic host with a habit of growth that makes it possible for upper and lower leaves to be kept at different temperatures.

6.2 PROTOPLASTS

Protoplasts are isolated plant cells that lack the rigid cellulose wall found in intact tissue. They are prepared by treating leaf tissue from which the lower epidermis has been stripped with a fungal pectinase, which gives rise to separated cells. These are then treated with a fungal cellulase to remove the cell wall (Fig. 6.6).

Protoplasts can be infected with intact virus or viral RNAs. Besides greatly improved synchrony of infection compared with intact leaves, protoplasts have several other advantages: (1) close control of experimental conditions; (2) uniform sampling can be carried out by pipetting; (3) the high proportion of infected cells (often 60–90%); (5) the relatively high efficiency of infection; (6) organelles, such as chloroplasts and nuclei, can be isolated in much better condition from protoplasts than from intact leaves.

However, a number of actual or potential limitations and difficulties must be borne in mind. (1) Protoplasts are very fragile—both mechanically and biochemically. Their fragility may vary markedly, depending on the growing conditions of the plants, season of the year, time of day, and the particular age of leaf chosen. (2) Under culture conditions that favor virus replication, protoplasts survive only for 2 to 3 days, and then

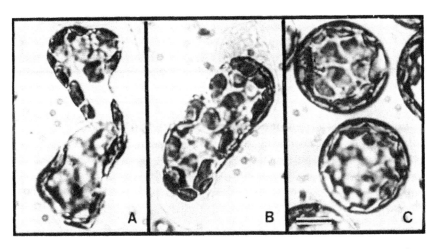

Figure 6.6 Isolation of protoplasts from chinese cabbage leaves. (A) A separated spongy mesophyll cell and (B) a separated palisade cell, following pectinase treatment. The cells still retain the cellulose wall. Following treatment of isolated cells with cellulase, spherical protoplasts are produced (C). Bar = 10 μm. Courtesy of Y. Sugimura.

decline and die. (3) To prevent growth of microorganisms during in-cubation, antibiotics may be added to the medium. These may have unexpected effects on virus replication. (4) Compared to intact tissue, relatively small quantities of cells are made available. (5) Cytological effects observed in thin sections of infected leaf tissue may not be re-produced in protoplasts, probably because of the effects of changed os-motic conditions on cell membranes. (6) The isolation procedure and the medium in which they are maintained may drastically affect the physiological state of the cells.

6.3 SITES OF SYNTHESIS AND ASSEMBLY

Two general kinds of procedures have been used in attempts to define the intracellular sites of virus synthesis and assembly: (1) fractionation of cell components from tissue extracts followed by assay for virus or virus components in the various fractions, and (2) light and electron microscopy.

We have learned the most about sites of synthesis and assembly where both kinds of technique have been applied. Immunocytochemical methods are being used increasingly to locate viral-coded proteins within cells and tissues.

7 ERRORS IN VIRUS REPLICATION

The process of virus replication gives rise to a proportion of progeny particles that are not identical to the infecting virus. We can distinguish three kinds of differences: (1) mutations, etc., that give rise to new viable strains of the virus, discussed in Chapter 12; (2) defective inter-fering virus particles, discussed in Chapter 9; and (3) mixed virus as-sembly, discussed here.

Mixed virus assembly can be shown to take place *in vitro* between the RNA of one strain of a virus and the coat protein of another; between RNA and protein from unrelated viruses; and between one kind of RNA and two different coat proteins. Mixed virus particles are also formed *in vivo*. When two viruses multiply together in the same tissue, some progeny particles may be formed that consist of the genome of one vi-rus housed in a particle made partially or completely from the structural components of the other virus. Barley yellow dwarf *Luteovirus* (BYDV) provides an interesting example.

Strains of BYDV show aphid vector specificity (Chapter 13, Section 3.2.4). When a strain of the virus normally transmitted in a particular vector was grown in oats in a double infection with a serologically unrelated strain not normally transmitted by the aphid, this latter strain was transmitted. This transmission was the result of the fact that some of the RNA of the second strain had been assembled into protein shells of the normally transmitted strain. Such phenotypic mixing can occur between two unrelated helical viruses with different dimensions. It has even been found between a helical virus (barley stripe mosaic *Hordeivirus*) and an icosahedral one, brome mosaic *Bromovirus* (BMV). With some viruses, significant amounts of host RNA may be incorporated into virus-like particles.

FURTHER READING

Davies, J. W., Wilson, T. M. A., and Covey, S. N. (1991). "The Molecular Biology of Plant Viruses." Chapman and Hall, in press.

Matthews, R. E. F. (1991). "Plant Virology," 3rd Ed. Academic Press, New York.

REPLICATION OF VIRUSES WITH ss-POSITIVE SENSE RNA GENOMES

7

As discussed in Chapter 6, the plant viruses with ss positive sense RNA genomes have developed five major strategies for efficient translation in a eukaryotic protein-synthesizing system. For the purposes of this chapter, I shall include translational frameshifting with the readthrough mechanism, as they have essentially the same result. Table 7.1 summarizes the strategies used by 18 groups of viruses. These can be arranged in seven categories as indicated in Table 7.1. In this chapter I shall consider one example from each category. The other groups are described by Matthews (1991). The arrangement shown in Table 7.1 is used for convenience, and should not be regarded as the basis for a proposed classification.

Organization and expression of the genome is summarized in a map for each virus considered. The conventions used in these maps are summarized in Table 7.2. There is no doubt that further work will lead to changes in particular aspects for most of these genome maps. Aspects of the replication of different viruses have been studied to widely differing extents, partly for technical reasons, and partly by chance. This situation is reflected in the very uneven coverage of topics for different viruses in this chapter and in Chapter 8.

1 THE *POTYVIRUS* GROUP

The potyviruses are a large and economically very important group of viruses infecting a wide range of species. They are transmitted mechanically, but insect vectors, especially aphids, are most important in

Table 7.1 Summary of Genome Strategies Adopted by 18 ss-Positive Sense RNA Plant Virus Groups

Number	Strategy (see Fig. 6.2)	Virus group	No. of ORFs	No. of proteins coded
I	One strategy. Polyprotein	*Potyvirus*	1	8
II	One strategy. Subgenomic RNA	*Potexvirus*	5	4–5
		Tombusvirus	≥3	≥3
III	Two strategies. Subgenomic RNA plus read-through or frameshift protein	*Tobamovirus*	5	4–5
		Luteovirus	6	6
		Carmovirus	5	7
IV	Two strategies. Subgenomic RNAs and polyprotein	*Tymovirus*	3	3–4
		Sobemovirus	4	4–5
V	Two strategies. Multipartite genome and polyprotein	*Comovirus*	2	9
		Nepovirus	2	5
VI	Two strategies. Subgenomic RNAs and multipartite genome	*Bromovirus*	4	4
		Cucumovirus	4	4
		Alfalfa mosaic virus	4	4
		Ilarvirus	4	4
		Hordeivirus	7	7
VII	Three strategies. Subgenomic RNAs, multipartite genome and readthrough protein (or frameshift)	*Tobravirus*	5	5–6
		Furovirus	9	6–9
		Dianthovirus		4

the field. In contrast to most other plant viruses, some stable, nonstructural gene products may accumulate in substantial amounts in infected cells, greatly facilitating their study. This group of viruses is the only known plant group in which the polyprotein strategy is the only one involved. Figure 7.1 illustrates this strategy for tobacco etch virus (TEV), the virus about which most is known.

Figure 7.1 shows a proposed scheme for the proteolytic processing of the initial long polyprotein. It is not certain whether any complete polyprotein molecules are synthesized or whether all are cleaved at least once before synthesis is complete. Two proteinase activities have been assigned to the 87-K and 49-K products and a third proteinase activity has been detected. Five sites that are cleaved by the 49-K protein have the following sequence.

$$\text{Glu-Xaa-Xaa-Tyr-Xaa-Gln} \overset{\downarrow}{-} \text{Ser (or Gly)}$$

Table 7.2   Conventions Used in the Genome Maps for RNA Viruses

Scale for genome length in 1000s of nucleotides

```
L_____L_____L_____L
0      1      2      3 kb
```

Genomic RNA
2200 nt

━━━━━━━━━━━━━━━━━━ Genome or subgenomic RNAs with length indicated in nucleotides

End groups, and structures near termini

5' cap = me^{7}Gppp

VPg = covalently bonded protein

5' = no special structure

? = not known

poly(A) 3' = a run of As usually of variable length (not included in quoted genome length)

tRNA^{tyr}3' = 3' terminal nucleotides can fold to give a tRNA like structure accepting the amino acid indicated

3' OH = no special structure at 3' termini

```
AUG              UAA
3495             4917
┌───────────────────┐
│    ORF1   54K     │
└───────────────────┘
```
= open reading frame 1, with a coding potential of Mr ≈ 54000
position of initiation and termination codons indicated by number of nucleotides from 5' terminus of the genomic RNA

ORFs are numbered in one series for the whole genome

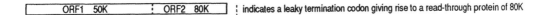

```
┌──────────────────┊──────────────────┐
│   ORF1   50K     ┊   ORF2   80K     │
└──────────────────┊──────────────────┘
```
┊ indicates a leaky termination codon giving rise to a read-through protein of 80K

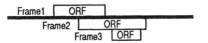

```
Frame1   ┌─ORF─┐
    Frame2  ┌──ORF──┐
       Frame3 ┌ORF┐
```
Where more than one reading frame is involved they are arranged as indicated

∿∿∿∿∿∿∿∿∿∿∿ = Protein product

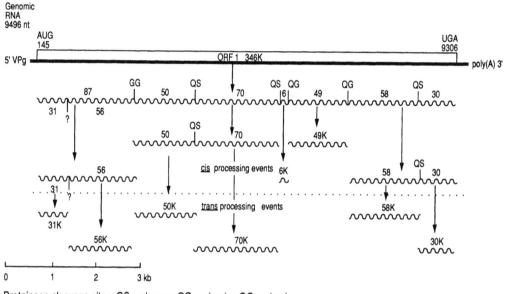

Proteinase cleavage sites QS = gln-ser ; QG = gln-gly ; GG = gly-gly

Protein functions :   87K , proteinase ; 31K , cell-to-cell movement ; 56K , insect transmission ; 50K , proteinase ;
                      70K , replication ; 6K , ? ; 49K , proteinase ; 50K polymerase ; 30K , coat protein.

· · · · · · · · · · · · ·   indicates *cis* or *trans* processing

Figure 7.1 Organization and Expression of a *Potyvirus* genome (TEV).

Xaa = any amino acid. Arrow indicates the bond that is cleaved. As
shown in Fig. 7.1, some proteolytic cleavage occurs while the enzyme is
still part of the polyprotein (cis cleavage). Other cuts are made inter-
molecularly by the free enzyme (trans cleavage).

About 2000 molecules of the 20-K coat protein are needed to make
one rod-shaped virus particle. Since the coat protein gene is at the 3'
end of the genome, a complete polyprotein must be translated for each
coat protein molecule. This may account for the fact that several other
viral-coded proteins accumulate in large quantities in infected cells.
These are (1) the cytoplasmic pinwheel inclusion protein of $M_r \simeq 70$ K,
probably involved in RNA replication. It aggregates to form pinwheel
inclusion bodies that are characteristic of the *Potyvirus* group; (2) the
cytoplasmic amorphous inclusion body protein, which is probably a
viral-coded protein necessary for insect transmission of the virus; (3)
nuclear inclusion proteins of $M_r \simeq 58$ K (polymerase) and 49 K (proteinase), which form the nuclear inclusions found in some *Potyvirus*

infections. Only full-length genomic RNA and its negative sense coun-
terpart have been found in infected tissues.

2 THE *POTEXVIRUS* GROUP

Individual viruses in this group have rather limited host ranges, but as a
group, members infect a wide range of species, and are geographically
very widespread. They are stable and transmitted by mechanical means.
The group provides an example of viruses that use only subgenomic
RNAs in order to have their genes translated in a eukaryotic system.

Various *Potexvirus* genomes have been fully sequenced. All show
the same genome organization, with five open reading frames (ORFs),
as illustrated in Fig. 7.2 for white clover mosaic virus (WCMV) (data

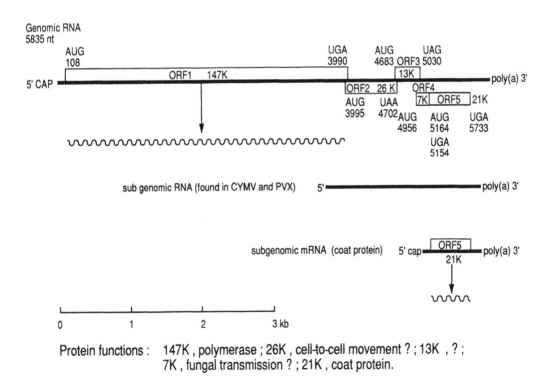

Figure 7.2 Organization and expression of a *Potexvirus* genome (white clover mosaic
virus).

from Beck *et al.* 1991). The protein encoded by the 147-K ORF has domains of sequence similarity with the presumed polymerase genes of other RNA viruses. ORFs 2, 3, and 4, which are all involved in cell-to-cell movement, are known as the *triple gene block*. Such an arrangement has been found for some other ssRNA plant viruses. There is a polyadenylation signal (AAUAAA) at various distances from the 3′ terminus in different potexviruses.

Subgenomic RNAs for coat protein synthesis have been found *in vivo*, sometimes in association with polyribosomes. A larger subgenomic RNA has also been isolated. Genomic RNA has also been found in the polyribosome fraction, indicating active translation *in vivo*. Genome-sized double-stranded RNAs (dsRNAs) have been found *in vivo*, probably representing a replicative form (RF) of the RNA.

Among the plant viruses with flexuous rod-shaped particles, most is known about the assembly *in vitro* of potexviruses, particularly papaya mosaic virus (PapMV). The process shows both similarities to and differences from the assembly of tobacco mosaic virus (TMV), but much less detail is understood. The site for nucleation of coat protein on the RNA is at the 5′ end. Coat protein forms aggregates of various sorts. In particular, at pH 7.0, a double disk is formed containing two layers of nine subunits. This double disk probably initiates virus assembly. Five nucleotides are associated with each protein subunit in the virus rod (compared with three for TMV).

3 THE *TOBAMOVIRUS* GROUP

TMV, the type member of this group, was the first virus to be isolated, and it remains one of the most studied among the plant viruses. Most tobamoviruses have wide host ranges among the angiosperms, and they are found wherever their crop and weed hosts grow. The replication of TMV has been studied more than any other plant virus. For this reason it is discussed in more detail than are the other groups. In particular, a great deal of experimental work has been carried out on the assembly of the virus *in vitro* from its components. Tobamoviruses combine two genome strategies—subgenomic RNAs, together with a read-through protein.

3.1 GENOME STRUCTURE

The nucleotide sequence of TMV revealed several closely packed ORFs. A m^7 Gppp cap is attached to the first nucleotide (guanylic acid). This is followed by an untranslated leader sequence of 69 nucleotides. The 5′ ORF codes for a 126-K protein. Experiments described below show that the termination codon for this protein (UAG) is leaky, and that a second, larger read-through protein is possible. This read-through protein has a MW of 183 K. The terminal five codons of this read-through protein overlap a third ORF coding for a protein of 30 K. This ORF terminates two nucleotides before the initiation codon of the fourth ORF that is closest to the 3′ terminus. It codes for a 17.6-K coat protein. As discussed below, the two smaller ORFs at the 3′ end of the genome are translated from subgenomic RNAs. A third subgenomic RNA termed I_1 RNA, representing approximately the 3′ half of the genome, has been isolated from TMV-infected tissue. S1 mapping showed that this RNA species had a distinct 5′ terminus. The 3′ untranslated sequences of these RNAs fold in the terminal region to give a transfer RNA (tRNA)-like structure accepting histidine. These relationships are summarized in Fig. 7.3. This genome structure is probably common to all members of the *Tobamovirus* group.

3.2 PROTEINS SYNTHESIZED *IN VITRO*

TMV genomic RNA has been translated in several cell-free systems. Two large polypeptides are produced, but no coat protein is made. The synthesis of the two larger proteins is initiated at the same AUG. The larger protein is generated by partial read-through of a termination codon. The two proteins are read in the same phase, so the amino acid sequence of the smaller protein is also contained within the larger one. The I_1 subgenomic RNA isolated from infected tissue is translated *in vitro* in the rabbit reticulocyte system to produce a polypeptide of M_r 54 K, but this product has not been detected in infected tissue.

The I_2 RNA isolated from infected tissue is translated to produce a protein of M_r 30 K. The I_2 RNA also contains the smaller 3′ coat protein gene (Fig. 7.3), but this is not translated in *in vitro* systems. The smallest TMV gene, (the 3′ coat protein gene) is translated *in vitro* only from the monocistronic subgenomic RNA (Fig. 7.3). This gene can also be

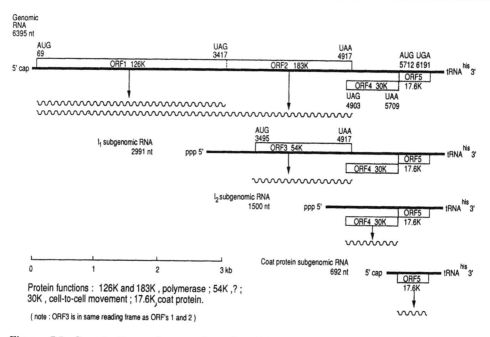

Figure 7.3 Organization and expression of a *Tobamovirus* genome (TMV, vulgare strain).

translated efficiently *in vitro* by prokaryotic protein-synthesizing machinery from *Escherichia coli*.

3.3 PROTEINS SYNTHESIZED *IN VIVO*

Proteins corresponding in size to the 183-K, 126-K, and 30-K proteins, and of course the 17.5-K coat protein, have been detected in infected cells and tissues. Their identity was confirmed in the following ways: determination of the nucleotide sequence in the 3' region of the TMV genome readily located the gene for coat protein, since its full amino acid sequence was already known. Peptide analysis on a large protein found *in vivo* showed it to be the same as an *in vitro* product of similar size. For the 30-K protein, a synthetic peptide was made with the sequence for the 16 C-terminal amino acids predicted from the nucleotide sequence. Antibodies raised against this peptide precipitated the 30-K protein, positively identifying it as the I_2 RNA product. No protein has yet been detected *in vivo* that corresponds to the 54-K *in vitro* translation product of the I_1 subgenomic RNA; thus, the TMV genome codes for four gene products, with a probable fifth yet to be established.

3.4 FUNCTIONS OF THE VIRAL-CODED PROTEINS

3.4.1 Coat Protein

Besides its structural function in the virus particle, the coat protein can influence disease in various ways. For example, some strains of TMV can cause necrotic local lesions rather than systemic mosaic disease. It has been shown that a substitution of a phenylalanine for a serine at position 148 in the coat protein is responsible for the resistant necrotic response in one tobacco host, but other hosts containing different resistance genes are not affected by this mutation. In addition, some coat protein mutants establish systemic infection less effectively than do wild-type viruses.

3.4.2 The 126-K and 183-K Proteins

The following lines of evidence show that one or both of these proteins are involved in viral RNA replication: (1) they have significant amino acid sequence homology with known viral RNA-dependent RNA polymerases; (2) mutation in the 126-K gene caused a reduction in the synthesis of the 30-K protein and its mRNA, suggesting that the 126-K (and/or 183-K) proteins are involved in the synthesis of I_2 RNA; (3) The most convincing evidence is the fact that a mutant in which both the 30-K and the coat-protein genes were deleted, replicated, in infected protoplasts to yield a shortened viral RNA; (4) *in vitro* mutagenesis at or near the leaky termination codon of the 126-K gene indicates that both proteins are necessary for normal TMV replication in tobacco leaves; (5) although the TMV replicase has not yet been characterized, a replication complex containing the 126-K and 183-K polypeptides has been solubilized from infected cells.

3.4.3 The 30-K Protein

The following evidence shows that the 30-K protein of TMV is required for cell-to-cell movement: (1) studies on the four proteins encoded by a

TMV mutant *ts* for cell-to-cell movement (LsI) showed that the mutant differed from normal virus only in the 30-K protein. The mutation leading to the *ts* state was shown to change a serine to a proline in the 30-K protein; (2) transgenic tobacco seedlings expressing the normal 30-K protein complemented the LsI mutant, allowing it to spread from cell to cell and move systemically; (3) in a reverse genetics approach, the point mutation found in the LsI mutant was introduced into the parental virus. The resultant virus had the LsI phenotype; (4) various frame-shift mutations in the 30-K gene yielded a defective phenotype. These mutations could replicate in protoplasts, but none showed infectivity for tobacco, i.e., local lesions or systemic infection. The 30-K protein has been shown to bind strongly to ssRNA in a nonsequence-specific manner. Such binding might be involved in moving viral RNA to the plasmodesmata.

3.5 CONTROLLING ELEMENTS IN THE VIRAL GENOME

Several controlling elements have been recognized or inferred in TMV RNA: (1) the nucleotide sequence involved in initiating assembly of virus rods. This is discussed in Section 3.9.1; (2) the replicase recognition site in the 3′ noncoding region (Chapter 6, Section 5.1); (3) the start codon sequence context may be one form of translational regulation. The context differs for each of the four known gene products. For example in strain U$_1$, the contexts are as follows:

$$126 \text{ K: ACA\underline{AUG}G} \qquad 54 \text{ K: GAU\underline{AUG}C}$$
$$30 \text{ K: UAG\underline{AUG}G} \qquad \text{coat protein: AAU\underline{AUG}U.}$$

The reasons for these differences are not known; (4) RNA promoters presumably have a role in regulating the amounts of subgenomic RNAs produced, but the exact sequences for these subgenomic promoters have not been identified. That for the coat protein lies within 100 nucleotides upstream of the ORF.

3.6 VIRAL RNA SYNTHESIS *IN VITRO*

In vitro RNA synthesis has been detected only in crude fractions from infected cells. A virus-specific, RNA-dependent RNA polymerase associated with the membrane fraction can catalyze the synthesis of both positive and negative sense strands. The major products labeled with radioactive nucleotides had the expected properties for TMV replicative form (RF) and replicative intermediate (RI) (Fig. 6.2). Much more label was incorporated into plus strands than into minus strands in the RF fraction. In the RI fraction, only labeled plus strands could be detected. The system has a number of limitations. Synthetic activity is short-lived. The system does not respond to added template RNA, and little or no free progeny viral RNA molecules are formed. RF of genomic size but not of coat protein subgenomic size has been detected. Furthermore, no minus-strand RNA of this size has been found. Thus, the subgenomic coat protein RNA is probably synthesized by internal initiation on the full-length negative-strand RNA rather than by transcription from a subgenomic negative sense strand.

3.7 EARLY EVENTS FOLLOWING INFECTION

3.7.1 Disassembly of the Virus *in Vitro*

To initiate infection, TMV RNA must be uncoated, at least to the extent of allowing the first cistron to be translated. Wilson (1984) found that treatment of TMV rods briefly at pH 8.0 allowed some polypeptide synthesis to occur when the treated virus was incubated in an mRNA-dependent rabbit reticulocyte lysate. He suggested that the alkali treatment destabilizes the 5' end of the rod sufficiently to allow a ribosome to attach to the 5' leader sequence and then to move down the RNA, displacing coat-protein subunits as it moves. He termed the ribosome–partially stripped rod complexes *striposomes,* and suggested that a similar uncoating mechanism may occur *in vivo*. Probably because it contains no G residues, the 5' leader sequence interacts weakly with coat protein subunits. Such microinstability may allow ribosomes to attach to the leader sequence, initiating cotranslational disassembly (Mundry *et al.,* 1991).

3.7.2 Experiments with Protoplasts

Protoplasts have been used in various studies aimed at elucidating early events in the infection process. However, in view of the abnormal state of the cells and particularly the nature of the suspension medium, the relevance of studies in protoplasts to the infection process in leaves following mechanical inoculation is open to question.

3.7.3 Early Events in Intact Leaves

The uncoating process has been examined directly by applying to intact leaves TMV radioactively labeled in the protein or the RNA or in both components. There is a fundamental difficulty with all such experiments. Concentrated inocula must be used to provide sufficient virus for analysis, but this means that large numbers of virus particles enter cells rapidly. It is impossible to know which among these particles actually establish an infection.

There is some evidence suggesting that TMV particles are uncoated *in vivo* in epidermal cells as an 80 S ribosome moves along the RNA from the 5′ end in the manner suggested from *in vitro* studies (Section 3.7.1). Translation complexes with the expected properties of striposomes have been isolated from the epidermis of tobacco leaves shortly after inoculation; however, more definitive experiments are needed to establish that this uncoating mechanism operates *in vivo*.

3.8 RNA AND PROTEIN SYNTHESIS *IN VIVO*

Viral protein synthesis does not suppress total host cell protein synthesis, but occurs in addition to normal synthesis. Two days after infection, viral coat-protein synthesis in leaves may account for about 7% of total protein synthesis. Synthesis of the 126-K protein was about 1.4%, and that of the 183-K protein about 0.3% as much as coat protein. Much of the TMV-induced RNA polymerase in tobacco leaves is in bound form. The TMV–RNA replication complex is found in a membranous complex bound in cytoplasmic ribosomes. The data in Fig. 7.4 illustrate the use of protoplasts to study the timing of production of various viral-coded products.

As might be expected, RI and RF appear early in infection, as does the 126-K protein (estimated as 140 K in Fig. 7.4). Coat-protein mRNA and genomic RNA are early products. At a later stage, virus production follows closely that of coat-protein synthesis. Thus, it appears that the amount of coat protein available may limit the rate at which progeny virus is produced.

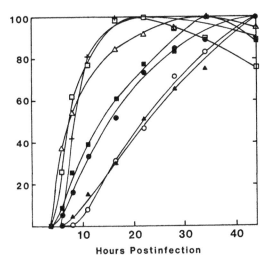

Figure 7.4 Time course of production of TMV-related RNAs, proteins, and progeny virus particles in synchronously infected protoplasts. One half of a batch of protoplasts was incubated with ^{14}C-uridine in the presence of actinomycin D from the time of inoculation. The other half was incubated with ^{14}C-leucine under the same conditions. Samples were taken for analysis at the times indicated. Data are expressed as the percentages of the maximal values attained for each component during the time course studies. (□) RI; (△) RF; (●) TMV-RNA; (■) coat protein mRNA; (+) 140-K protein; (▲) coat protein; (0) progeny virus particles. From Ogawa and Sakai (1984).

3.9 ASSEMBLY OF THE TMV ROD

3.9.1 Assembly *in Vitro*

In their classical experiments, Fraenkel-Conrat and Williams (1955) showed that it was possible to prepare TMV coat protein and TMV RNA, and to reassemble these into intact virus particles. TMV RNA alone had an infectivity about 0.1% that of intact virus. Reconstitution of virus rods gave greatly increased specific infectivity (about 10 to 80% that of the native virus), and the infectivity was resistant to RNAase attack. Since these early experiments, many workers have studied the mechanism of assembly of the virus rod. The three-dimensional structure of the coat protein is known in atomic detail (Fig. 5.7), and the complete nucleotide sequence of several strains of the virus and related viruses is known. The system therefore provides a useful model for studying interactions during the formation of a macromolecular assembly from protein and RNA.

There are four aspects of rod assembly to be considered: the site on the RNA where rod formation begins; the initial nucleating event that begins rod formation; rod extension in the 5′ direction; and rod extension in the 3′ direction. There is a general consensus concerning most of the details of the initiation site and the initial event, but the nature of the elongation processes remains controversial. A central problem has been the fact that the coat-protein monomer can exist in a variety of aggregation states, the existence of which is closely dependent on conditions in the medium. Equilibria exists between different aggregates, so that whereas one species may dominate under a given set of experimental conditions, others may be present in smaller amounts.

The Assembly Origin in the RNA

The experiments of many workers agree in showing that the origin of assembly of TMV is at a specific internal sequence of nucleotides. In the type virus, this sequence lies outside the coat-protein gene in a 5′ direction. The initiating sequence consists of three triplets— AAGAAGUCG—in the form of a loop at the end of a base-paired stem containing about 14 base pairs. The nine-base loop combines first with a 20 S aggregate of coat. Binding is the result mainly of the regularly spaced G residues, as shown by site-directed mutagenesis. The origin of assembly is located elsewhere in some tobamoviruses.

The Initial Nucleating Event

The structure of the 20 S double disk is known to atomic resolution (Fig. 5.7). For many years it has been assumed that this was the configuration that initiated assembly. Some recent work suggests that the double disk may already be in protohelical form under reconstitution conditions. However, on balance, the disk structure seems most probable. It certainly makes possible the model for TMV assembly in which the inner part of the two-layered disk acts as a pair of jaws (Fig. 5.7) to bind specifically to the origin of assembly loop in the RNA, in the process converting to a protohelix (Fig. 7.5).

Rod Extension in the 5′ Direction

All workers agree that following the initiation of rod assembly, rod extension is faster in the 5′ direction than in the 3′, but disagreement

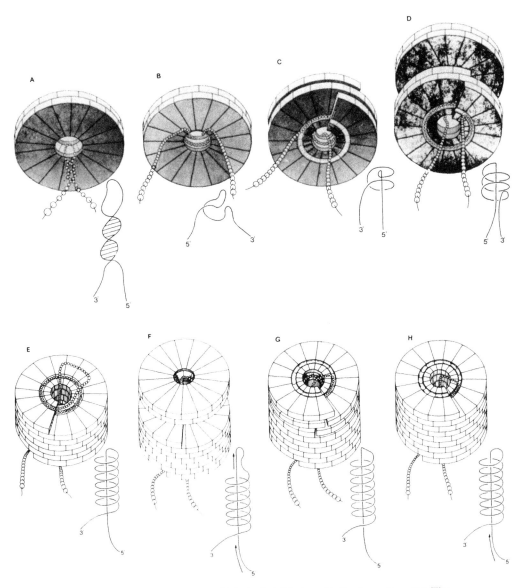

Figure 7.5 Model for the assembly of TMV; A–C, initiation; D–H, elongation. (A) The hair-pin loop inserts into the central hole of the 20 S disk. This insertion is from the lower side of the disk as viewed in Fig. 5.8. It is not yet apparent how the correct side for entry is chosen. (B) The loop opens up as it intercalates between the two layers of subunits. (C) This protein–RNA interaction causes the disk to switch to the helical lockwasher form (a protohelix). Both RNA tails protrude from the same end. The lockwasher–RNA complex is the beginning of the helical rod. (D) A second double disk can add to the first on the side away from the RNA tails. As it does so, it switches to helical form, and two more turns of the RNA become entrapped. (E–H) Growth of the helical rod continues in the 5′ direction as the loop of RNA receives successive disks, and the 5′ tail of the RNA is drawn through the axial hole. In each drawing, the three-dimensional state of the RNA strand is indicated. Courtesy of P. J. G. Butler, Copyright, Medical Research Council.

remains as to other aspects. Butler's group believed that the 20 S aggregate is used in 5' extension, with complete rods being formed in 5 to 7 min. Okada's group claimed that 5' extension uses 4 S or A protein with complete rods taking 40–60 min to form (A protein consists of aggregates of up to about five coat-protein monomers). Some of the earlier relevant experiments are discussed in detail by Butler (1984) and Okada (1986).

Rod Extension in the 3' Direction

As with extension in the 5' direction, there is disagreement between Butler's and Okada's groups as to whether A protein or the 20 S aggregate is involved in extension of the rod in the 3' direction.

3.9.2 Assembly *in Vivo*

The 20 S aggregate has not been definitively established as occurring *in vivo*. Nevertheless, the following evidence shows that the process involved in the initiation of assembly outlined in the preceding sections is almost certainly used *in vivo*. (1) The conditions under which assembly occurs most efficiently *in vitro* (pH 7.0, 0.1 M ionic strength, and 20°C) can be regarded as reasonably physiological. (2) In some tobamoviruses, the origin of assembly sequence lies within the coat-protein gene. In these viruses, the coat-protein subgenomic RNA is assembled into a short rod *in vivo*. (3) The TMV mutant Ni2519, which is *ts* for viral assembly *in vivo*, has a single base change that weakens the secondary structure near the origin of assembly loop. (4) Foreign genes that include the TMV origin of assembly have been introduced into the nuclear DNA to make transgenic plants. RNA transcripts of these genes can be assembled into virus-like rods with the TMV coat protein when the transgenic plants are systemically infected with TMV.

3.10 INTRACELLULAR LOCATION OF VIRUS SYNTHESIS AND ACCUMULATION

Many years ago, light microscopy revealed the existence of amorphous inclusions named X bodies in cells infected with TMV (Fig. 7.6). In to-

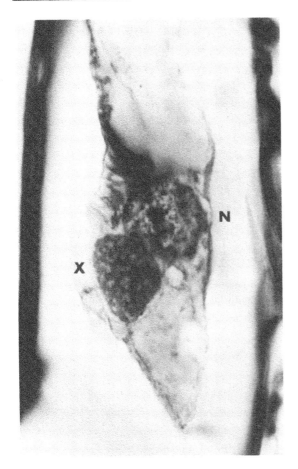

Figure 7.6 Viroplasm induced by TMV in a tobacco cortical parenchyma cell. Light microscope (×1500). N, nucleus; X, X-body. From Esau and Cronshaw (1967).

bacco cells, electron microscopy has shown that these structures consist of an assemblage of endoplasmic reticulum, ribosomes, virus rods, and wide filaments that may be bundles of tubules (Fig. 7.7). They occur in all types of cells infected with TMV.

The weight of evidence suggests that TMV components are synthesized and assembled into virus in or near these cytoplasmic viroplasms. In particular, various studies using immunocytochemical labeling have localized the large TMV-coded polypeptides in the viroplasms (Fig. 7.8). This is strong indication that the viral RNA is synthesized at these sites.

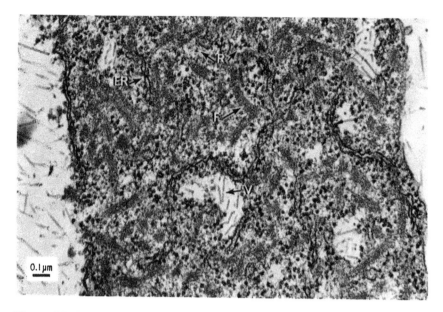

Figure 7.7 Viroplasm induced by TMV in a tobacco parenchyma cell showing endoplasmic reticulum (ER), ribosomes (R), virus rods (V), and wide filaments (F); (×53,000). The inclusion has no delimiting membrane. From Esau and Cronshaw (1967).

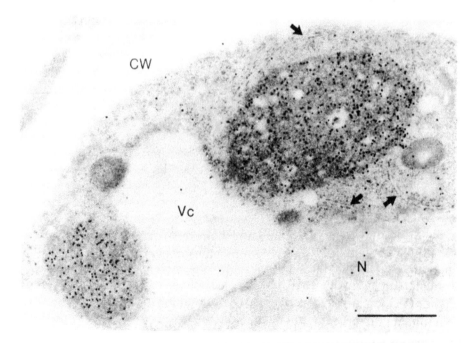

Figure 7.8 Immunocytochemical localization of the 130-K–180-K TMV proteins in a cell from a young infected tobacco leaf. Leaf sections were treated with an anti-130-K–anti-180-K antiserum, and then with a protein A–gold complex. Gold label is strongly localized in the viroplasm regions of the cytoplasm. CW, cell wall; Vc, vacuole; N, nucleus; arrows indicate ER; Bar = 1 μm. From Saito *et al.* (1987).

4 THE *TYMOVIRUS* GROUP

Except for a few crops, members of the *Tymovirus* group are not of major importance. They are transmitted mechanically and by beetles.

4.1 GENOME STRUCTURE

The genomic organization is very compact. For example, with the European turnip yellow mosaic virus (TYMV), only 192 (3%) of the 6318 nucleotides are noncoding. The largest ORF initiates at position 95 and ends at position 5627 with a UAG (to give a protein of 206 K, 1844 amino acids). A second ORF runs from the first AUG on the RNA (beginning nucleotide 88) and terminates with a UGA at 1972 (69 K; 628 amino acids). This ORF overlaps the 206-K gene in a different reading frame, but is in the same reading frame as the coat-protein gene. The coat-protein gene is in reading frame -1 compared with the 206-K gene for the virus, illustrated in Fig. 7.9. A 105-nucleotide noncoding 3′ region, which can form a tRNA-like structure (Fig. 7.9), appears downstream of the coat-protein cistron.

4.2 SUBGENOMIC RNA

A small subgenomic RNA, which is the mRNA for coat protein, is packaged with the genomic RNA and in a series of partially filled particles. No double-stranded RNA (dsRNA) corresponding in length to this subgenomic RNA appears to be present *in vivo*. The coat-protein mRNA is synthesized *in vivo* by internal initiation on minus strands of genomic length. Comparison of various *Tymovirus* nucleotide sequences around the initiation site for the subgenomic mRNA revealed two conserved regions, one at the initiation site, and the other, a 16-nucleotide sequence on the 5′ side of it. This longer sequence has been called the *tymobox* and may be an important component of the promoter for subgenomic RNA synthesis.

4.3 *IN VITRO* TRANSLATION STUDIES

The coat-protein gene in the genomic RNA is not translated *in vitro*, but the small subgenomic RNA is very efficiently translated in *in vitro* sys-

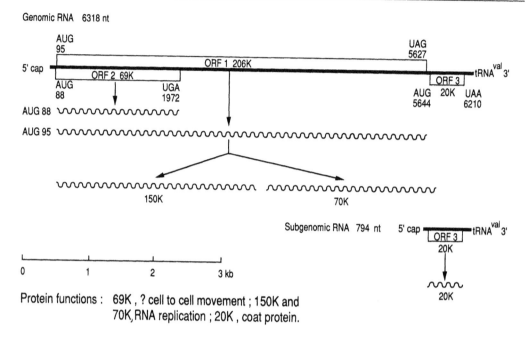

Figure 7.9 Organization and expression of a *Tymovirus* genome (TYMV, European strain).

tems to give coat protein. Both the large ORFs are expressed in *in vitro* systems. The 206-K polypeptide is cleaved *in vitro* to give smaller products, but the proteinase involved has not been identified. The scheme for proteolysis shown in Fig. 7.9 may be modified by further studies.

4.4 FUNCTIONS OF THE PROTEINS

The 206-K polypeptide contains sequences conserved in a number of RNA viral gene products that are involved in nucleotide binding, and in replicase function. The latter sequence is located toward the 3′ terminus of the 206-K structure in a 70-K cleavage product. However, the TYMV replicase isolated from infected plants contains a 115-K (=120 K) viral-coded polypeptide. This situation is not yet resolved.

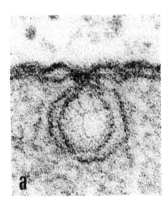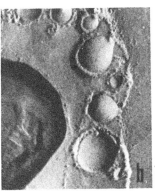

Figure 7.10 Fine structure of TYMV-induced peripheral vesicles in the chloroplasts of infected chinese cabbage cells. (a) Thin section showing continuity of inner chloroplast and outer vesicle membranes and stranded material inside the vesicle with the staining properties of ds nucleic acid (×235,000). Courtesy of S. Bullivant. (b) Fine structure of vesicle membranes revealed by freeze fracturing of isolated chloroplasts (×92,000). From Hatta *et al.* (1973).

4.5 RNA SYNTHESIS

Various lines of evidence have shown that the TYMV RNA polymerase activity is associated with the small virus-induced vesicles formed in large numbers at the periphery of chloroplasts in infected cells (Fig. 7.10).

In vivo, the replicating form of the RNA is mainly in a single-stranded (ss) state. TYMV infection stimulates nucleic acid synthesis in the nucleus, but the significance of this observation is not established. The subgenomic RNA containing the coat-protein gene becomes labeled *in vivo* with ^{32}P more rapidly than does genomic RNA, presumably reflecting more rapid synthesis of the coat-protein mRNA. However, the site of production of this mRNA is not yet established.

In spite of the close association of TYMV with the chloroplasts, the presence of chlorophyll is not necessary for TYMV replication, since the virus can multiply in chlorophyll-less protoplasts from etiolated Chinese cabbage hypocotyls.

4.6 VIRAL PROTEINS SYNTHESIZED *IN VIVO*

Viral protein synthesis in Chinese cabbage protoplasts is inhibited by cycloheximide but not by chloramphenicol, indicating that coat protein is synthesized on 80 S cytoplasmic ribosomes. The cytoplasmic site for viral protein synthesis has been confirmed by cytological evidence (Section 4.8, following), but viral protein also accumulates in the nucleus.

The 69-K protein has been detected *in vivo*. It is expressed at a 500 times lower level than coat protein, and appears to be an early nonstructural protein. Other large polypeptides have been detected, but their relationship to the scheme shown in Fig. 7.9 has not been established.

4.7 EARLY EVENTS FOLLOWING INOCULATION WITH TYMV

Following mechanical inoculation of leaves, a significant proportion of the RNA in the retained inoculum is uncoated within 45 sec, and the process is complete within 2 min. At least 80–90% of this uncoating takes place in the epidermis. With high inoculum concentrations, approximately 10^6 particles per epidermal cell can be uncoated. The process gives rise to empty shells that have lost about five to six protein subunits, and to low-MW protein. This uncoating process also occurs with plant species not known to be hosts of TYMV.

4.8 A MODEL FOR TYMV SYNTHESIS AND ASSEMBLY

The small vesicles at the margins of the chloroplasts, illustrated in Fig. 7.10, are formed by invagination of the chloroplast membranes. They are a consistent feature of all *Tymovirus*-infected species and all types of infected cells and tissues that have been examined. These vesicles are the first observable cytological change to be brought about by infection. A series of subsequent cytological effects has been established, including the fact that, at an early stage, coat protein accumulates in the cytoplasm over clusters of vesicles in the chloroplasts. From these various observations, a model for the assembly of TYMV has been proposed (Fig. 7.11). *Tymovirus* coat proteins in the form of empty virus shells accumulate in large quantities in the nuclei of infected cells. The significance of this phenomenon is not understood.

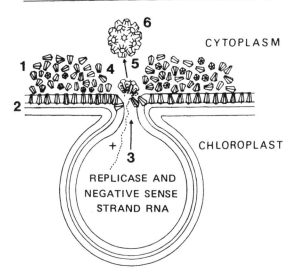

Figure 7.11 A model for the assembly of TYMV. (1) Pentamer and hexamer clusters of coat-protein subunits are synthesized by the ER and accumulate in the cytoplasm overlying clustered vesicles in the chloroplast. (2) These become inserted into the outer chloroplast membrane in an oriented fashion, i.e., with the hydrophobic sides that are normally buried in the complete protein shell lying within the lipid bilayer, with the end of the cluster that is normally inside the virus particle at the membrane surface. (3) An RNA strand synthesized or being synthesized within a vesicle begins to emerge through the vesicle neck. (4) At this site, a specific nucleotide sequence in the RNA recognizes and binds a surface feature of a pentamer cluster lying in the outer chloroplast membrane near the vesicle neck, thus initiating virus assembly. (5) Assembly proceeds by the addition of pentamers and hexamers from the uniformly oriented supply in the membrane. (6) The completed virus particle is released into the cytoplasm. From Matthews (1981).

5 THE *COMOVIRUS* GROUP

Members of this group usually have rather narrow host ranges and restricted distribution. Most are transmitted in the field by beetles, and some, through the seed. They are not responsible for many diseases of major significance.

Cowpea mosaic virus (CPMV) is the type member of the *Comovirus* group. The following discussion centers on this virus, since it has been used for most experimental work. Structure and organization of the bipartite CPMV genome is remarkably similar to that of the animal picornaviruses, which have a monopartite genome. This aspect is discussed in Chapter 16, Section 5.

5.1 STRUCTURE OF THE GENOME

The genome of CPMV consists of two strands of positive sense ssRNA called M- and B-RNAs, with M_r of 1.22×10^6 and 2.04×10^6. Both RNAs have a small VPg covalently linked at the 5′ end, and a poly(A) sequence at the 3′ end. There is a single, long, ORF in both RNAs. In M-RNA there are three AUGs, beginning at nucleotides 161, 512, and 524. Studies using site-directed mutagenesis have shown that the AUG at position 161 is used *in vitro* to direct the synthesis of a 105-K product. Both the 105-K protein initiating at nucleotide 161, and a 95-K product initiating at nucleotide 512 have been detected in infected protoplasts, as have their cleavage products illustrated in Fig. 7.12.

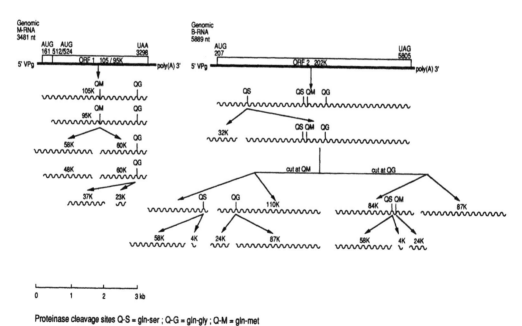

Proteinase cleavage sites Q-S = gln-ser ; Q-G = gln-gly ; Q-M = gln-met

Protein functions : M encoded , 58K and 48K , cell-to-cell movement ; 37K and 23K , coat proteins
　　　　　　　　　B encoded , 32K , Modulation of protease ; 4K , VPg ; 24K , protease ; 58K , RNA replication complex ; 110K, replicase.

Figure 7.12　Organization and expression of a *Comovirus* genome (CPMV).

The first 44 nucleotides in the 5′ leader sequences of M- and B-RNA show 86% homology. The last 65 nucleotides preceding the poly(A) sequence show 83% homology. The seven nucleotides before the poly(A) sequence are the same in the two RNAs. This conservation suggests that these sequences are involved in recognition signals for viral or host proteins, and studies with artificial mutants confirm their importance. Neither RNA has an AAUAAA polyadenylation signal preceding the 3′ poly(A) sequence. The presence of poly(U) stretches at the 5′ end of negative sense strands in RF molecules indicates that the poly(A) sequences are transcribed in the RNA-replication process.

5.2 ROLES OF THE M- AND B-RNAs

Both the B and M virus particles, or the RNA they contain, are necessary for an infection producing progeny virus particles. However, inoculation of protoplasts with B component leads to production of B-encoded polypeptides and to replication of B component RNA. M-RNA is not replicated in the absence of B-RNA. Thus, B-RNA must code for proteins involved in replication, while M-RNA codes for the structural proteins of the virus. B-RNA inoculated into intact leaves cannot move to surrounding cells in the absence of M-RNA.

5.2.1 Proteins Coded by B-RNA

B-RNA is translated both *in vitro* and *in vivo* into a 200-K polyprotein, which is subsequently proteolytically cleaved at specific sites, in a series of steps, to give the functional proteins. The relations between the various polypeptides synthesized in protoplasts have been established by comparison of the proteolytic digestion patterns of the isolated proteins and by immunological methods. More recently, radioactively labeled CPMV proteins synthesized in infected protoplasts were isolated, and partial amino acid sequences determined. This allowed precise location of the coding sequence for each protein on the B-RNA, precise determination of molecular weights, and determination of the cleavage sites at which they are released from the polyprotein precursor (Fig. 7.12).

5.2.2 Proteins Coded by M-RNA

A scheme for the translation of M-RNA is shown in Fig. 7.12. It is based on *in vitro* translation and processing studies and the use of antisera raised against synthetic peptides to detect the corresponding proteins *in vivo*. However, not all the polypeptides shown in Fig. 7.12 have been positively identified as occurring *in vivo*.

5.2.3 Functions of the Viral-Coded Proteins

Coat Proteins

In vitro studies demonstrated directly that the two-coat proteins are coded on M-RNA. This virus can spread through the plant only if the RNAs are in the form of virus particles.

Viral Protease

Various experiments, including studies with deletion mutants made using copy DNA (cDNA), have shown that the 24-K protein is the protease responsible for all cleavages in both B- and M-polyproteins. For efficient cleavage of the glutamine-methionine site in the M-RNA polyprotein, a second B-encoded protein (32 K) is essential, although this protein does not itself have proteolytic activity.

Replicase

Two lines of evidence indicate that the 110-K protein encoded by B-RNA is involved in RNA replication. First, it is part of the replicase complex isolated from infected leaves, and cosediments precisely with RNA-polymerase activity. Second, the polypeptide shows significant sequence similarity with the replicase of picornaviruses (see Chapter 16, Section 5).

Membrane Attachment

The B-RNA-encoded 58-K polypeptides may be involved in membrane attachment in the membrane-bound replication complex.

The VPg

Partial amino acid sequence of the VPg allowed it to be precisely mapped on the B-RNA.

Transport Function

As noted in Section 5.1, B-RNA is dependent on a function of M-RNA for cell-to-cell movement. Besides being a requirement for the coat proteins, experiments using insertion and deletion mutants have shown that both the 58-K and 48-K proteins encoded by M-RNA are needed for cell-to-cell movement.

5.3 REPLICATION *IN VIVO*

A CPMV RNA replication complex, containing a 110-K viral-coded polypeptide, has been isolated from infected cowpea leaves. This replication complex is capable of elongating nascent viral RNA strands *in vitro* to full-length RNA. The initiation of viral RNA replication may be linked to processing of the polyprotein, and the VPg may play some role, but further work is needed. The position of the 110-K polypeptide in the polyprotein (Fig. 7.12) requires that for every molecule of this protein produced, the other B-RNA-encoded polypeptides must also be formed, if not fully cleaved. CPMV-infected cells develop quite large pathological structures that contain the RF form of the viral RNA and virus particles. They are, therefore, probably viroplasms—the sites of virus synthesis and assembly.

6 THE *BROMOVIRUS* GROUP

The *Bromovirus* group contains four definitive members. The type member, brome mosaic virus (BMV), has a limited host range in the Poaceae, whereas the others have hosts mainly in the Fabaceae. Application of molecular biological techniques has led to a rapid development in our understanding of the genome of BMV. Bromoviruses are representative of several groups that combine two strategies: a tripartite genome, and translation from a subgenomic RNA (Table 7.1).

6.1 GENOME STRUCTURE

BMV has a tripartite genome totalling 8243 nucleotides. In addition, a subgenomic RNA containing the coat-protein gene is found in infected plants and in virus particles. This coat-protein mRNA is encoded in the sequence toward the 3′ end of RNA3. Each of the four RNAs has a 5′ cap and a highly conserved 3′ terminal sequence of about 200 nucleotides. The terminal 135 nucleotides of this sequence can be folded into a three-dimensional tRNA-like structure that accepts tyrosine, in a reaction similar to the aminoacylation of tRNAs.

RNAs 1 and 2 encode single proteins, as indicated in Fig. 7.13. RNA3 encodes a protein of predicted MW of 32,480. Between this ORF and the coat-protein ORF an intercistronic noncoding region approximately 250 nucleotides long is present. An internal poly(A) sequence of heterogeneous length (16–22 nucleotides) occurs in this intercistronic region,

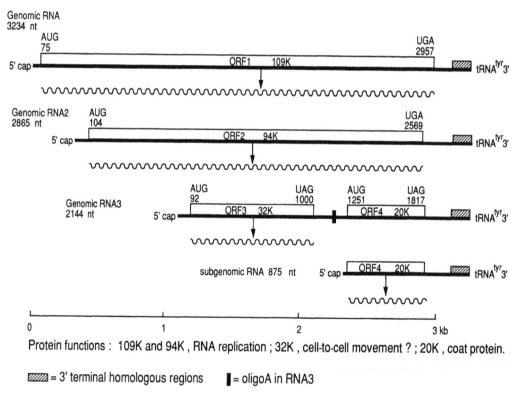

Protein functions : 109K and 94K , RNA replication ; 32K , cell-to-cell movement ? ; 20K , coat protein.

▨▨▨ = 3′ terminal homologous regions   ▮ = oligoA in RNA3

Figure 7.13 Organization and expression of a *Bromovirus* genome (BMV).

which ends 20 bases 5′ to the start of the coat-protein gene. The first 9 bases of the subgenomic RNA (RNA4) consist of the last 9 bases of the intercistronic region.

6.2 PROTEINS ENCODED AND THEIR FUNCTIONS

BMV RNAs are efficient mRNAs in *in vitro* systems, especially in the wheat germ system, which is derived from a host plant of the virus. In this system, RNA1 directs the synthesis of a single polypeptide of M_r 110 K (109 K in Fig. 7.13), and RNA2, a single polypeptide of M_r 150 K (94 K in Fig. 7.13). RNA3 directs the synthesis of a 35 K protein (32 K in Fig. 7.13). The coat protein ORF in RNA3 is not translated. Four new proteins corresponding in size to the genome products indicated in Fig. 7.13 have been found in protoplasts infected with BMV.

The gene products of RNAs 1 and 2 are involved in viral RNA replication because (1) these two RNAs can replicate in the absence of RNA3; (2) they have amino acid sequence similarities with proteins of other viruses known to function as RNA polymerases; (3) two proteins isolated from an active replicase fraction from infected tissue had the same tryptic peptide pattern as polypeptides translated *in vitro* from RNAs 1 and 2; (4) antibodies raised against synthetic peptides corresponding to the C-terminal regions of the gene products of RNAs 1 and 2 recognized the native proteins in replicase preparations; (5) defined mutations introduced into RNA2 blocked RNA synthesis.

On the basis of amino acid sequence similarities, it has been suggested that the protein translated from RNA3 may have a role in cell-to-cell movement analogous to that of the 30-K protein of TMV (Section 3.4). This possible function is supported by the fact that RNAs 1 and 2 will replicate in protoplasts in the absence of RNA3, but this RNA is needed for systemic plant infection. Thus, a function has been assigned or tentatively indicated for all four proteins known to be translated from the BMV genome.

6.3 ELEMENTS CONTROLLING RNA REPLICATION
IN THE BMV GENOME

In Chapter 6, the roles of 5′, 3′, and internal nucleotide sequences in regulating RNA replication were discussed. Many of the experiments on which these ideas are based were carried out with BMV.

6.4 REPLICATION OF BROMOVIRUSES

In vitro experiments with cowpea chlorotic mottle virus (CCMV) indicate that cotranslational disassembly may occur. In translational mixes with CCMV, up to four ribosomes were associated with each virus particle. However, cotranslational disassembly could not be demonstrated to occur *in vivo*.

In young cowpea leaves in which infection with CCMV was synchronized by differential temperature treatment, the three largest RNAs were synthesized at relatively constant ratios throughout the infection, but very little RNA4 was produced early in infection. As infection progressed, the proportion of RNA4 continued to increase. The RFs of components 1, 2, and 3 were produced with kinetics similar to that for the corresponding ssRNAs.

In vitro, CCMV coat protein can assemble with RNA under mild conditions to produce infectious virus that cannot be distinguished from native virus by physicochemical or structural means.

Various *in vitro* studies with BMV and CCMV indicate that the highly basic N-terminal region of the coat protein is involved in reactions with the viral RNA. Experiments with BMV variants containing known deletions in the coat-protein gene confirmed this view. The first seven amino acids are completely dispensible for packaging of the RNA *in vivo*, but if 25 N-terminal amino acids are missing, no virus particles are produced.

7 THE *TOBRAVIRUS* GROUP

This group contains three distinct viruses [tobacco rattle virus (TRV), pea early browning virus, and pepper ringspot virus]. They have a bipartite genome. In addition, there is a read-through protein, and three subgenomic RNAs are used.

7.1 GENOME STRUCTURE

7.1.1 RNA1

The RNA1 of TRV and other tobraviruses is about 6800 nucleotides long. There are four ORFs arranged as shown in Fig. 7.14. There is a 1-base

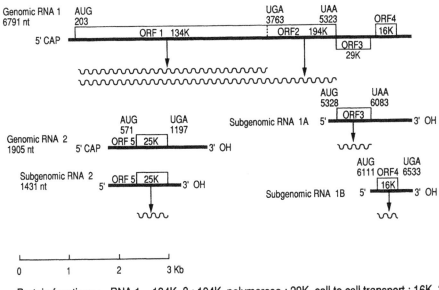

Protein functions :   RNA 1   134K, ? ; 194K, polymerase ; 29K, cell to cell transport ; 16K, ?
                      RNA 2   25K, coat protein.

Figure 7.14   Organization and expression of a *Tobravirus* genome (TRV). (RNA1, strain SYM; RNA2, strain PSG.)

intergenic region between the 194-K and 29-K ORFs in TRV. There is a 3′ noncoding region of 255 nucleotides. Although TRV RNA1 cannot be aminoacylated, there is a tRNA-like feature in the 3′ terminal region.

7.1.2 RNA2

The RNA2 of tobraviruses has widely different lengths in the range of 1800 to 4000 nucleotides. Different strains of TRV show substantial diversity in the length and in sequences of their RNA2. This wide variability is owing to (1) a variation in the length of the 3′-terminal sequences that RNA2s have in common with RNA1, and (2) the presence or absence of an additional RNA2-specific ORF (Fig. 12.2). No protein product has yet been found for this ORF.

All the natural TRV isolates have a 3′ region of identical nucleotide sequence in RNAs 1 and 2. However, experiments with pseudo-recombinants showed that this perfect 3′ homology is not a prerequisite for a stable genotype.

7.2 PROTEINS ENCODED

7.2.1 RNA1

The RNA1 of the SYM strain of TRV encodes a protein of $M_r \simeq 134$ K, together with a read-through protein of $\simeq 194$ K. The 194 K protein shows sequence similarities with the putative replicase genes of TMV, AMV, BMV, and CMV. The similarities are particularly strong with TMV. The 29-K protein shows some sequence similarity to the TMV 30-K protein and thus may be involved in virus movement from cell to cell. No similarities were found between the 16-K ORF and proteins of other viruses.

The gene products indicated for RNA1 and subgenomic RNAs 1A and 1B have been synthesized in *in vitro* protein-synthesizing systems. A 16-K protein has been found in infected chloroplasts.

7.2.2 RNA2

No messenger activity has been detected for TRV strain SYM RNA2 in *in vitro* tests. However, a subgenomic mRNA derived from RNA2 was shown to be the mRNA for coat protein. RNA2 has one known gene product— the coat protein. However, if coat protein is the only gene product, it is not clear why a subgenomic RNA should be required. Certain types of disease symptoms are specified by the short rods, even when these give rise to identical coat proteins. Thus there may be a second protein, as yet unidentified, coded for by RNA2.

7.3 ASSEMBLY OF VIRUS RODS

Like TMV, TRV coat protein can form a series of stable aggregates. In particular, there is a 40 S disk consisting of three layers of protein subunits arranged in a helix. *Tobraviruses* can be reassembled from protein and RNA *in vitro* under appropriate conditions. The binding site for the disk on RNA2 is at or near the 5′ terminus.

FURTHER READING

Matthews, R. E. F. (1991). "Plant Virology." Academic Press, New York.

Potyvirus group

Dougherty, W. G., and Carrington, J. C. (1988). Expression and function of potyvirus gene products. *Annu. Rev. Phytopathol.* **26**, 123–143.

Potexvirus group

Sit, T. L., AbouHaidar, M. G., and Holy, S. (1989). Nucleotide sequence of papaya mosaic virus RNA. *J. Gen. Virol.* **70**, 2325–2331.

Tobamovirus group

Solis, I., and Garcia-Arenal, F. (1990). The complete nucleotide sequence of the genomic RNA of the tobamovirus tobacco mild green mosaic virus. *Virology* **177**, 553–558.

Tymovirus group

Ding, S., Howe, J., Keese, P., Mackenzie, A., Meek, D., Osorio-Keese, M., Skotnicki, M., Srifah, P., Torronen, M., and Gibbs, A. (1990). The tymobox, a sequence shared by most tymoviruses: Its use in molecular studies of tymoviruses. *Nucleic Acids Res.* **18**, 1181–1187.

Comovirus group

van Kammen, A., and Eggen, H. I. L. (1986). The replication of cowpea mosaic virus. *Bio-Essays* **5**, 261–266.

Bromovirus group

Ahlquist, P., Allison, R., Dejong, W., Janda, M., Kroner, P., Pacha, R., and Traynor, P. (1990). Molecular biology of *Bromovirus* replication and host specificity. *In* "Viral Genes and Plant Pathogenesis" (T. P. Pirone and J. G. Shaw, eds.), pp. 144–155. Springer-Verlag, New York.

Tobravirus group

Angenent, G. C., van den Ouwéland, J. M. W., and Bol, J. F. (1990). Susceptibility to virus infection of transgenic tobacco plants expressing structural and non-structural genes of tobacco rattle virus. *Virology* **174**, 191–198.

REPLICATION OF OTHER VIRUS GROUPS AND FAMILIES

This chapter summarizes knowledge concerning five kinds of plant viruses that do not have positive sense single-stranded RNA (ssRNA) genomes. These are the *Caulimovirus* group with double-stranded (dsDNA); the *Geminivirus* group with ssDNA; the Reoviridae family with dsRNA; the Rhabdoviridae family with negative sense ssRNA; and the Bunyaviridae family, represented by tomato spotted wilt virus (TSWV) which is in the genus *Tospovirus* with ambisense ssRNA. A final section deals with the use of plant viruses as gene vectors.

1 *CAULIMOVIRUS* GROUP

Some members of this group such as cauliflower mosaic virus (CaMV) are very widespread, and are found wherever their host plants are grown. Individual viruses have a limited host range. Apart from the newly established commelina yellow mottle virus group, the caulimoviruses are the only plant viruses known to have a dsDNA genome. In 1979 very little was known about the replication of this group, but since then progress has been very rapid. There have been two main motivating factors. First it was hoped that these viruses, because of their dsDNA genomes, might be effective gene vectors in plants. This aspect is discussed in Section 6. Second, the realization that the DNA is replicated by a process of reverse transcription, similar to that of the animal retroviruses, made their study a matter of wide interest. Most experimental work has been carried out on CaMV. During virus replication, viroplasms, which have an appearance characteristic for the group, appear in infected cells.

1.1 STRUCTURE OF THE GENOME

The circular dsDNA of CaMV (about 8 kb) has a single gap in one strand
and two in the complementary strand. Gap 1, the single discontinuity in
the α strand, which is the strand used for transcription, corresponds to
the absence of only one or two nucleotides compared with the complementary strand. The two discontinuities in the complementary strand
have no missing nucleotides. They are regions of overlap where a short
sequence is displaced from the double helix by an identical sequence at
the other boundary of the discontinuity. In CaMV, the terminal 5′ deoxy-
ribonucleotide is at a fixed position and often has one or more ribonucleotides attached. The DNA encodes six and possibly eight genes.
These genes are close-spaced, but have very little overlap except for
the possible gene VIII. The arrangement of the open reading frames
(ORFs) in relation to the dsDNA is shown in Fig. 8.1.

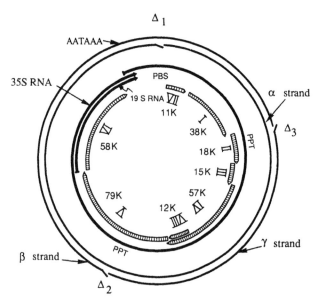

Figure 8.1 Genetic organization of CaMV. The 8-kbp viral DNA features three single-
stranded interruptions, one (△1) in the α (or coding) strand, and two (△2 and △3) in the
noncoding strand, defining the β and γ DNA species. The DNA encodes 8 potential ORFs.
The M_r of the predicted proteins are indicated. The capped and polyadenylated 19 S and
35 S RNAs have different promoters, but share the same 3′ termini. The two mRNAs are
translated from a fully ds form of the DNA and not from the gapped form shown here.
From Pfeiffer *et al.* (1987).

1.2 PROTEINS CODED FOR AND THEIR FUNCTIONS

The eight ORFs illustrated in Fig. 8.1 have been cloned, enabling *in vitro* transcription and translation experiments to be carried out. All eight ORFs could be translated *in vitro*. However, not all the protein products have been detected *in vivo*, and not all have had functions unequivocally assigned to them.

1.2.1 ORF I

The gene product of ORF I has a close association with the cell wall matrix associated with modified plasmodesmata, suggesting a role in cell-to-cell movement. Such a role is supported by some degree of sequence similarity between the ORF I protein and the 30K tobacco mosaic virus (TMV) protein, which is involved in viral movement (Chapter 7, Section 3.4). Purified ORF I protein binds strongly to RNA with no nucleotide sequence specificity. Since 35 S CaMV RNA is homologous to the entire genome, Citovsky *et al.* (1991) proposed that CaMV might move through plasmodesmata as a 35 S RNA–ORF I protein complex.

1.2.2 ORF II

Some naturally occurring strains of CaMV are transmitted by aphids; others are not. Experiments in which various recombinant genotypes were produced between various cloned CaMV strains showed that aphid transmissibility mapped to gene II. These studies and molecular genetic experiments have clearly demonstrated that the product of gene II functions in aphid transmission. The gene II protein also appears to increase the extent to which virus particles are held within the viroplasms.

1.2.3 ORF III

The gene III product is a non-sequence-specific DNA-binding protein with a preference for dsDNA. It may be a structural protein within the virus particle.

1.2.4 ORF IV

The product of ORF IV is a 57-K precursor of the 42-K protein subunit of the icosahedral shell of the virus. The 42-K protein is assumed to be derived from the 57-K molecule by proteolysis after formation of the virus shell. The coat protein is phosphorylated at serine and threonine residues by a protein kinase that is firmly bound to the virus. The coat protein is also glycosylated to a limited extent.

1.2.5 ORF V

This is the largest ORF in the genome. The mRNA derived from cloned DNA is translated *in vitro* to give a protein of the expected MW of the viral reverse transcriptase gene: (1) a protein of the expected size is present both in replication complexes and in virus particles. The 80-K polypeptide associated with the replication complexes is recognized by antibodies against a gene V *in vitro* translation product; (2) there are regions of significant sequence similarity between gene V of CaMV and the reverse transcriptase gene of retroviruses; (3) yeast expressing the cloned gene V gene accumulated significant levels of reverse transcriptase activity.

1.2.6 ORF VI

The product of gene VI (58 K) has been identified as the major protein found in CaMV viroplasms. An RNA transcribed from cloned DNA was translated to give a polypeptide of appropriate size. The gene VI product was detected immunologically in a cell fraction enriched for viroplasms. Various studies show that the gene VI product plays a major role in disease induction, in symptom expression, and in controlling host range; but that other genes may also play some part. Gene VI also functions in *trans* to activate translation of other viral genes.

1.2.7 ORFs VII and VIII

The two smallest ORFs (VII and VIII) are not present in another *caulimovirus* (carnation etched ring virus), so their significance is doubtful.

1.3 RNA AND DNA SYNTHESIS

By 1983 various aspects of CaMV nucleic acid replication led three groups to propose that CaMV DNA is replicated by a process of reverse transcription involving an RNA intermediate. Some of the observations that led to the model were (1) the fact that a full-length RNA transcript with terminal repeats is produced; (2) the fact that DNA in virus particles has discontinuities (Fig. 8.1), whereas that found in the nucleus does not, but is supercoiled and is associated with histones as a minichromosome; (3) the existence of dsDNA in knotted forms; (4) the existence of other forms of CaMV DNA in the cell that are not encapsidated, such as an unusual ss molecule consisting of 625 nucleotides with the same polarity as the α-strand, which is covalently linked to about 200 ribonucleotides. Since 1983, much further evidence has confirmed the reverse transcription model, including the partial characterization of the viral-coded reverse transcriptase discussed in the previous section.

There are two phases in the nucleic acid replication cycle of CaMV (Fig. 8.2). In the first, the dsDNA of the infecting particle moves to the cell nucleus, where the overlapping nucleotides at the gaps are removed, and the gaps are covalently closed to form a circular dsDNA. These minichromosomes form the template used by the host enzyme, RNA polymerase II, to transcribe two RNAs of 19 S and 35 S. At 31 nucleotides upstream from the initiation site of the 35 S RNA, there is a TATATAA sequence, and a similar TAT$_A$TAAA is upstream of the 19 S RNA initiation site. The two promoters are very active, particularly that for the 35 S RNA which is highly active when used in constructs to express genes in a variety of plant cells (Section 6.2). An enhancer sequence of 338 base pairs has been identified in the region upstream from the 35 S TATA sequence.

The two polyadenylated RNA species migrate to the cytoplasm for the second phase of the replication cycle, which takes place in the viroplasms. The 19 S RNA is the mRNA for the viroplasm protein, which is produced in large amounts. The 35 S RNA must be used to produce the other gene products, since no functional smaller transcripts for most of the ORFs have been found.

There is an AUG at the beginning of every ORF, and the ORFs are close together in the genome. These observations led to a *relay race* model for the translation of the 35 S RNA. In this model, a ribosome binds first to the 5′ end of the RNA, translates to the first termination codon. At this point, it does not completely leave the RNA but reinitiates

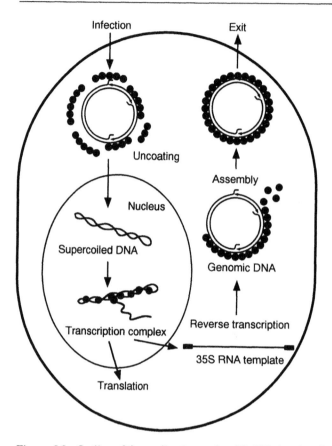

Figure 8.2   Outline of the replication cycle of CaMV, showing the initial uncoating phase in the cytoplasm; the formation of minichromosomes of viral DNA and host protein in the nucleus; and reverse transcription of viral RNA to give progeny DNA copies in the cytoplasm. For details see text.

protein synthesis at the nearest AUG, whether just downstream or upstream from the termination. Support for the model came from site-directed mutagenesis studies in ORF VII and the region between ORF VII and ORF I, regions which are not essential for infectivity under laboratory conditions. Insertion of an AUG into either of these regions rendered the viral DNA noninfectious, unless the inserted AUG was followed by an in-frame termination codon.

Gene VI is the only *Caulimovirus* gene to be transcribed as a separate transcript from its own promoter, suggesting that it may have an important role at an early stage following infection. Reporter genes in various plasmid constructions have been used to show gene VI functions in *trans* in the posttranscriptional expression of the closely spaced genes of the full-length RNA transcript. Mutagenesis of the coding part of gene VI showed that the protein product rather than the mRNA is responsible for transactivation. The mechanism by which this protein product enhances substantially the translation of the other viral genes on a polycistronic mRNA is not understood.

To commence viral DNA synthesis on the 35 S RNA template, a plant methionyl tRNA molecule base pairs over 14 nucleotides at its 3' end with a site on the 35 S RNA corresponding to a position immediately downstream from the △1 discontinuity in the α strand DNA. The viral reverse transcriptase commences synthesis of a DNA minus strand and continues until it reaches the 5' end of the 35 S RNA. At this point, a switch of the enzyme to the 3' end of the 35 S RNA is needed to complete the copying. The switch is made possible by the 180 nucleotide direct-repeat sequence at each end of the 35 S RNA. When the template switch is completed, reverse transcription of the 35 S RNA continues up to the site of the tRNA primer, which is displaced and degraded. The △1 gap is present in the newly synthesized DNA.

The used 35 S template is then removed by an RNase H activity. It is not certain whether this activity is a function of the reverse transcriptase or a host enzyme. In this process, two purine-rich tracts of the RNA are left near the position of gaps △2 and △3 in the second DNA strand (plus strand). Synthesis of the second (plus) strand of the DNA then occurs, initiating at these two RNA primers. The growing plus strand has to pass the △1 gap in the minus strand, which again involves a template switch.

1.4 RECOMBINATION IN CaMV DNA

The fact that CaMV DNA is converted to a covalently closed ds circle to allow transcription shows that there must be an early involvement of host-plant DNA-repair enzymes following infection. This idea is reinforced by the fact that cloned DNA, excised from the plasmid in linear form, is infectious, and that the progeny DNA is circular. The following evidence shows that recombination can occur *in vivo*. Coinfection of plants with nonoverlapping defective deletion mutants usually leads to the production of viable virus particles. Analysis of the progeny DNA showed that the rescue was by recombination rather than complementation. Pairs of noninfectious recombinant full-length CaMV genomes,

integrated with a plasmid at different sites, can regain infectivity on inoculation to an appropriate host. In the progeny virus DNA, all the plasmid DNA is eliminated, and the viral DNA has a normal structure.

1.5 REPLICATION *IN VIVO*

The characteristic viroplasms induced by caulimoviruses are the site
for progeny viral DNA synthesis and for the assembly of virus particles
(Fig. 8.3). Viral coat protein appears to be confined to them. Most virus
particles are retained within the viroplasms.

At an early stage in their development, the viroplasms appear as
very small patches of electron-dense matrix material in the cytoplasm,
surrounded by numerous ribosomes. Larger viroplasms are probably
formed by the growth and coalescence of the smaller bodies. The mature viroplasms vary quite widely in size from about 0.2 to 20 μm in diameter. They are usually spherical, and are not membrane bound. They
often have ribosomes at the periphery and consist of a fine granular
matrix with some electron-lucent areas not bounded by membranes.

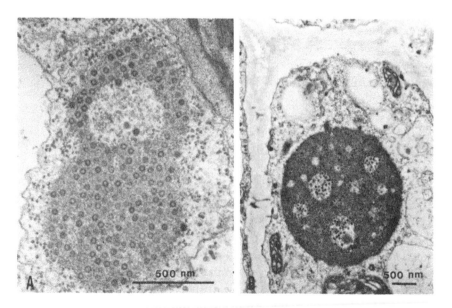

Figure 8.3 *Caulimovirus* viroplasms. (A) A *Brassica rapa* cell with a CaMV viroplasm,
containing virus particles. Courtesy of G. P. Martelli. (B) Cell of infected *Brassica perviri-
dis* with a viroplasm (V) containing virus particles. From Martelli and Castellano (1971),
by permission from Cambridge University Press.

Virus particles are present scattered or in irregular clusters in the lucent areas and the matrix.

Little is known about the way CaMV particles are assembled. No empty virus shells and very little unencapsidated DNA of the sort found in virus particles are found in infected tissue. These observations suggest that encapsidation may be closely linked to DNA synthesis. The role of glycosylation and phosphorylation of the coat protein remains to be determined.

2 *GEMINIVIRUS* GROUP

The geminiviruses are the only recognized group of plant viruses with a genome consisting of ssDNA. They are responsible for a number of important diseases, especially in the tropics. Like that of the caulimoviruses, interest in members of this group was stimulated by the possibility that, because they contained DNA, they might be developed as gene vectors for plants.

2.1 GENOME STRUCTURE

Three subgroups within the *Geminivirus* group have been approved by the ICTV as follows:

Sub-group	Type member	Hosts	Vectors	Genome structure
I	Maize streak virus (MSV)	Monocotyledons	Leafhoppers	One circular ssDNA
II	Beet curly top virus (BCTV)	Dicotyledons	Leafhoppers	One circular ssDNA
III	Bean golden mosaic virus (BGMV)	Dicotyledons	Whiteflies	Two circular ssDNAs

Thus subgroup II combines some properties of subgroups I and III.

All geminiviruses have a conserved genome sequence in common.

There is a large (~200 base) noncoding intergenic region in the two-component geminiviruses. This region has sequences capable of forming a hairpin loop. Within this loop is a conserved sequence found in all geminiviruses. Because it is so highly conserved, it has been considered to be involved in RNA transcription (see Section 2.4) and in DNA replication. African cassava mosaic virus (ACMV) will serve as an example of the *Geminivirus* group (but see also recent work with this virus, Etessami *et al.*, 1991).

The genome of ACMV consists of two ssDNA circles of similar size, both of which are needed for infectivity. The nucleotide sequences of the two DNAs are very different except for a 200 nucleotide noncoding region, called the common region, which is almost identical on each DNA. The two DNAs have been arbitrarily labeled 1 and 2 (some workers use the terminology A and B). The strands that are found in virus particles are designated virus (or plus), and the complementary strands, minus. For these viruses, this is an arbitrary distinction because both plus and minus strands contain coding sequences. Comparison of the sequences of ACMV with other bipartite geminiviruses revealed that six of these ORFs are in a conserved arrangement (Fig. 8.4).

2.2 FUNCTIONS OF THE ORF PRODUCTS

2.2.1 Coat Protein

The coat-protein gene has been mapped to the 30.2 K ORF of DNA1. It might be expected that the properties of the coat protein play a role in insect vector specificity. This role was demonstrated by the following experiment: DNA clones were constructed in which the coat-protein gene in DNA1 of ACMV (whitefly transmitted) was replaced with the coat-protein gene of beet curly top virus (BCTV) (leafhopper transmitted). The BCTV gene was expressed in plants and gave rise to particles containing the ACMV DNAs. These chimeric particles were transmitted by the BCTV leafhopper vector when injected into the insects, whereas normal ACMV was not.

2.2.2 ORFs Related to DNA Replication

In vitro mutagenesis has shown that the largest gene in the minus strand of DNA1 (=40.3 K in Fig. 8.4) is required for DNA synthesis. This

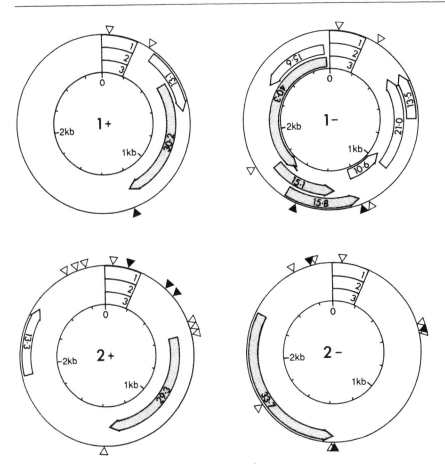

Figure 8.4   Genome organization of ACMV. Arrows indicate ORFs and direction of transcription. Coding capacity of each ORF is indicated. The shaded ORFs are the six found in other bipartite geminiviruses, and those for which RNA transcripts have been identified. The boxed region to the right of the 0-kb position, showing reading frames 1, 2, and 3, is the 200-base common intergenic region. △, possible promoter sequences (TATA boxes); ▲, possible polyadenylation signals (AATAAA sequences). From Townsend *et al.* (1985).

is the only viral gene product that is absolutely required for ss and dsDNA replication, but accumulation of ssDNA requires the 15.1-K ORF in 1− DNA of Fig. 8.4. Maximal replication of both ss and dsDNA depends on the presence of a functional gene product from the 15.8-K ORF in 1− DNA of Fig. 8.4.

2.2.3 Virus Movement

Since DNA1 of ACMV can replicate in protoplasts, but does not infect plants, DNA2 must encode a function for cell-to-cell movement of the virus.

2.2.4 Insect Transmission

Since DNA2 is found only with whitefly-transmitted geminiviruses, products of this DNA may be involved in insect transmission. Insertion and deletion mutagenesis experiments show that both coding regions of ACMV DNA2 must have other functions in addition to a possible role in insect transmission. The experiments described in Section 2.2.1 show that the coat protein, a product of a DNA1 gene, controls transmission by insect vectors.

2.3 mRNAs

Five virus-specific polyadenylated RNA transcripts found in ACMV-infected plants were mapped to either the plus or minus DNA strands, demonstrating that transcription is bidirectional in both DNAs. A sixth transcript presumably exists, as has been found for tomato golden mosaic virus (TgMV). Recent experiments suggest, however, that the genome organization of geminiviruses may be more complex, and the number of encoded proteins greater, than is indicated in Fig. 8.4. Wheat dwarf *Geminivirus* contains a functional intron, an unusual feature for a plant virus. The presence of an intron may be a feature of geminiviruses infecting the Poaceae.

2.4 CONTROL OF TRANSCRIPTION

The genome of ACMV has consensus promoter sequences (TATAA/TAA/T) lying outside or just within the common untranslated region.

There is a 122-bp sequence upstream from the start site for transcription of coat-protein mRNA, which enhances promoter activity. This *upstream activating sequence* lies in the large intergenic region and includes a region common to all geminiviruses.

2.5 DNA REPLICATION

The details of *Geminivirus* DNA replication are not well established. *Geminivirus* particles usually accumulate in the nucleus, and with some, such as maize streak virus (MSV), large amounts of virus accumulate there. In some infections, fibrillar rings, which must be part of a spherical structure, appear in the nucleus, but their composition and significance are not known.

The presence of the same 200-nucleotide sequence in the intergenic regions of both DNAs 1 and 2 of ACMV indicates that it plays an important part in viral replication. A common region with the capacity to form a stable loop containing a conserved nononucleotide (TAATATTAC) sequence has been found in all geminiviruses. This loop structure is necessary for DNA replication and has been proposed as a binding site for the host enzymes responsible for priming complementary strand synthesis.

Extracts of *Nicotiana* leaves infected with ACMV contain a variety of viral DNA forms in addition to ssDNA. Three forms of unit-length dsDNA have been found: closed circular supercoiled, relaxed circular, and linear. Other high-MW species appeared to be concatamers consisting of two or more unit-length genomes. The existence of these ds forms and the concatamers indicates that the virus replicates via a circular ds replicative form, possibly by a rolling-circle mechanism. Support for this mechanism comes from the observation that there is strong similarity between the highly conserved sequence TAATATTAC found in the DNA components of all geminiviruses that have been sequenced, and the cleavage site for the gene A protein of the ssDNA bacterial virus ΦX174, which acts as an origin for rolling-circle replication of the DNA.

Recombination is known to occur during ACMV DNA replication. Insertion or deletion mutagenesis of the two large ORFs of ACMV DNA2 destroyed infectivity, but infectivity was restored by coinoculation of constructs that contained single mutations in different ORFs.

3 PLANT REOVIRIDAE

Plant members of the Reoviridae family are placed in two genera: *Phytoreovirus* with 12 dsRNA genome segments, the type member being wound tumor virus (WTV); and *Fijivirus*, with 10 dsRNA genome segments and Fiji disease virus (FDV) as the type member. Some members are responsible for important diseases, especially in the humid tropics and subtropics.

3.1 GENOME STRUCTURE

Nine of the 12 genome segments of WTV have been fully sequenced. Each segment contains a single functional ORF and all have conserved nucleotide sequences at their termini. Inverted terminal repeats also occur just inside these conserved termini (Fig. 8.5).

3.2 RNA TRANSCRIPTION AND TRANSLATION

Particles of the plant reoviruses, like their counterparts infecting vertebrates, contain a transcriptase, which can transcribe ssRNA using the RNA in the particle as template. WTV has also been shown to contain a methylase that catalyzes the incorporation of methyl groups from S-adenosyl-L-methionine into the RNA strands synthesized *in vitro*, giving the 5′ terminal structure $^7mG(5')ppp(5')Ap^m$——.

The transcriptase in purified WTV synthesizes *in vitro* 12 ssRNA products corresponding to the 12 dsRNA segments of the genome. The existence of this active transcriptase within the virus particles has facilitated the identification of the protein products of the 12 genes.

The proteins coded by segments 1, 3, and 6 are found in the nucleoprotein core of the virus. Those coded by segments 2, 5, 8, and 9 form part of the protein shell, whereas those from segments 4, 7, 10, 11, and 12 are nonstructural proteins. The products of segments 2 and 5 are essential for insect transmission.

3.3 INTRACELLULAR SITE OF REPLICATION

Plant reoviruses replicate in the cytoplasm. Following infection, densely staining viroplasms appear in the cytoplasm of infected plants and in various tissues of infected leafhopper vectors. Detailed electron microscopic studies support the view that the viroplasms caused by FDV in sugarcane are the sites of virus-component synthesis and assembly.

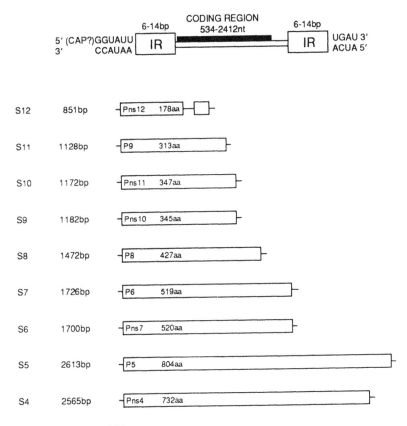

S12	851bp	Pns12 178aa
S11	1128bp	P9 313aa
S10	1172bp	Pns11 347aa
S9	1182bp	Pns10 345aa
S8	1472bp	P8 427aa
S7	1726bp	P6 519aa
S6	1700bp	Pns7 520aa
S5	2613bp	P5 804aa
S4	2565bp	Pns4 732aa
S1–S3:	in excess of 3kbp	

Figure 8.5 Organization of the WTV genome. The general organization of WTV genomic segments is illustrated at the top of the figure. The terminal conserved hexanucleotide and tetranucleotide sequences are indicated in large letters, whereas the positions of the 6–14 base pair (bp) segment-specific, terminal inverted repeats (IR) are indicated by the large, open boxes. The solid black rectangle indicates the general position of the coding region (534–2412 nt). The remainder of the figure illustrates specific information regarding the size of individual genomic segments (bp) and their encoded polypeptides (aa) as well as the coding assignments. There is no evidence for expression of the second small coding region present in genomic segment S12. From Nuss and Dall (1990).

They are composed mostly of protein and dsRNA. Some areas contain numerous isometric particles 50–60 nm in diameter. Some appear to be empty shells, whereas others contain densely staining centers of dsRNA. These particle types appear to be incomplete virus particles or cores. Complete virus particles have been seen only in the cytoplasm.

3.4 RNA SELECTION DURING VIRUS ASSEMBLY

Every WTV particle appears to contain one copy of each genome segment because (1) RNA isolated from virus has equimolar amounts of each segment, and (2) an infection can be initiated by a single particle. Thus, there is a significant problem with understanding the assembly of this kind of virus. What are the macromolecular recognition signals that allow one, and one only, of each of 10 or 12 genome segments to appear in each particle during virus assembly? For example, the packaging of the 12 segments of WTV presumably involves 12 different and specific protein–RNA and/or RNA–RNA interactions.

Various experiments indicate that the fully conserved hexanucleotide sequence at the 5′ termini and a fully conserved tetranucleotide sequence at the 3′ termini of viral RNAs shown in Fig. 8.5 might form the recognition signals for viral as opposed to host RNA. The segment-specific inverted repeats of variable length located just inside the conserved segments (Fig. 8.5) may represent the specific recognition sequence for each individual genome segment.

4 PLANT RHABDOVIRIDAE

Rhabdoviruses are very widespread, but individual viruses often have a limited host range. The rhabdoviruses have the largest particles among the plant viruses. They are one of only two families of plant viruses whose particles are bounded by a lipoprotein membrane. They have been placed in two groups depending on their site of maturation and accumulation within the cell.

4.1 GENOME STRUCTURE

Plant rhabdoviruses, like those infecting vertebrates, possess a genome consisting of a single piece of ss negative sense RNA, with a length in the range of 11,000 to 13,000 nucleotides. Six viral-coded proteins are known. A discrete mRNA is transcribed from the negative sense genome for each encoded protein. Gene-sequencing techniques have

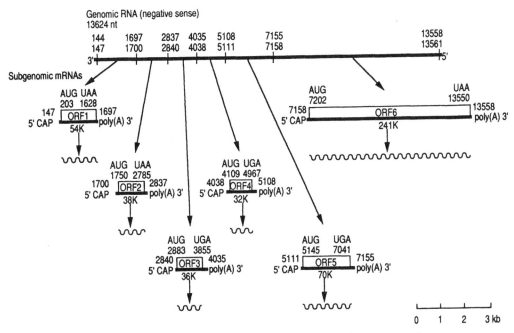

Protein products (in order from the 3' end of the genome) are   N, M₂, NV (nonstructural), M₁, G, and L. Data courtesy of K. B. Goldberg and A. O. Jackson.

Figure 8.6   Structure and expression of a plant *Rhabdovirus* genome (SYNV).

only recently been applied to one plant member of the family—Sonchus yellow net virus (SYNV). Current knowledge is summarized in Fig. 8.6.

Hybridization, using a library of cDNA clones, has established that the order of the six genes in the SYNV genome is: 3'N-M2-sc4-M1-G-L-5'. The intergenic and flanking sequences are highly conserved. The transcription site for all of the SYNV mRNAs begins with the sequence AACA or AACU (positive sense).

4.2 ENCODED PROTEINS

Five viral-coded proteins are found in SYNV virus particles. The proteins coded by the first two mRNAs to be sequenced have been identified by raising antibodies against the product of the cloned gene and showing that they react with a structural component of the purified virus. On this basis, the first gene from the 3' end of the genomic RNA

codes for the nucleocapsid (N) protein, whereas the second gene encodes another structural protein, M_2.

4.3 THE VIRAL TRANSCRIPTASE

Variable amounts of transcriptase activity have been found in different
plant rhabdoviruses, but the enzyme is not yet well characterized. Lettuce necrotic yellows virus (LNYV) transcriptase is located in the internal nucleoprotein core of the virus and uses the viral RNA as a template.

4.4 GENOME REPLICATION

Replication of the genome of plant rhabdoviruses is assumed to use a
full-length positive sense RNA as a template. The detailed mechanism
for genome replication has not been established for any plant rhabdoviruses. In particular the question remains as to how the RNA polymerase enzyme in producing a complete positive sense strand, can ignore
the intergenic junctions that delineate the mRNAs.

4.5 mRNA AND PROTEIN SYNTHESIS

From patterns of hybridization with cDNA clones, it has been established
that the SYNV genome is transcribed into a short 3′ terminal *leader* RNA
and six mRNAs. Four major proteins (G, N, M_1, and M_2) have been detected by immunoprecipitation in protoplasts infected with SYNV.

4.6 CYTOLOGICAL OBSERVATIONS ON REPLICATION

Because of their large size and distinctive morphology, the rhabdoviruses are particularly amenable to study in thin sections of infected
cells. They appear to fall into three groups: (1) those that accumulate in
the perinuclear space with some particles scattered in the cytoplasm.
With some viruses of this group, structures resembling the inner nucleoprotein cores have been seen within the nucleus. The envelopes of
some particles in the perinuclear space can be seen to be continuous
with the inner lamella of the nuclear membrane. (2) In the second
group, for example LNYV, maturation of virus particles occurs in asso-

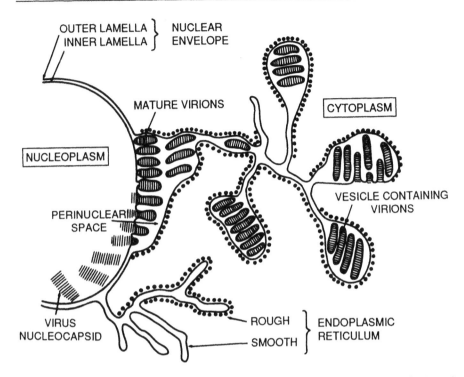

Figure 8.7 Diagrammatic representation of morphogenesis and cellular distribution of a rhabdovirus such as LNYV. From Francki *et al.* (1985).

ciation with the endoplasmic reticulum, and particles accumulate in vesicles in the reticulum. Figure 8.7 gives a graphical representation for this group.

Biochemical evidence suggests that the nucleus is involved in the early stages of infection by members of this group. (3) In this group, exemplified by barley yellow striate mosaic virus, infection leads to the formation of large viroplasms in the cytoplasm of infected cells. These consist of electron-dense granular or fibrous material. Mature virus particles appeared in membrane-bound sacs within the viroplasm, particularly near its surface. There is probably not a nuclear phase associated with the replication of this group.

5 PLANT BUNYAVIRIDAE

The genome of tomato spotted wilt *Tosposvirus* (TSWV) consists of three RNA segments: (L ≃ 8200 nucleotides); M (≃ 5400); and S (≃

3000). TSWV is the only virus in this genus. It is widespread throughout the subtropical and temperate zones, has a very wide host range, and is the cause of some important diseases. There are four structural polypeptides. Two are glycosylated proteins (G1 and G2) and are at the surface. The N protein binds to the RNA, and there is a large protein present in minor amounts. On the basis of these and other properties noted later, TSWV has been placed in a new genus, *Tospovirus*, in the Bunyaviridae, a large family of viruses replicating in vertebrates and invertebrates.

The M and S RNAs have been sequenced. Figure 8.8 summarizes main features of these two genome segments. Both RNAs contain inverted complementary repeats at their termini. These are probably involved in RNA replication and in the formation of the circular configuration found in virus particles.

The M RNA is negative sense and has nine nucleotide complementary termini. The complementary strand of mRNA contains a long ORF with a leaky UGA termination codon. Thus, a 35-K protein and a read-

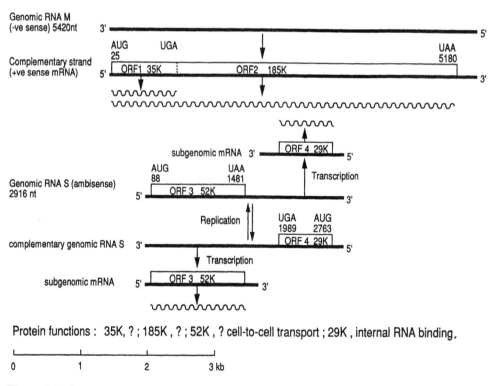

Protein functions : 35K, ? ; 185K , ? ; 52K , ? cell-to-cell transport ; 29K , internal RNA binding.

0   1   2   3 kb

Figure 8.8 Organization and expression of TSWV genomic RNAs M and S.

through protein of 185 K can be produced. The N-terminal part of these polypeptides shows sequence similarity to the nonstructural protein coded on the M RNA of Rift Valley Fever virus, a member of the *Phlebovirus* genus in the Bunyaviridae.

The S RNA has an ambisense coding strategy. In other words, one gene is present in the positive sense, and the other occurs in the negative sense. The positive sense ORF encodes a protein with a predicted size of 52 K. The ORF in the complementary sense encodes a 29-K protein. Both proteins are translated from subgenomic RNAs, which have been identified in extracts from TSWV-infected cells. There is a stable hairpin loop structure in the intergenic region of S RNA. The two subgenomic RNAs probably terminate at this loop.

Little is known about the replication of TSWV *in vivo*, except that it almost certainly takes place in the cytoplasm.

6 POSSIBLE USES OF VIRUSES FOR GENE TRANSFER

In the early 1980s there was considerable interest in the possibility of developing plant viruses as vectors for introducing foreign genes into plants. At first, interest centered on the caulimoviruses, the only plant viruses with dsDNA genomes, because cloned DNA of the viruses was shown to be infectious. Interest later extended to the ssDNA geminiviruses, and then to RNA viruses when it became possible to reverse transcribe these into dsDNA, which could produce infectious RNA transcripts.

The main potential advantages of a plant virus as a gene vector were seen to be (1) the virus or infectious nucleic acid could be applied directly to leaves, thus avoiding the need to use protoplasts and the consequent difficulties in plant regeneration; (2) it could replicate to high copy number; and (3) the virus could move throughout the plant, thus offering the potential to introduce a gene into an existing perennial crop such as orchard trees.

Such a virus vector would have to be able to carry a nonviral gene (or genes) in a way that did not interfere with replication or movement of the genomic viral nucleic acid. Ideally, it would also have the following properties: (1) inability to spread from plant to plant in the field, providing a natural containment system; (2) induction of very mild or no disease; (3) a broad host range; this would allow one vector to be used for many species, but would be a potential disadvantage in terms of safety; and (4) maintenance of continuous infection for the lifetime of the host plant.

The major general limitations in the use of plant viruses as gene vectors are as follows: (1) they are not inherited in the DNA of the host plant, and therefore genes introduced by viruses cannot be used in conventional breeding programs; (2) annual crops would have to be inoculated every season, unless there were a very high rate of seed transmission; (3) by recombination or other means, the foreign gene introduced with the viral genome might be lost quite rapidly, with the virus reverting to wild type; and (4) it would be necessary to use a virus that caused minimal disease in the crop cultivar. The virus used as vector might mutate to produce significant disease, or be transmitted to other susceptible crops.

In recent years technological advances in the use of the modified Ti plasmid of *Agrobacterium tumefaciens* for gene transfer have made this the method of choice for most dicotyledon plant systems. For monocotyledons, methods involving DNA introduction into protoplasts, or direct introduction of DNA into leaf cells by a particle gun have been developed. The present status of plant viruses as useful gene vectors can be gauged from the fact that they are not mentioned in a recent review of genetic engineering applied to crop improvement (Gasser and Fraley, 1989). However, plant viruses have been very useful in one respect. They have acted as a source of control elements for use in other vector systems.

6.1 TRANSIENT EXPRESSION VECTORS

In principle, viral DNA might be altered to provide plasmid type vectors for high copy number, rapid expression of modified or foreign genes.

Transient expression systems allow for the rapid screening of DNA constructs designed to study the activities of promoter sequences, RNA processing signals, etc., in cells, and as a preliminary screen in the construction of transgenic plants. TGMV DNA A contains the genes necessary for viral DNA replication. Constructs have been made in which the coat protein ORF of DNA A was replaced by the bacterial CAT or beta glucuronidase (GUS) genes, under the control of the coat-protein promoter. Following inoculation using *Agrobacterium tumefaciens* (see Chapter 2, Section 6.1), these foreign genes were transiently expressed in petunia leaf discs.

6.2 PROMOTERS

The 19 S and 35 S promoters of CaMV are both strong constitutive promoters that have found wide application for the expression of a range of heterologous genes. The 35 S promoter has been found to be much more effective than the 19 S in several systems. For example, expression of the α-subunit of β-conglycinin in petunia plants under control of the 35 S promoter was 10–50 times greater than from the 19 S promoter. The 35 S promoter is also 10–30 times more effective than the nopaline synthase promoter from *Agrobacterium tumefaciens*.

6.3 UNTRANSLATED LEADER SEQUENCES AS ENHANCERS OF TRANSLATION

Untranslated leader sequences of several viruses have been shown to act as very efficient enhancers of mRNA translational efficiency, both *in vitro* and *in vivo* and in prokaryotic and eukaryotic systems. AMV RNA4 is known to be a well-translated message for AMV coat protein. The natural leader sequences of a barley and a human gene have been replaced with AMV RNA4 leader sequence. These constructs showed up to a 35-fold increase in mRNA translational efficiency in the rabbit reticulocyte and wheat germ systems. Similar results in a wide range of systems have been obtained using constructs containing the untranslated 5' leader sequence of TMV RNA. These untranslated viral leader sequences probably reduce RNA secondary structure, making the 5' terminus more accessible to scanning by ribosomal subunits or to interaction with initiation factors.

FURTHER READING

Davies, J. W., and Stanley, J. (1989). Geminivirus genes and vectors. *Trends Genet.* 5, 77–81.

de Haan, P., Wagemakers, L., Peters, D., and Goldbach, R. (1990). The SRNA segment of tomato spotted wilt virus has an ambisense character. *J. Gen. Virol.* 71, 1001–1007.

Jackson, A. O., Francki, R. I. B., and Zuidema, D. (1987). Biology, structure, and replication of plant rhabdoviruses. *In* "The Rhabdoviruses" (R. R. Wagner, ed.), pp. 427–508. Plenum, New York.

Matthews, R. E. F. (1991). "Plant Virology." 3rd Ed. Academic Press, New York.

Nuss, D. L., and Dall, D. J. (1990). Structural and functional properties of plant reovirus genomes. *Adv. Virus Res.* **38**, 249–369.

Pfeiffer, P., Gordon, K., Fütterer, J., and Hohn, T. (1987). The life cycle of cauliflower mosaic virus. *In* "Plant Molecular Biology" (D. von Wettstein and N. H. Chua, eds.), pp. 443–458. Plenum, New York.

Wilson, T. M. A. (1989). Plant viruses: A tool-box for genetic engineering and crop protection. *BioEssays* **10**, 179–186.

SMALL NUCLEIC ACID MOLECULES THAT CAUSE OR MODIFY DISEASES

9

Several different kinds of small nucleic acid molecules may be found in plants showing viruslike disease symptoms. Some of these are actually the cause of the disease, whereas others modify the disease produced by a virus.

1. *Viroids* are very small circular ssRNAs that cause disease and replicate independent of any virus. They do not code for any polypeptides.

2. *Satellite viruses* are very small viruses that encode their own coat protein, but are otherwise dependent on a helper virus for replication. They usually have very little nucleotide sequence similarity with the helper virus, but affect disease symptoms.

3. *Satellite RNAs* are small RNAs dependent on a helper virus both

for replication and for packaging into virus particles made of helper virus protein. Most have only limited sequence similarity with the helper virus, but some exceptions exist. They modify disease symptoms.

4. Defective interfering (DI) RNAs and particles These molecules are derived from a viral genome by substantial deletions of internal nucleotide sequences. They depend on the presence of the intact virus for replication and ameliorate disease symptoms.

1 VIROIDS

A variety of viruslike diseases in plants have been shown to be caused by pathogenic RNAs known as viroids. They are small circular ssRNA molecules a few hundred nucleotides long, with a high degree of secondary structure. They do not code for any polypeptides, and replicate independent of any associated plant virus. Indeed there is no evidence to show that viroids are related to viruses in an evolutionary sense. Viroids are of practical importance as the cause of several economically significant diseases, and are of general biological interest as the smallest known agents of infectious disease. The most-studied viroid is potato spindle tuber viroid (PSTVd). Viroid names are abbreviated to the initials with d added to distinguish viroid from abbreviations for virus names (Appendix 1, Table 2).

1.1 STRUCTURE OF VIROIDS

1.1.1 Circular Nature of Viroid RNA

The circular nature of many viroids was first shown directly by electron microscopy. Under nondenaturing conditions, the molecules appear as small rods with an axial ratio of about 20:1, and for PSTVd, an average length of about 37 nm. Spread under denaturing conditions, the molecules can be seen to be covalently closed circles. All viroid preparations also contain a variable proportion of linear molecules.

1.1.2 Nucleotide Sequences

The nucleotide sequences of about 50 viroids and viroid sequence variants are now known. These sequences have confirmed the circular nature of the molecules (Fig. 9.1). All viroids have some degree of sequence similarity. On the basis of degree of overall sequence similarity, 16 distinct viroids have been recognized, and other distinct viroids continue to be discovered. The use of copy DNA (cDNA) clones to sequence field isolates of viroids has revealed that a single isolate may contain a range of closely related sequence variants.

1.1.3 Secondary Structure

From the primary sequence, it is possible to predict a secondary structure that maximizes the number of base pairs. This gives rise to rod-like molecules with base-paired regions interspersed with unpaired loops (Fig. 9.1).

A

B

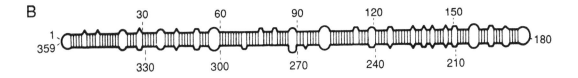

C

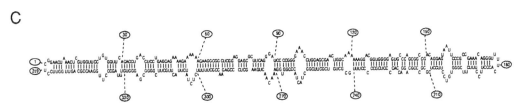

Figure 9.1 Potato spindle tuber viroid. (A) Nucleotide sequence as determined by Gross *et al.* (1978). (B) Proposed secondary structure in outline. (C) Three-dimensional representation of the viroid molecule. Courtesy of H. Sanger.

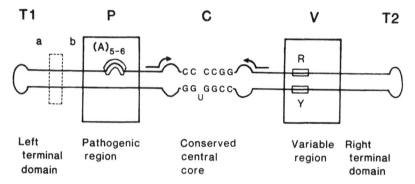

Figure 9.2   Model of the five viroid domains, based on nucleotide sequences of over 40 PSTVd-like viroids. The *conserved central domain* of about 95 nucleotides has a strictly conserved bulged helix, flanked by inverted repeat sequences indicated by the arrows. The *pathogenic domain* contains an adenine-dominated, purine-rich sequence in one strand. This domain has been implicated in pathogenesis. The *variable domain* may show less than 50% homology between closely related viroids. R, Y, delineate a short oligopurine oligopyrimidine helix. The two *terminal domains* have been implicated in viroid replication. The main sequence homologies between PSTVd-like viroids in these two domains are indicated. From Keese and Symons (1985).

1.1.4 Structure Domains

Viroids in the PSTVd group have been the most studied, and a model of secondary structure involving five structural domains has been developed (Fig. 9.2).

1.2 VIROID REPLICATION

1.2.1 A Model

It is now generally accepted that viroid RNAs do not code for any polypeptides. This being so, they must use preexisting host nucleic acid-synthesizing enzymes. RNA strands with sequences complementary to viroid RNA (defined for viroids as minus strands) have been found in infected tissues. Minus-strand PSTVd RNA can exist as a tandem multimer of several unit-length monomers. Monomeric PSTVd plus strands have been found complexed with long multimeric minus strands. These experiments and many others have demonstrated that viroids replicate via an RNA template, and that replication almost certainly involves a rolling-circle mechanism. Several models have been put forward. One is shown in Fig. 9.3. None of the models can be taken as proven in every detail, and different viroids may replicate in somewhat different ways.

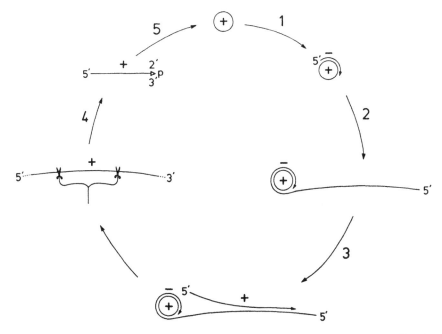

Figure 9.3 Asymmetric model of rolling-circle viroid replication. The model begins at the top of the diagram, with an infecting circular plus strand (+), which becomes a template for minus-strand (−) synthesis (steps 1 and 2). A multimeric linear minus strand is then copied into a multimeric plus-strand precursor (step 3), and the plus-strand precursor is cleaved (step 4) and circularized (step 5). From Branch *et al.* (1988).

1.2.2 Enzymes Involved

Enzymes are required for at least three steps in viroid replication.

Copying of Plus and Minus Strands

Four possible host enzymes are known: DNA-dependent RNA polymerases I, II, and III, and an RNA-dependent RNA polymerase. The enzyme (or enzymes) has not been identified beyond doubt, but present evidence favors DNA polymerase II.

Cleavage of Multimers

Avocado Sunblotch Viroid (ASBVd) Dimers or multimers of ASBVd plus- or minus-strand RNA can self-cleave *in vitro* at specific sites, giving rise to plus- or minus-strand monomers. As with satellite RNAs (Section 2.2) self-cleavage of ASBVd RNAs requires a divalent ion, and gives rise to 5′ OH and 2′–3′ cyclic phosphate 3′ termini. Single or double hammerheadlike structures have been proposed for the self-cleavage of ASBVd RNAs (Fig. 9.4).

The sequences required to form the plus- and minus-strand self-cleavage structures are situated side by side in the middle of the molecule, and about one third of the viroid sequence is involved.

PSTVd and Other Viroids   Secondary structures such as those illustrated in Fig. 9.4 have not been found in other viroids. All viroid multimer RNAs except those of ASBVd are processed by a host nuclear

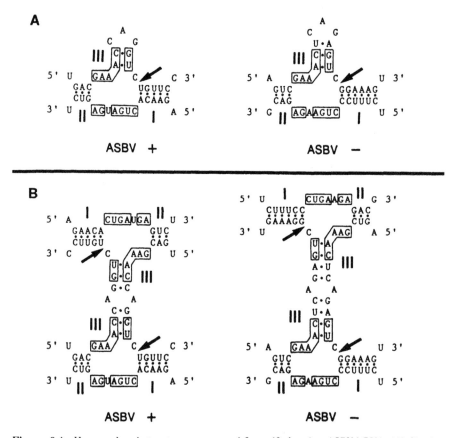

Figure 9.4 Hammerhead structures proposed for self-cleaving ASBVd RNA. (A) Single-hammerhead structures proposed by Hutchins *et al.* (1986), which are similar to those proposed for self-cleaving satellite RNAs (Fig. 9.10). (B) Double-hammerhead structure proposed for ASBVd RNA by Forster *et al.* (1988). Sites of cleavage are indicated by arrows. Nucleotides conserved between ASBVd, newt self-cleaving RNA, and plant satellite RNAs are boxed. From Forster *et al.* (1988).

enzyme to give rise to monomers, but the enzyme has not yet been identified. The central conserved domain plays an important role in the generation of monomers from multimers.

Ligation of Linear Monomers

The last step in viroid replication is the ligation of linear monomers to give the circular molecules. A host enzyme must be involved, and it has been shown that an RNA ligase isolated from wheat germ will efficiently convert PSTVd linear monomers to circular form.

1.3 CLASSIFICATION OF VIROIDS

Since viroids cause some economically important diseases, it is necessary to be able to identify a viroid isolated from a particular crop, and this requires some form of classification for viroids, but there are difficulties in arriving at a satisfactory classification.

1.3.1 General Difficulties

The only structural data available for viroid classification are the nucleotide sequences. Several classifications have been proposed, on the basis of either overall sequence similarities, or the sequences of defined areas in the molecule. Arbitrary decisions must be made. For example, it has been proposed that an overall sequence similarity of 65% or more would place two viroids in the same group, and that more than 90% sequence similarity would indicate strains of a single viroid. Such arbitrary distinctions are not particularly satisfactory. One classification is shown in Fig. 9.5.

1.3.2 Recombination between Viroids

The available nucleotide sequence data make it almost certain that recombination events have occurred quite frequently between different viroids, presumably during replication in mixed infections. For example, Australian grapevine viroid (AGVd) appears to have originated from extensive recombination between several other viroids (Fig. 9.6). Such events complicate the question of viroid classification.

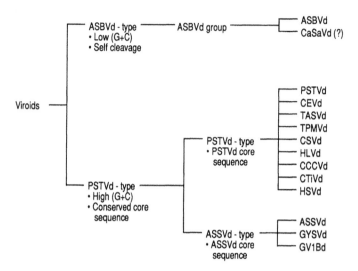

Figure 9.5 A proposed classification of viroids into three groups based on differences in the conserved core sequences of PSTVd and apple scarskin viroid (ASSVd) and on the self-cleavage site of ASBVd. Additional name abbreviations are CaSaVd, carnation stunt-associated viroid; CEVd, citrus exocortis viroid; TASVd, tomato apical stunt viroid; TPMVd, tomato planta macho viroid; CSVd, chrysanthemum stunt viroid; HLVd, hop latent viroid; CCCVd, coconut cadang-cadang viroid; CTiVd, coconut tinangaja viroid; HSVd, hop stunt viroid; GYSVd, grapevine yellow speckle viroid; GV1Bd, grapevine viroid 1B. From Koltunow and Rezaian (1989).

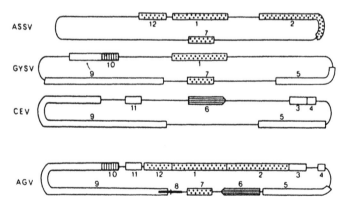

Figure 9.6 Evidence for the origin of Australian grapevine viroid (AGVd) by recombination events with three other viroids. AGVd has 369 nucleotides and has less than 50% sequence similarity with any known single viroid. Nevertheless, its sequence can be divided into regions showing high sequence similarity with several other viroids. Boxes denote these regions. Regions of high similarity are shown by shading and by numbers. From Rezaian (1990).

1.3.3 Circular Subviral RNA Pathogens

A circular RNA containing about 1700 nucleotides called the delta agent is a satellite RNA of hepatitis B virus (HBV). It is the only agent with viroidlike properties known to exist in mammals and to cause human disease. It replicates by a rolling-circle mechanism, and can carry out a self-cleavage reaction like ASBVd and the plant viral satellite RNAs that replicate by a rolling-circle mechanism. Branch *et al.* (1990) suggested that because these are unique and distinctive features not found in other agents, the viroids, the viral satellite RNAs that replicate by a rolling-circle mechanism, and the delta agent should be placed in a group called *circular subviral RNA pathogens*.

1.4 BIOLOGICAL ASPECTS OF VIROIDS

1.4.1 Macroscopic Disease Symptoms

As a group, there is nothing that distinguishes disease symptoms produced by viroids from those caused by viruses. These symptoms include stunting, mottling, leaf distortion, and necrosis. From an agricultural crop point of view, symptoms cover a whole range from the slowly developing lethal disease in coconut palms caused by coconut cadang-cadang viroid (CCCVd) to the worldwide symptomless infection of hops with hop latent viroid (HLVd). It is probable that many more symptomless viroid infections remain to be discovered.

1.4.2 Cytopathic Effects

Various effects of viroid infection on cellular structures have been reported. For example, in some infections, changes have been observed in membranous structures called *plasmalemmasomes*. Several workers have described pronounced corrugations and irregular thickness in cell walls of viroid-infected tissue. A variety of degenerative abnormalities have been found in the chloroplasts of viroid-infected cells.

1.4.3 Biochemical Changes

Viroid infection appears to cause no gross changes in host nucleic acid metabolism. By contrast, marked changes in the amounts of various host proteins have been described in infected tissue. Perhaps the most dramatic effect is the increase in a 140-K host protein in tomatoes infected with several different viroids. Induction of this protein was not specific to viroid infection.

1.4.4 Movement in the Plant

Viruses that have coat proteins defective for virus assembly have naked RNAs. Most of these move slowly through the plant by cell-to-cell movement. By contrast, viroids move rapidly through a host plant at a rate like that of competent viruses, almost certainly through the phloem. The relative resistance of viroid RNA to nuclease attack probably facilitates their long-distance movement. It is also possible that viroid particles undergo translocation while bound to some host protein.

1.4.5 Transmission

Viroids are readily transmitted by mechanical means in most of their hosts. Transmission in the field is probably mainly by contaminated tools and similar means. This ease of transmission in the presence of nucleases is probably owing to viroid secondary structure and to the complexing of viroids to host components during the transmission process.

PSTVd is transmitted through the pollen and true seed of potato plants and can survive in infected seed for long periods. However, this route of transmission is not likely to be of great commercial importance in potato.

1.4.6 Interference between Viroids

Inoculation of tomato plants with a very mild strain of PSTVd gives substantial protection against a second inoculation with a severe strain applied 2 weeks after the first. Cross-protection occurs not only between strains of a particular viroid, but also between different viroids. Protection is usually only temporary. The molecular basis for this interference is not understood.

1.4.7 Epidemiology

The main methods by which viroids are spread through crops are vegetative propagation, mechanical contamination, and through pollen and seed. The relative importance of these methods varies with different viroids and hosts. For example, vegetative propagation is dominant for PSTVd in potatoes and chrysanthemum stunt viroid (CSVd) in chrysanthemums. Mechanical transmission is a significant factor for others such as citrus exocortis viroid (CEVd) in citrus and HSVd in hops. Seed and pollen transmission are factors in the spread of ASBVd in avocados.

For most viroid diseases, the reservoir of inoculum appears to be within the crop itself, which raises the question as to where the viroid diseases came from. Although the evolution of viroids has been the subject of much speculation, the sudden appearance and rapid spread of most "new" viroid diseases can probably be accounted for by the following factors. Viroids are readily transmitted by mechanical means. Many modern crops are grown as large-scale monocultures. Thus from time to time, a viroid present in a natural host, and probably causing no disease, might escape into a nearby susceptible commercial crop and spread rapidly within it. If the viroid and crop plant had not evolved together, disease would be a likely outcome. There is direct evidence for such a sequence of events with the tomato planta macho disease in Mexico.

1.5 MOLECULAR BASIS FOR BIOLOGICAL ACTIVITY

Because of their very small size, their autonomous replication, the known structure of many variants, and the lack of any viroid-specific polypeptides, it has been a hope of many workers that viroids might provide a simple model system that would provide insights as to how variations in the structure of a pathogen modulate disease expression. This hope has not yet been realized. Very small changes in nucleotide sequence may give rise to dramatic changes in the kind of disease induced by a viroid. Therefore, disease induction must involve specific recognition of the viroid sequence by some host macromolecule. Until the nature of this host macromolecule (or molecules) is known, the interpretation of correlations between nucleotide sequence and biological properties of viroids will remain speculative.

One approach to the problem is to compare the nucleotide sequences of naturally occurring variants of a viroid and relate this to differences in the disease produced. For example, variants of PSTVd could be placed in four classes with respect to disease severity in tomato: mild, intermediate, severe, and lethal. With these variants, nucleotide differences all lie within the P domain (Fig. 9.2). On the other hand, pathogenicity of CEVd variants was associated with nucleotide changes in both the P and variable domains.

Attempts to apply site-directed mutagenesis to study the effects of single nucleotide changes on the disease produced have not been successful. The mutants were either noninfectious or reverted rapidly to wild type. Similarly, the construction of chimeric viroids has met with little success.

Sequence similarities have been noted between domains in viroid RNAs and certain host cell RNAs, but the significance of these similarities remains to be established.

1.6 DIAGNOSTIC PROCEDURES FOR VIROIDS

Since viroids produce no specific proteins, the immunological methods applied so successfully to viruses cannot be used for the diagnosis of the diseases they cause. Similarly, because no characteristic particles can be readily detected, electron microscopic techniques are inappropriate. For these reasons diagnostic procedures have been confined to biological tests, gel electrophoresis, and more recently to nucleic acid hybridization tests.

1.6.1 Biological Tests

Biological tests for viroid detection and diagnosis have been important where suitable diagnostic test plants have been identified, and they remain important for some viroids, for example, where strains of different severity exist. However, no suitable indicator hosts have been found for some viroids such as CCCVd. Mild isolates of viroids may produce barely detectable symptoms. Environmental conditions also may markedly affect the disease produced by other isolates. For such reasons, *in vitro* tests based on the properties of the viroid RNA have assumed considerable importance.

1.6.2 Gel Electrophoresis

Viroids generally occur in very low concentration in the infected host; thus, partial purification and concentration must be used before the nucleic acids are run in an appropriate polyacrylamide gel electrophoresis (PAGE) system. Various procedures have dramatically increased the sensitivity of viroid detection by electrophoresis.

1.6.3 Nucleic Acid Hybridization

Methods for detection and diagnosis based on nucleic acid hybridization are becoming increasingly important. For application in routine testing, it was necessary to develop methods for the large-scale production of cloned, highly labeled viroid cDNA, and to develop procedures by which clarified tissue extracts could be used in the tests. The dot-blot hybridization procedure is now being widely used.

1.6.4 General

With present procedures, it is relatively simple to verify a positive test result for viroid infection. It is much more difficult to ensure that a negative result means a viroid-free plant. This may be an important practical issue where vegetatively propagated crops are concerned.

2 SATELLITE VIRUSES AND SATELLITE RNAs

Purified virus preparations isolated from infected plants may contain a variety of RNAs other than the genomic RNAs. Some of these, such as subgenomic RNAs, have already been discussed in Chapters 6 through 8. In addition, some isolates of certain plant viruses contain satellite agents. Two classes of these agents can be distinguished according to the source of the coat protein used to encapsidate the RNA. In *satellite viruses*, the satellite RNA codes for its own coat protein. In *satellite RNAs*, the RNA becomes packaged in protein shells made from coat protein of the helper virus. Satellite viruses and satellite RNAs have the following properties in common:

1. Their genetic material is an ssRNA molecule of small size. The RNA is not part of the helper virus genome, and usually has only limited sequence similarity to the helper virus RNA.
2. Replication of the RNA is dependent on a specific helper virus.
3. The agent affects disease symptoms, at least in some hosts.
4. Replication of the satellite interferes to some degree with replication of the helper.
5. Satellites are replicated in the cytoplasm on their own RNA template.

In recent years, several satellite RNAs associated with a particular group of viruses have been shown to have viroidlike structural properties. These agents have been termed virusoids, but this term is no longer favored.

2.1 SATELLITE PLANT VIRUSES

Three definite satellite plant viruses have been described, together with two probable examples. They are the smallest known viruses. Satellite tobacco necrosis virus (STNV) has been the most studied, and is the example discussed here.

The helper virus, tobacco necrosis virus (TNV), is a typical small icosahedral virus with a diameter of about 30 nm. It replicates independent of other viruses, and normally infects plant roots in the field. Certain cultures of TNV contain substantial amounts of a smaller viruslike particle with a diameter of about 18 nm, depending for its replication on the larger virus. There is significant specificity in the relationship between satellite and helper. Strains of both viruses have been isolated. Only certain strains of the helper virus will activate particular strains of the satellite.

Both STNV and TNV are transmitted by the zoospores of the fungus *Olpidium brassicae* (Chapter 10, Section 2.2). Transmission depends on an appropriate combination of four factors: satellite and helper virus strains, race of fungus, and species of host plant.

The complete nucleotide sequence of STNV RNA was one of the first viral sequences to be determined. The amino acid sequence of the coat protein was deduced from the nucleotide sequence and later confirmed by direct sequencing of the STNV coat protein. These results showed that STNV codes for only one gene product—its coat protein.

STNV RNA has no significant sequence similarity with the TNV genome. STNV RNA is remarkably stable *in vivo,* having been shown to survive in inoculated leaves for at least 10 days in the absence of helper virus. This stability may have evolved to allow the satellite to survive a period within a cell after uncoating, but before the cell becomes infected with helper virus. Little is known about the replication of STNV *in vivo,* but it is widely assumed that STNV RNA replication must be carried out by an RNA-dependent RNA polymerase coded for, at least in part, by the helper virus. Replication of STNV substantially suppresses TNV replication, and it is possible that this may involve competition for the replicase. The presence of STNV in the inoculum reduces the size of the local lesions produced by the helper virus. This could be owing to the reduction in TNV replication.

STNV is readily translated *in vitro* in both prokaryotic and eukaryotic systems, the only product being the coat protein. In the wheat germ system, where both TNV and STNV RNAs were translated in a mixture, STNV RNA was preferentially translated even in the presence of an excess of TNV RNA.

2.2 SATELLITE RNAs

Cucumber mosaic *Cucumovirus* (CMV) normally causes a fern–leaf–like disease in tomatoes in the field. The first satellite RNA was isolated from tomatoes infected with CMV but showing a lethal necrotic disease. It was the presence of the satellite RNA known as CARNA5 that increased the severity of disease.

Satellite RNAs have since been found in preparations of various viruses belonging to the *Cucumovirus, Nepovirus, Tombusvirus* and *Sobemovirus* groups. Thirty-three satellite RNAs have been found, associated with 29 viruses (Francki *et al,* 1991). They do not always increase severity of disease. Satellite RNAs differ from satellite viruses in that they do not code for a coat protein. The RNA is encapsidated in particles of the helper virus.

Satellite RNAs associated with four sobemoviruses discovered in Australia occur in both circular and linear forms. In their biological properties (dependence on the helper virus), they are undoubtedly satellite RNAs, but in some properties, particularly their small size (325–

390 nucleotides), circularity, high degree of base pairing, and lack of mRNA activity, they are like viroids. A sequence GAAAC is found in the central region of viroids and also in a similar position in some of these satellites.

Satellite RNAs can be divided into two groups: those that replicate via a negative sense monomer template, as does the helper virus, and those that have a rolling-circle method of replication.

2.2.1 Satellite RNAs Replicating via a Negative Sense Monomer Template

This group is typified by the satellite RNAs of CMV and tomato black ring virus (TBRV). Their properties with respect to replication may be summarized as follows:

1. *Terminal structures*   CMV satellite RNAs have a 5′ cap structure and a 3′ hydroxyl group, as does CMV. STBRV has a 5′ VPg and 3′ poly-adenylation, as does TBRV. These similarities to the corresponding helper virus suggest a common method for replication of the satellite and helper RNA.

2. *The presence of ORFs*   Satellite RNAs in this group that have been sequenced contain one or two ORFs. For example, five variants of STBRV have been sequenced. All contained a single large ORF, coding for a putative protein containing 419–424 amino acids with several regions of identical amino acid sequence. These features strongly suggest a functional role for the ORF.

3. *Translation of the ORFs*   Translation of satellite RNAs of this group in an *in vitro* protein-synthesizing system give rise to the expected polypeptide products. These have not been positively identified *in vivo* and their role in replication is unknown.

4. *RNA replication via a unit-length negative sense template*   The following lines of evidence strongly suggest that CMV satellite RNAs replicate via a unit-length negative sense template: (a) A dsRNA species of MW 220 K corresponding to the double-stranded form of CARNA5 has been isolated from infected tobacco. (b) Kinetics of labeling of the two strands with ^{32}P fit with the production of an excess of positive sense strands early in infection. (c) The ds forms of both CARNA5 and genomic CMV RNA3 contain an unpaired guanosine at the 3′ end of the minus strand. This is a feature of the RFs of some other viruses, and

suggests that the viral and satellite RNAs share a common replicative mechanism. (d) CARNA5 does not replicate in protoplasts unless the helper virus is present. This result suggests that the satellite depends on the helper for some replication function, rather than just for encapsidation or movement through the plant.

This evidence strongly suggests that replication of satellite RNAs of this type depends on a replicase function of the helper virus. This raises the question of the function of the polypeptide potentially coded for by these satellites. This polypeptide might modulate the activity of the host replicase so that it preferentially replicated the satellite RNA.

2.2.2 Satellite RNAs with a Viroidlike Replication

This group is typified by the satellite RNA of tobacco ringspot *Nepovirus* (STRSV), the satellite RNA of arabis mosaic virus, and by the viroid-like satellite RNAs.

STRSV has no detectable mRNA activity, nor do the nucleotide sequences indicate ORFs of significant length. There is no clear evidence as to why satellite RNAs of this group are dependent for their replication on a helper virus. However, there is quite strong evidence that they are replicated by a viroidlike, rolling-circle mechanism involving intermediates that are multimeric tandem repeats of the satellite RNA. Circular and multimer forms of STRSV RNA are found in infected tissue. Large amounts of the circular monomer form may be present in nucleic acids extracted from infected tissue, but only linear molecules are packaged in virus particles.

In vitro, multimeric forms of STRSV plus sense RNA can be processed autolytically in the presence of divalent ions to give biologically active monomers. Under appropriate conditions, this reaction is reversible, and ligation occurs to restore the original bonds. Similar processes occur with multimeric minus sense STRSV RNAs.

2.2.3 The Development of Simple and Specific RNA Enzymes

The idea that enzymes always involve proteins as part or all of their structure has been biochemical dogma for more than 60 years. The fact that satellite RNAs can cleave phosphodiester bonds and religate them in the absence of any protein has aroused considerable interest among biologists.

The self-cleavage activity of circular satellite RNAs is confined to relatively short and conserved parts of the molecule. They lie at the 5′ and 3′ termini of the linear forms. Hammerhead structures, like those proposed for ASBVd (Fig. 9.4), operate in the satellite RNA self-cleavage and ligation process. Three essential minimal sequences have been defined for the enzymatic activity of STRSV RNA (Fig. 9.7). The structure labeled B and C in Fig. 9.7 is known as a *ribozyme*. Haseloff and Gerlach (1989) proposed that the highly conserved B region in the ribozyme may provide a metal ion-binding site which precisely positions such an ion close to the 2′ OH group next to the phosphodiester bond to be cleaved.

The ability to synthesize tailor-made ribozymes with C sequences complementary to the RNA site of interest makes it possible in principle to cleave *in vivo* any RNA of known sequence, at a predetermined site. This opens up many possibilities, including a novel approach to the control of virus infection. It has long been assumed that proteins must have been one of the earliest macromolecules to develop in the precellular stages of evolution. The fact that RNA alone can cleave and ligate bonds lends support to the view that the macromolecules present in the earliest stage of evolution of life on earth may have consisted only of RNA.

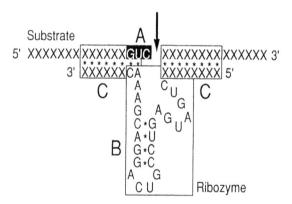

Figure 9.7 Essential sequences in the self-cleaving domain derived from STRSV RNA. (A) GUC triplet in the substrate RNA immediately 5′ to the bond to be cleaved; (B) Catalytic region consisting of a highly conserved sequence with a base-paired stem and a loop; (C) Flanking regions without necessary sequence conservation, but base-paired to the substrate RNA and allowing precise positioning of the ribozyme relative to the cleavage site. From Haseloff and Gerlach (1989).

Figure 9.8 Modulation of CMV symptoms by its satellite CARNA 5. Upper row of plants was infected by virus alone; lower row of plants, by virus plus CARNA 5. In tabasco pepper plants on the left, CARNA 5 attenuates disease symptoms; in tomato plants on the right, CARNA 5 induces lethal necrosis. From Kaper and Collnaer (1988).

2.2.4 Symptom Modulation by Satellite RNAs

The biological effects of the presence of a satellite RNA along with the helper virus may depend markedly on the host species. This is illustrated in Fig. 9.8 for CMV and its satellite CARNA5. Many such effects are known.

Various experiments have shown that where satellite RNAs occur, the disease outcome depends on interactions between (1) the strain of helper virus; (2) the strain of satellite RNA; (3) the species and cultivar of host plant; (4) environmental conditions; and (5) the presence of a related satellite. The presence of a satellite causing only mild symptoms may protect plants against a second inoculation with a satellite causing severe disease.

The sequence domains in satellite RNAs that are responsible for these effects on disease have been investigated using (1) comparisons between naturally occurring variants of known sequence, (2) site-directed mutagenesis, and (3) recombinant RNA genomes constructed via cDNAs.

An RNA of particular interest in relation to disease modulation, and also to the origin and evolution of satellite RNAs, is the virulent satellite RNA of turnip crinkle *Carmovirus* (TCV) shown in Fig. 9.9. Other related avirulent satellites lack the 3' sequence derived from the helper TCV genome.

Another interesting satellite RNA is involved in groundnut rosette, which is an important disease of groundnuts in Africa. Groundnut rosette virus (GRV) is transmitted by an aphid vector only in the presence of groundnut rosette assistor *Luteovirus*, which on its own causes no disease. Various forms of the rosette disease are known. Murant and Kumar (1990) have shown that a satellite RNA of GRV is largely responsible for the kind of disease that occurs. Different variants of the satellite cause the different forms of the disease.

The only general conclusion to be drawn from these various studies is that changes in disease induced by the presence of a satellite RNA depend on changes in the nucleotide sequence in the RNA. There is no convincing evidence that such changes are mediated by a polypeptide translated from the satellite RNA. Disease modulation is probably brought about by specific macromolecular interactions between the satellite RNA and (1) helper virus RNAs; (2) host RNAs; (3) host proteins; (4) viral-coded proteins; or (5) any combination of 1 to 4. Until the interactions involved have been established on a molecular basis, differences in nucleotide sequence between related satellites will remain largely uninterpretable with respect to disease modulation.

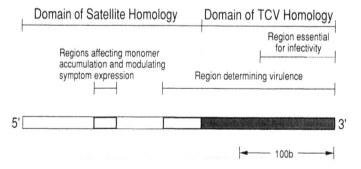

Figure 9.9 Domains in the virulent composite satellite RNA C of TCV. Modified from Simon *et al.* (1988).

3 DEFECTIVE INTERFERING PARTICLES

Defective interfering RNAs (DI RNAs) and DI particles are widespread in animal virus infections. These DI particles have the following properties: (1) they are derived from the viral genome by deletion of nucleotide sequences. Thus, they consist mainly or entirely of genomic nucleotide sequences; (2) they reduce the yield of helper virus; and (3) their presence causes milder disease symptoms.

These kinds of RNAs or DNAs may be present in several plant virus groups, including rhabdoviruses, reoviruses, tombusviruses, and geminiviruses. A good example is the abnormal RNA found in a culture of tomato bushy stunt virus (TBSV) that meets all the criteria noted above for a DI RNA (Morris and Hillman, 1989). The RNA was about 396 nucleotides long. It was derived from the genomic RNA by six internal deletions, the 5' and 3' sequences being conserved. Two of the deletions were large (1180 nt and 3000 nt), whereas the others were much smaller (Fig. 9.10). Coinoculation of the small RNA depressed virus synthesis in whole plants and attenuated disease symptoms. Although the DI RNA could represent 60% of viral-specific RNA in leaf extracts, only about 3

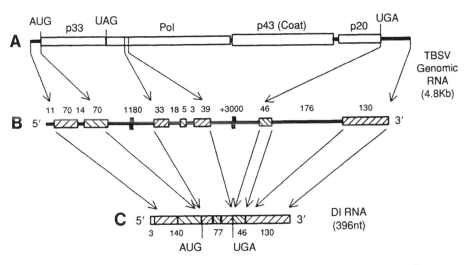

Figure 9.10 Origin of a DI RNA (C) from TBSV genomic RNA. (B) shows the size (in nucleotides) and the positions of the regions of genomic sequence deleted to produce the DI RNA. From Morris and Hillman (1989).

to 4% of the encapsidated RNA was DI RNA. The DI RNA probably replicates by the same mechanism as the viral RNA.

FURTHER READING

Diener, T. O. (ed.) (1987). "The Viroids." Plenum Press, New York.

Francki, R. I. B. (1985). Plant virus satellites. *Annu. Rev. Microbiol.* **39**, 151–174.

Matthews, R. E. F. (1991). "Plant Virology." 3rd Ed. Academic Press, New York.

Morris, T. J., and Knorr, D. A. (1990). Defective interfering viruses associated with plant virus infections. *In* "New Aspects of Positive Strand RNA Viruses" (M. A. Brinton and F. X. Heinz, eds.), pp. 123–127. Amer. Soc. Microbiology, Washington, D.C.

Semancik, J. S. (ed.) (1987). "Viroids and Viroid-like Pathogens." CRC Press, Boca Raton, Florida.

Symons, R. H. (1989). Self-cleavage of RNA in the replication of small pathogens of plants and animals. *Trends Biochem. Sci.* **14**, 445–450.

TRANSMISSION, MOVEMENT, AND HOST RANGE

10

Being obligate parasites, viruses depend for survival on being able to infect new individual host plants reasonably frequently in relation to the life span of the host species. However, there is a barrier to infection, because most plant surfaces are covered by layers of inert material, which viruses cannot penetrate on their own (Fig. 10.1). Different viruses overcome this problem in one or more of the following ways: (1) by avoiding the need to invade through the plant surface, as in transmission through the seed or in vegetative propagation; or (2) by some method that involves penetration through a wound in the surface layers, as in mechanical transmission or transmission by invertebrates.

The study of virus transmission is important for several reasons: (1) a knowledge of the way a virus is being transmitted in the field is an essential prerequisite for developing methods to control the disease it is causing; (2) the relationships between viruses and their invertebrate and fungal vectors are of considerable general biological interest; and (3) certain methods, particularly mechanical transmission, are very important for the study of viruses in the laboratory.

1 DIRECT PASSAGE IN LIVING HIGHER PLANT MATERIAL

1.1 THROUGH THE SEED

About one fifth of the known plant viruses are transmitted through the seed of infected host plants. Seed transmission provides a very effective means of introducing virus into a crop at an early stage, giving randomized foci of infection throughout the planting. Thus, when some other method of transmission can operate to spread the virus within the growing crop, seed transmission may be of very considerable economic importance. Viruses may persist in seed for long periods so that commercial distribution of a seed-borne virus over long distances may readily occur.

Two general types of seed transmission can be distinguished. With tomato mosaic virus (ToMV) in tomato, seed transmission is largely the result of contamination of the seed coat with virus, resulting in subsequent infection of the germinating seedling by mechanical means. This type of transmission may occur with other tobamoviruses. The external virus can be readily inactivated by certain treatment eliminating all, or almost all seed-borne infection.

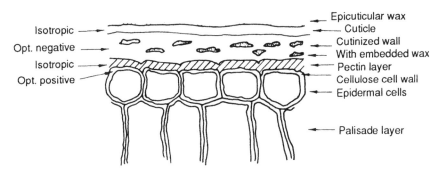

Figure 10.1   The barrier to virus infection. Diagrammatic representation of the epicuticle of the plant seen in cross section. The lines dividing the layers above the epidermal cells indicate regions of major change in the construction of components rather than sharp boundaries. Individual plant species may depart greatly from this general arrangement. B. E. Juniper; from Eglinton and Hamilton (1967).

In the second and more common type of seed transmission, the virus is found within the tissues of the embryo, for example with nepoviruses. The embryo may become infected through the ovary or via the pollen.

1.1.1 Factors Affecting the Proportion of Seed Infected Internally

Various factors influence the proportion of seed in which the embryo is infected:

1. *Virus and virus strain*   For viruses in a particular group, the proportion of infected seed may range from 0 to 100% for different viruses and virus strains.

2. *Host plant*   Different host species, and cultivars within a single species, may differ widely in the rate of seed transmission.

3. *Time at which the seed-bearing plant is infected*   For most hosts in which a virus is seed transmitted, the earlier the plant is infected, the higher the percentage of seed that will subsequently transmit the virus.

4. *Age of seed when sown*   Some viruses are lost quite rapidly from seed on storage, whereas others may persist for years.

5. *Temperature*   Viruses may survive in seed stored at high temperatures for longer periods than the ability of the seed to germinate.

1.1.2 Transmission through Infected Pollen

Most seed-transmitted viruses are probably also transmitted through pollen from infected plants. Conversely, there appears to be no example of a pollen-transmitted virus that is not also seed transmitted. Self-pollination of infected plants presumably can result in a higher percentage of infected seed than when only one of the gametes comes from an infected individual. With certain viruses, when a healthy plant is pollinated with infected pollen, only the resulting seed may become infected. With others, however, the plant itself may also become infected.

There are three mechanisms by which viruses may be transmitted by pollen: (1) By virus within the sperm cell nucleus or cytoplasm. (2) On the exine of the pollen grains. The germ tubes growing from such pollen may then pick up virus particles and carry active virus to the ovule. (3) By species of thrips that infect flowers and can transfer pollen to a second plant and introduce a virus from the pollen while feeding. The efficiency of pollen transmission varies widely. Members of the *Cryptovirus* group are unusual in that they are transmitted with high efficiency through pollen and seed, but not by mechanical transmission, grafting, or invertebrate vectors.

1.2 BY VEGETATIVE PROPAGATION

Vegetative propagation is an important horticultural practice, but it is unfortunately also a very effective method for perpetuating and spreading viruses. Economically important viruses spread systemically through most vegetative parts of the plant. A plant once systemically infected with a virus usually remains infected for its lifetime. Thus, any vegetative parts taken for propagation, such as tubers, bulbs, corms, runners, and cuttings, will normally be infected. There are many instances in which every individual of a particular cultivar tested has been found infected with a particular virus, for example, some potato cultivars infected with potato virus X (PVX).

1.3 BY GRAFTING

Grafting is essentially a form of vegetative propagation in which part of one plant grows on the roots of another individual. Once organic union has been established, the stock and scion become effectively a single plant. Where either the rootstock or the individual from which the scion is taken is infected systemically with a virus, the grafted plant as a whole will become infected if both partners in the graft are susceptible. Since the early days of work with plant viruses, the demonstration that a disease was transmissible by grafting, together with the absence of a pathogen visible by light microscopy, has been taken as an indication that the disease is attributable to a virus. Because these criteria were inadequate, some diseases once thought to be caused by a virus are now known to be caused by mycoplasmas or similar organisms.

Grafting transmission may lead to a different disease from that appearing after, say, mechanical inoculation. For example, *Nicotiana glutinosa* normally gives necrotic local lesions with no systemic movement of virus following mechanical inoculation with tobacco mosaic virus (TMV). However, healthy plants grafted with tobacco plants infected systemically with TMV die of a systemic necrotic disease. This effect of grafting is probably owing to the introduction of virus into the vascular elements of the hypersensitive host.

2 TRANSMISSION BY ORGANISMS OTHER THAN HIGHER PLANTS

2.1 INVERTEBRATES

Many plant viruses are transmitted from plant to plant in nature by invertebrate vectors. This important topic is considered in Chapter 13 of this volume.

2.2 FUNGI

Species in the chytrid genus *Olpidium* transmit four viruses with isometric particles, whereas species in two plasmodiophorus genera (*Polymyxa* and *Spongospora*) transmit about 11 rod-shaped or filamentous viruses.

The most-studied fungal vector is the chytrid *Olpidium brassicae* (Wor.) Dang., which is a soilborne obligate parasite infecting the roots of many plants. In root cells, the fungus forms resting spores, which are released into the soil when the root disintegrates. The resting spores, under appropriate conditions in the root or soil, release numerous zoospores into the soil water. These can then infect fresh roots. In the first phase of the infection process, the zoospore forms a cyst. In the second phase, the host cell wall is breached, and the cyst cytoplasm enters the host cell, leaving behind it the ectoplast and the tonoplast. The fungal cytoplasm becomes surrounded by a new tonoplast inside the host cell. A virus may be carried to the root on the surface of the zoospore (e.g., tobacco necrosis virus (TNV), satellite TNV (STNV), cucumber necrosis virus) or within the zoospores (e.g., lettuce big vein agent and tobacco stunt virus).

2.3 EXPERIMENTAL TRANSMISSION
BY *AGROBACTERIUM TUMEFACIENS*

DNA copies of viral genomes can be inserted into the Ti plasmid carried by *A. tumefaciens*. Infection of a plant by *Agrobacterium* carrying such an engineered plasmid offers a novel experimental method for transmission of plant viruses, some of which have proved impossible to transmit by mechanical inoculation.

3 MECHANICAL TRANSMISSION

Mechanical inoculation involves the introduction of infective virus or viral RNA into a wound made through the plant surface. When virus establishes itself successfully in the cell, infection occurs. This method of transmission is of great importance for many aspects of experimental plant virology, particularly for the assay of viruses (Chapter 4) and in the study of the early events in the interaction between a virus and susceptible cells (Chapter 7). Viral RNA as inoculum was discussed in Chapter 6, Section 1.1. When intact virus is used as inoculum, the viral nucleic acid must be partly or entirely uncoated at an early stage. This process is discussed in Chapter 7, for TMV in Section 3.7 and for turnip yellow mosaic virus (TYMV) in Section 4.7.

3.1 FACTORS AFFECTING EFFICIENCY
OF MECHANICAL TRANSMISSION

There are many factors affecting the efficiency of mechanical transmission. Factors affecting plant susceptibility are discussed in Chapter 11, Section 6. Here factors in the inoculation process are considered.

3.1.1 Method of Applying the Inoculum

In most circumstances, gently rubbing the leaf surface with some suitable object wetted with the inoculum gives efficient transmission. The objective in mechanical inoculation is to make numerous small wounds in the leaf surface without causing death of the cells. The pressure required to do this depends on many factors such as plant species, age and condition of the leaf, and additives present in the inoculum. With a few viruses and hosts, severe abrasion or cutting with a knife is more effective.

3.1.2 Additives That Increase Efficiency

The efficiency of mechanical inoculation is greatly increased when some abrasive material such as carborundum or celite is added to the inoculum or sprinkled over the leaves before inoculation. The increase in the number of local lesions obtained by the use of abrasives varies with different hosts and viruses but may be 100-fold or more. Other substances such as dipotassium phosphate, potassium sulfite, and divalent metal ions may also give substantial increases in inoculation efficiency, depending on the virus, the host plant used as source of inoculum, and the host species being inoculated.

3.1.3 Frozen and Freeze-Dried Inoculum

Frozen or freeze-dried leaf material ground with a pestle and mortar may form an effective inoculum for some viruses and hosts.

3.1.4 Host Components in the Inoculum

When crude leaf extracts are used as inoculum, host components such as polyphenols, polysaccharides, etc., may have an inhibitory effect on infection. Additives to the inoculum may be used to counteract such substances.

3.1.5 Washing Leaves

Washing leaves immediately after inoculation or dipping them in water is a common practice. It may improve efficiency of the inoculation if inhibitors are present in the inoculum. In other circumstances, it may reduce efficiency.

3.1.6 The Plant Part Inoculated

Sometimes a virus can be transmitted mechanically by inoculating cotyledon leaves but not the first true leaves.

3.1.7 The Plant Part Used as Inoculum

Virus inhibitors are often present in highest concentration in leaves, and may be present in low concentrations or absent in other organs, e.g., petals. Thus, extracts from petals have been used successfully to transmit several viruses mechanically.

3.1.8 Viral RNA as Inoculum

For any virus that has positive sense ssRNA as its genetic material, it should be possible, in principle, to prepare an extract of total nucleic acids from infected tissue with a phenol water system and use this to inoculate healthy plants after removal of the phenol. This procedure may allow mechanical transmission when whole-leaf extracts are ineffective.

3.1.9 Plant Factors

There are many plant factors, such as age and genotype, that affect the efficiency of mechanical transmission. These are discussed in Chapter 11, Section 6.

3.2 NATURE AND NUMBER OF INFECTIBLE SITES

The upper surface of leaves commonly used for mechanical inoculation usually contains about 1 to 5×10^6 cells of various types. Most will be epidermal cells, with a smaller number of stomatal guard cells. Many leaves possess trichomes. There are numerous kinds of trichomes, and more than one kind may occur on the same leaf. It seems probable that all the cell types making up the epidermis are potentially capable of being infected by mechanical inoculation. The cuticle is probably the major barrier to infection. Wounds that penetrate right through the cuticle and the cell wall are probably effective in allowing virus to enter. Experiments with several viruses indicate that, following mechanical inoculation, some underlying mesophyll cells may be infected directly with the inoculum.

Compared with the plaque assays for bacterial viruses on lawns of growing bacteria, or for animal viruses on cell monolayers, the mechanical inoculation of plant viruses is a highly inefficient process. As usually carried out, about 10^4 to 10^7 virus particles may be applied to the leaf for each local lesion produced. By using very small volumes of concentrated inoculum, this number may be reduced to 10 to 100. The reasons for this inefficiency have not been established, but it is likely that a high proportion of virus particles in the applied inoculum never reach an infectible wound. Experiments suggest that the infectibility of wounds falls off very rapidly after they are made.

3.3 MECHANICAL TRANSMISSION IN THE FIELD

Compared to transmission by invertebrate vectors or vegetative propagation, field spread by mechanical means is usually of minor importance. With some viruses, however, it is of considerable practical significance. TMV can readily contaminate hands, clothing, and implements and be spread by workers in tobacco and tomato crops. This is particularly important during the early growth of the crop, for example, during the setting out of plants. Plants infected early act as sources of infection for further spread, either during cultural operations such as disbudding, or by rubbing together of healthy and infected leaves by wind. TMV may be spread mechanically by tobacco smokers, because the virus is commonly present in cigarettes.

PVX can also be readily transmitted by contaminated implements or machines and by workers or animals that have been in contact with diseased plants. On some materials, the virus persists for several weeks. Several viruses, such as tomato bushy stunt virus (TBSV) and Southern bean mosaic virus (SBMV), may infect roots from virus-contaminated soil, apparently without fungi or arthropods being involved. In glasshouses, TMV can be transmitted by soil containing the virus coming into contact with leaves.

4 MOVEMENT AND FINAL DISTRIBUTION IN THE PLANT

Some viruses are confined to the inoculated leaf, whereas others may move systematically through the plant. If a virus replicates in the initially infected cell but cannot move to neighboring cells, its replication may remain undetected. This has implications in considering the host range of viruses (Section 5). In many host–virus combinations, cell-to-cell movement is limited, giving rise to a relatively small zone of infected tissue that may be visible as a local lesion, often consisting of dead cells (Fig. 11.1). In still others, movement within the inoculated leaf is not limited, but systemic spread may not occur.

Viruses spread through the plant in two ways—a slow cell-to-cell spread through the plasmodesmata from the site of inoculation, and a much more rapid movement through the vascular tissues, usually through the phloem. In these processes, there are three possible barriers to movement: (1) movement from the first infected cell; (2) movement out of parenchyma cells into vascular tissues; and (3) movement out of vascular tissue into the parenchyma cells of an invaded leaf.

The final distribution of virus through tissues and organs may be very uneven. Many factors, including host genes, viral genes, and environmental conditions, affect the rate and extent of virus movement through the plant. The effects of viral genes are discussed in Section 4.4; host genes, in Chapter 11, Section 6.2; and environmental factors, in Chapter 11, Section 6.2.

4.1 THE FORM IN WHICH VIRUS MOVES

Since both intact virus and isolated positive sense ssRNA can infect cells, both of these structures may be involved in cell-to-cell spread. Virus particles have been visualized in plasmodesmata and sieve pores. Such observations by no means demonstrate, however, that intact virus particles are the form, or the only form in which a virus moves from cell to cell. There is clear evidence that viral RNA can move into and infect neighboring cells. For example, defective mutant strains of certain viruses (e.g., TMV) are unable to produce complete virus particles. Infective viral RNA and defective coat protein are produced, and yet the infection moves from cell to cell.

Russian virologists have suggested that TMV and some other viruses move over long distances through the plant in complexes that they called virus-specific ribonucleoprotein (vRNP), containing viral RNA, viral coat protein, and various other proteins. The evidence, including binding of movement protein to nucleic acids, suggests, but by no means proves, that vRNPs are the form in which TMV and other viruses move both from cell to cell and over long distances. The presence of characteristic virus particles in sieve tubes or phloem exudate suggests that, in some circumstances, the intact virus may be translocated.

4.2 THE ROLE OF PLASMODESMATA

Plasmodesmata consist of two concentric cylinders of plasma membrane and endoplasmic reticulum that traverse the cellulose walls between adjacent plant cells. The annulus between the two membrane cylinders gives continuity of the cytosol between cells. This continuity is probably regulated and therefore, not permanent. Even small molecules like sucrose are transported via the plasmodesmata. Virus particles have been observed by electron microscopy in the plasmodesmata of infected tissue. Thus, it is possible that this is the route by which viruses move from cell to cell. However, it is known that ribosomes, which are roughly the same size as small viruses, do not pass through the plasmodesmata. Therefore we must conclude that viruses alter the properties of plasmodesmata in a manner that allows their passage through them. Evidence for such changes has recently been obtained with TMV-infected plants, and with cells infected with caulimoviruses. With TMV, slightly enlarged plasmodesmata can be observed by electron microscopy. Movement of large dye particles is also facilitated in TMV-infected tissue. In CaMV-infected cells, plasmodesmata become greatly enlarged, and this may also facilitate virus movement. The rates of virus movement from cell to cell through the plasmodesmata are of course dependent on temperature. Movement of TMV out of the inoculated epidermis takes about 4 hr at 27°C. Movement from cell to cell in the mesophyll, as infection spreads, occurs at the rate of about one cell every few hours.

4.3 LONG-DISTANCE MOVEMENT

Most viruses move longer distances through the plant in the phloem, but viruses transmitted by beetles may move in the xylem. Once virus enters the phloem, movement may be very rapid. A rate of 2.5 cm/min

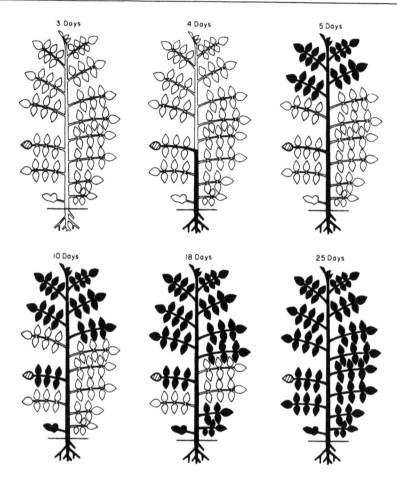

Figure 10.2 Diagram showing the spread of TMV through a medium young tomato plant. The inoculated leaf is shaded, and systemically infected tissues are shown in black. From Samuel (1934).

has been recorded for beet curly top *Geminivirus*. The pattern of movement of TMV from an inoculated leaflet through a tomato plant is shown in Fig. 10.2. This pattern of movement is fairly typical for viruses infecting herbaceous plants, at least as far as stem and leaves are concerned.

4.4 THE ROLE OF VIRAL GENE PRODUCTS IN VIRUS MOVEMENT WITHIN THE PLANT

Various lines of evidence indicate that the coat protein is not required for cell-to-cell or long-distance systemic movement of some viruses. Experiments with designed mutations in the coat-protein gene show that effective packaging of the RNA into virus particles is essential for systemic infection with some viruses such as brome mosaic virus (BMV), but others such as barley stripe mosaic virus (BSMV) are able to move very efficiently when the entire coat-protein gene has been deleted.

Experiments involving the use of recombinant DNA technology, discussed in Chapter 7, Section 3.4, have shown beyond doubt that the 30-K protein coded for by TMV is involved in localized cell-to-cell movement of this virus. Genes with a cell-to-cell movement function have now been identified positively or tentatively for many groups of RNA viruses, and for cauliflower mosaic virus (CaMV). The cell-to-cell movement proteins of several viruses have amino acid sequence similarities. Cell-to-cell movement proteins may not be found in all virus groups. Luteoviruses are almost entirely confined to the phloem, and the nucleotide sequence of potato leafroll virus (PLRV) shows no open reading frame (ORF) with a sequence related to the TMV movement protein.

More than one gene product, other than coat protein, may be involved in the movement of some viruses. For example, while the ORF1 product of CaMV is involved in cell-to-cell movement, gene VI is also necessary for systemic spread. Viral-coded movement proteins may act in one, or more than one way. The proteins encoded by different viruses may act by different mechanisms; but the fact that there are amino acid sequence similarities between proteins encoded by different viruses would favor some similarity in mechanisms involved in movement. These are most likely to involve structural and functional alterations to the plasmodesmata, which allow virus particles to be transported, as well as some *molecular chaperone* activity involving virus nucleic acid transport from cell to cell.

4.5 FINAL DISTRIBUTION OF VIRUS IN THE PLANT

Some viruses in certain hosts are confined to the local lesions that form in inoculated leaves (Fig. 11.1). However, even with viruses that move systemically through the plant, the final distribution of virus is usually far from uniform. There are several reasons for this:

1. *Escaping infection* Certain viruses move with difficulty into particular leaves. With viruses infecting woody perennials, distribution may be very uneven, even in plants that have been infected for several years.

2. Rise and fall of virus concentration with age   The concentrations of some viruses initially rise rapidly in infected leaves and then decrease as the infection progresses. Considering the plant as a whole, this can lead to a very uneven distribution of virus in different leaves at any given time during growth.

3. Uneven concentration in different organs and tissues   In a plant infected for some time with a systemic virus, the concentration of virus may not be uniform in different organs. With most mosaic-type viruses, the virus reaches a much higher concentration in the leaf lamina than in other parts of the plant.

4. Viruses confined to certain tissues   Members of the *Luteovirus* group, such as barley yellow dwarf virus (BYDV), are found in the phloem parenchyma, companion cells, and sieve elements. A few other viruses, such as the plant reoviruses, also appear to be confined to the phloem.

5. Uneven distribution in leaves showing mosaic patterns   Dark green areas in the mosaic patterns of diseased leaves usually contain very little virus compared with yellow or yellow-green areas. This has been found for viruses that differ widely in structure, and among those that infect mono- or dicotyledonous hosts. The phenomenon may, therefore, be a fairly general one for diseases of the mosaic type. This uneven distribution of virus in mosaic diseases may also extend to the petals of flowers showing color-breaking.

6. Distribution of viruses near apical meristems   Electron microscopic observations show that some viruses, for example tobacco rattle virus (TRV), can invade the apical meristems of both shoots and roots. However, with many virus–host combinations, there appears to be a zone of variable length (usually about 100 μm but up to 1000 μm) near the shoot or root tip that is free of, or contains very little virus.

5 THE MOLECULAR BASIS FOR HOST RANGE

Since the early years of this century, plant virologists have used host range as a criterion for attempting to identify and classify viruses. In a typical experiment, the virus under study would be inoculated by mechanical means into a range of plant species. These plants would then be observed for the development of viruslike disease symptoms. Back-inoculation to a host known to develop disease might be used to check for symptomless infections. In retrospect, we can see that reliance on such a procedure gives an oversimplified view of the problem of virus host ranges. Over the past few years our ideas of what we might mean by *infection* have been considerably refined, and some possible molecular mechanisms that might make a plant a host or a nonhost for a particular virus have emerged.

On the basis of present knowledge, there are four possible stages in which a virus might be blocked from infecting a plant and causing systemic disease: (1) during initial events—the uncoating stages; (2) during attempted replication in the initially infected cell; (3) during movement from the first cell in which the virus replicated; and (4) by stimulation of the host's cellular defences in the region of the initial infection. These stages will be considered in turn.

5.1 INITIAL EVENTS

5.1.1 Recognition of a Suitable Host Cell or Organelle

Bacterial viruses and most of those infecting vertebrates have specific proteins on their surface, which act to recognize a protein receptor on the surface of a susceptible host cell. The surface proteins on plant rhabdoviruses and reoviruses that replicate in their insect vectors may have such a cell-recognition function in their animal hosts.

However, there is no evidence for such receptors on the surface of any of the positive sense ssRNA plant viruses. For such viruses, coat proteins appear to play little if any positive role in cell recognition. Specialized surface recognition proteins may be of little use to plant viruses because of the requirement that they enter cells through wounds on the plant surface. Most plant viruses may have evolved a recognition system basically different from that of viruses, which normally encounter and recognize their host cells in a liquid medium or at a plasma membrane surface. The evidence available for the small plant viruses suggests that host range is usually a property of the viral RNA rather than of the protein coat.

5.1.2 The Uncoating Process

Experiments with several ss positive sense RNA plant viruses, including TMV and TYMV, show that uncoating of the RNA from virus particles introduced by mechanical inoculation processed about as efficiently in nonhost species as in hosts. Thus, there is no evidence that the uncoating process is involved in host specificity.

5.2 REPLICATION

There are no recorded examples among the ss positive strand RNA viruses in which particular steps in the viral replication cycle have been proved to be a determinant of host range among different plant species. Following inoculation of TMV to plant species considered not to be hosts for the virus, viral RNA has been found in polyribosomes. Furthermore, TMV particles can uncoat and express their RNA in *Xenopus* oocytes. However, a block in the virus-replication cycle is the basis for resistance to a virus in one variety of a particular host species (Chapter 11, Section 6.2.2).

Most strains of CaMV infect only members of the Brassicaceae, but a few have solanaceous hosts. Tests with recombinant viruses made by exchanging DNA segments between cloned strains have demonstrated that the first half of gene VI, the gene for the inclusion body protein (Fig. 8.1), determines whether the virus can systemically infect solanaceous species.

Some viruses synthesize read-through proteins (Chapters 6 and 7). Successful read-through depends on the presence of an appropriate suppressor transfer RNA (tRNA) in the host plant. The presence or absence of such a tRNA might also be a factor that could influence host range for such viruses.

5.3 CELL-TO-CELL MOVEMENT

Two lines of evidence strongly support the view that possession of a compatible and functional cell-to-cell movement protein is one of the factors determining whether a particular virus can give rise to readily detectable virus replication in a given host species or cultivar.

First, the experiments summarized earlier in Section 4.4 show that many viruses contain a gene coding for a cell-to-cell movement protein. Viruses cannot usually infect leaves through the cut end of petioles or stems. However, if the upper leaves are already infected with a helper virus, a virus in solution may be able to pass into and infect the upper leaves. This can happen with mutants defective in a transport-function gene, and also with viruses that are not normally able to infect the species of plant involved. Thus, the resistance of an apparent nonhost species due to a block in the transport function may be overcome by pre-infection with a virus that has a transport protein compatible with the particular host. This appears to be a fairly general phenomenon.

Second, a number of viruses have been shown to be able to infect and replicate in protoplasts derived from species in which they show no macroscopically detectable sign of infection following mechanical inoculation of intact leaves. Other evidence shows that, for some viruses at least, gene products in addition to the movement protein may participate in determining host range.

5.4 HOST GENES AFFECTING HOST RANGE

Host genes must be involved in all interactions between a virus and the plant cell following inoculation, irrespective of the outcome of an inoculation. The effects on virus infection of certain host genes involved in generalized host-defense mechanisms have been studied in some detail. These are discussed in Chapter 11, Section 3.

6 DISCUSSION AND SUMMARY

The ability to be transferred to healthy host plant individuals is crucial for the survival of all plant viruses. Viruses cannot, on their own, penetrate the undamaged plant surface. For this reason, individual viruses have evolved ways to bypass or overcome this barrier. Viruses that are transmitted from generation to generation via pollen and seed, or in plant parts in vegetatively reproducing species, can circumvent the need to penetrate the plant surface.

Many groups of plant viruses are transmitted either by invertebrate or fungal vectors, which penetrate the cell during feeding or infection, and carry infectious virus into the plant cells at the same time.

Mechanical transmission is a process whereby small wounds are made on the plant surface in the presence of infectious virus. For many groups of viruses, if conditions are favorable, infection will follow in infectible hosts. Mechanical inoculation is a very important procedure in experimental virology. This process is not of great importance for infection in the field except for a few groups such as tobamoviruses, potexviruses, and hordeiviruses, which appear to have evolved no other effective means of plant-to-plant transmission. These viruses are, however, well adapted to mechanical transmission, since they are relatively stable and occur in high concentration in infected leaves. These features make transmission possible when leaves of neighboring plants abrade one another.

It is probable that successful entry of a single virus particle into a cell is sufficient to infect a plant, but in the normal course of infection, many virus particles often enter a single cell. Following initial replication in the first infected cell, the virus may move by a slow process to neighboring cells via the plasmodesmata. The virus may then enter the vascular tissue, usually the phloem, giving rise to rapid movement. Some viruses code for specific gene products that are necessary for the virus to move out of the initially infected cell and through the plant.

The final distribution of a virus through the plant may be quite uneven. In some host–virus combinations, virus movement is limited to local lesions. In others, some leaves may escape infection, whereas in mosaic diseases, dark green islands of tissue may contain little or no virus. Some viruses are confined to certain tissues. For example, luteoviruses are confined mainly to the phloem. Some viruses penetrate to the dividing cells of apical meristems. Others appear not to do so.

There is clear evidence that both virus particles and infectious viral RNA can move from cell to cell, but this may not be the form in which all viruses move. Evidence suggests that TMV may move in the form of nucleoprotein complexes composed of viral RNA, viral-encoded proteins, and some host components. Viral movement proteins may exert their effect by altering the properties of the plasmodesmata to facilitate viral passage as well as by binding RNA, but neither of these processes has been unequivocally proven to have a direct role in virus movement.

The molecular basis for the host range of viruses is not understood. Most plant viruses do not have a mechanism by which they can recognize a host cell that is suitable for replication. Uncoating of the viral nucleic acid appears to occur in hosts and nonhosts alike. No particular step in the replication of a virus has yet been implicated in limiting host range. However, there is substantial evidence that in apparent nonhosts of some viruses, the virus can replicate in the first infected cell but cannot move to neighboring cells. This presumably is because the viral-coded cell-to-cell movement protein fails to function in the particular plant. For practical purposes, such a plant is resistant to the virus in question.

In other virus–host combinations, movement may be limited by host responses to cells in the vicinity of the initially infected cell. Such responses often give rise to a *local lesion host,* which from a practical point of view is *field resistant.* Using these ideas, the different kinds of host response to inoculation with a virus are defined in Table 11.1.

FURTHER READING

Atabekov, J. G., and Taliansky, M. E. (1990). Expression of a plant virus-coded transport function by different viral genomes. *Adv. Virus Res.* **38,** 201–248.

Hull, R. (1989). The movement of viruses in plants. *Annu. Rev. Phytopathol.* **27,** 213–240.

Matthews, R. E. F. (1991). "Plant Virology," 3rd Ed. Academic Press, New York.

Maule, A. J. (1991). Virus movement in infected plants. *Crit. Rev. Plant Sci.* **9,** 459–473.

11

HOST PLANT RESPONSES TO VIRUS INFECTION

1 THE KINDS OF HOST RESPONSE TO INOCULATION WITH A VIRUS

The terms for describing the various kinds of response made by plants to inoculation with a virus have been used in ambiguous and sometimes inconsistent ways. Thus, over many years, there has been confusion about the meaning of certain terms. To a significant degree, the confusion in terminology has been the result of a lack of molecular biological knowledge concerning the different kinds of virus–plant interactions. However, recent work allows some of the terms to be defined with greater precision. In Table 11.1, I have defined the relevant terms in the light of current knowledge. The use of some of these terms differs from that in other branches of virology. For example, *latent* used in reference to bacterial or vertebrate viruses usually indicates that the viral genome is integrated into the host genome. Table 11.1 shows that there are two kinds of infectible host plants—susceptible and resistant. The responses of these two kinds of host are considered next.

2 THE RESPONSES OF SUSCEPTIBLE HOSTS

Viruses are economically important only when they cause some significant deviation from normal in the growth of a susceptible host plant. For experimental studies, we are usually dependent on the production of disease in some form to demonstrate biological activity. Symptomology was particularly important in the early days of virus research, before any of the viruses themselves had been isolated and characterized. Dependence on disease symptoms for identification and classification led to much confusion, because it was not generally recognized that many factors can have a marked effect on the disease produced by a given virus.

Most virus names in common use include terms that describe an important symptom in a major host, or the host from which the virus was first described. Some viruses, under appropriate conditions, may infect a plant without producing any obvious signs of disease. Others may lead to rapid death of the whole plant. Between these extremes, a wide variety of diseases can be produced.

Table 11.1   Types of Response by Plants to Inoculation with a Virus

---

Immune (nonhost). Virus does not replicate in protoplasts; nor in cells of the intact plant, even in the initially inoculated cells. Inoculum virus may be uncoated, but no progeny viral genomes are produced.

Infectible (host): Virus can infect and replicate in protoplasts.

     Resistant (*Expreme hypersensitivity*). Virus multiplication is limited to initially infected cells because of an ineffective viral-coded movement protein, giving rise to a *subliminal infection*. Plants are *field resistant*.

     Resistant (*Hypersensitivity*). Infection limited by a host response to a zone of cells around the initially infected cell, usually with the formation of visible necrotic local lesions. Plants are *field resistant*.

     Susceptible (*Systemic movement and replication*)

         Sensitive. Plants react with more or less severe disease.

         Tolerant. There is little or no apparent effect on the plant, giving rise to *latent* infections.

Virus infection does not necessarily cause disease at all times in all parts of an infected plant. We can distinguish six situations in which obvious disease may be absent: (1) infection with a very mild strain of the virus; (2) a tolerant host; (3) nonsterile *recovery* from disease symptoms in newly formed leaves; (4) leaves that escape infection because of their age and position on the plant; (5) dark green areas in a mosaic pattern; and (6) plants infected with cryptic viruses.

2.1 MACROSCOPIC SYMPTOMS

2.1.1 Local Symptoms

Localized lesions that develop near the site of entry on leaves are not usually of any economic significance but are important for biological assay and sometimes for diagnosis. Infected cells may lose chlorophyll and other pigments, giving rise to chlorotic local lesions (Fig. 11.1A). For many host–virus combinations, the infected cells die, giving rise to necrotic lesions. These vary from small pinpoint areas to large, irregular, spreading necrotic patches (Fig. 11.1C). In a third type, ringspot lesions appear. Typically these consist of a central group of dead cells. Beyond this group, one or more superficial concentric rings of dead cells develop that have normal green tissue between them (Fig. 11.1B). Some ringspot local lesions consist of chlorotic rings rather than necrotic ones. Some viruses elicit no visible lesions in the intact leaves of certain hosts, but when the chlorophyll is removed by ethanol, and the leaf is stained with iodine, *starch lesions* may become apparent.

Figure 11.1 Types of local lesion: (A) Chlorotic lesions caused by beet mosaic *Potyvirus* in *Chenopodium amaranticolor.* Courtesy of P. R. Fry. (B) Ringspot lesions due to AMV in tobacco. Courtesy of S. A. Rumsey. (C) Necrotic lesions due to TSWV in tobacco. Courtesy of E. E. Chamberlain.

2.1.2 Systemic Symptoms

The following sections summarize the major kinds of effects produced by systemic virus invasion. It should be borne in mind that these various symptoms often appear in combination in particular diseases, and that the pattern of disease development for a particular host−virus combination often involves a sequential development of different kinds of symptoms.

Effects on Plant Size

Reduction in plant size is the most general symptom induced by virus infection. There is probably some slight general stunting of growth even with *masked* or *latent* infections, in which the systemically infected plant shows no obvious sign of disease. In vegetatively propagated plants, stunting is often a progressive process. For example, virus-infected strawberry plants and tulip bulbs may become smaller in each successive year. A reduction in total yield of fruit or other usable product of the plant is a common feature, and an important economic aspect of virus disease.

Mosaic Patterns and Related Symptoms

One of the most common obvious effects of virus infection is the development of a pattern of light and dark green areas giving a mosaic effect in infected leaves. The detailed nature of the pattern varies widely for different host–virus combinations. In dicotyledons, the areas making up the mosaic are generally irregular in outline. There may be only two shades of color involved—dark green and a pale or yellow-green, or there may be many different shades of green and yellow (Fig. 11.2).

In mosaic diseases infecting herbaceous plants, there is usually a fairly well-defined sequence in the development of systemic symptoms. The virus moves up from the inoculated leaf into the growing shoot and into partly expanded leaves. In these leaves, the first symptoms are a *clearing* or yellowing of the veins. These symptoms may be very faint or may give striking emphasis to the pattern of veins. Vein-clearing may persist as a major feature of the disease.

Mosaic patterns are laid down at a very early stage of leaf development and may remain unchanged, except for general enlargement, for most of the life of the leaf. In some mosaic diseases, the dark green areas are associated mainly with the veins to give a vein-banding pattern. In leaves that are past the cell-division stage of leaf expansion when they become infected (about 4 to 6 cm long for leaves such as tobacco and chinese cabbage), no mosaic pattern develops. The leaves become uniformly paler than normal.

Figure 11.2 Mosaic disease in chinese cabbage caused by TYMV. Leaf from a plant infected with the Cambridge stock culture. Inoculation with extracts from different-colored blocks of tissue in this leaf gave plants with different symptoms. Courtesy of J. Endt.

In monocotyledons, a common result of virus infection is the production of stripes or streaks of tissue lighter in color than the rest of the leaf. The shades of color vary from pale green to yellow or white, and the more-or-less angular streaks or stripes run parallel to the length of the leaf (Fig. 11.3).

A variegation or *breaking* in the color of petals commonly accompanies mosaic symptoms in leaves. The breaking usually consists of flecks, streaks, or sectors of tissue with a color different from normal (Fig. 11.4). The breaking of petal color is frequently the result of loss of anthocyanin pigments, which reveals any underlying coloration due to plastid pigments. Flower breaking is usually a good indication that the plant is infected with a virus.

Fruits formed on plants showing mosaic disease in the leaves may

Figure 11.3 Narcissus mosaic *Potexvirus* in daffodil. Healthy leaf on right. Courtesy of S. A. Rumsey.

show a mottling, for example, cucumbers infected with cucumber mosaic virus (CMV) (Fig. 11.5A). In other diseases, severe stunting and distortion of fruit may occur (Fig. 11.5B).

Yellows Diseases

Viruses that cause a general yellowing of the leaves are not so numerous as those causing mosaic diseases, but some, such as the viruses causing yellows in sugar beet, are of considerable economic importance.

Ringspot Diseases

The major symptom in many virus diseases is a pattern of concentric rings and irregular lines on the leaves and sometimes also on the fruit. The lines may consist of yellowed tissue or may be owing to

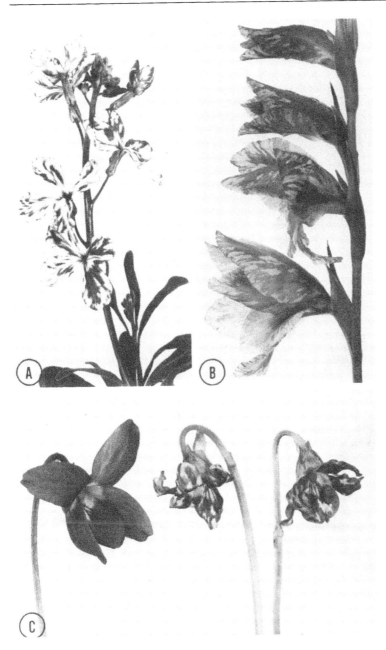

Figure 11.4   Flower symptoms. (A) Turnip mosaic *Potyvirus* in stock. Courtesy of R. I. Hughes. (B) CMV in gladiolus. Courtesy of S. A. Rumsey. (C) CMV in violet. Healthy flower on left. Courtesy of S. A. Rumsey.

Figure 11.5   Fruit symptoms. (A) CMV in cucumber. Healthy fruit on left. Courtesy of H. Drake. (B) Pear stony pit virus in pear. Healthy fruit on left. Courtesy of L. H. Wright. (C) Apple, var. Granny Smith, with concentric ring and russett patterns due to apple ringspot virus. Courtesy of S. A. Rumsey.

death of superficial layers of cells, giving an etched appearance. In severe diseases, complete necrosis through the full thickness of the leaf lamina may occur. With the ringspot viruses, such as tobacco ringspot *Nepovirus* (TRSV), there is a marked tendency for plants to recover from the disease after an initial shock period. Leaves that have developed symptoms do not lose these, but younger growth may show no obvious symptoms in spite of the fact that they contain virus.

Necrotic Diseases

Death of tissues, organs, or the whole plant is the main feature of some diseases. Necrotic patterns may follow the veins as the virus moves into the leaf. In some diseases, the whole leaf is killed. Necrosis may extend fairly rapidly throughout the plant.

Developmental Abnormalities

Besides being generally smaller than normal, virus-infected plants may show a wide range of developmental abnormalities. Such changes may be the major feature of the disease or may accompany other symptoms. For example, uneven growth of the leaf lamina is often found in mosaic diseases. Dark green areas may be raised to give a blistering effect, and the margin of the leaf may be irregular and twisted.

Some viruses cause swellings in the stem, which may be substantial in woody plants, for example, in cocoa swollen shoot disease. Another group of growth abnormalities are known as enations. These are outgrowths from the upper or lower surface of the leaf, usually associated with veins. They may be small ridges of tissue or larger, irregularly shaped leaflike structures, or long filiform outgrowths.

Tumorlike growths are caused by a few viruses. The tumor tissue is less organized than that of enations. Some consist of wartlike outgrowths on stems or fruits. The most-studied tumors are those produced by wound tumor Phytoreovirus (WTV). In a systemically infected plant, external tumors appear on leaves or stems where wounds are made. In infected roots they appear spontaneously, probably as a consequence of damage during root growth.

Genetic Effects

Infection with barley stripe mosaic *Hordeivirus* (BSMV) induces an increase in mutation rate in *Zea mays* and also a genetic abnormality known as an aberrant ratio. This effect is inherited in a stable manner, with a low frequency of reversion to normal ratios. It is inherited in plants in which virus can no longer be detected. The way in which virus infection causes these abnormalities is not well understood.

Recovery from Disease

Not uncommonly a plant shows disease symptoms for a period and then new growth appears in which symptoms are milder or absent, although virus is still present. This commonly occurs with *Nepovirus* infections. Many factors, such as age of plant when infected, influence this recovery phenomenon.

The Cryptovirus Group

The viruses now placed in the *Cryptovirus* group escaped detection for many years because most of them cause no visible symptoms, or in a few situations elicited only very mild symptoms. These viruses are not transmissible mechanically or by vectors, but are transmitted efficiently in pollen and seed. They occur in very low concentrations in infected plants. Nevertheless, they have molecular characteristics that might be expected of disease-producing viruses. They share some properties with the reoviruses.

2.1.3 Agents Inducing Viruslike Symptoms

Disease symptoms similar to those produced by viruses can be caused by a range of physical, chemical, and biological agents. These include the following:

1. *Very small cellular parasites such as mycoplasmas, spiroplasmas, and rickettsialike organisms*   Diseases caused by these organisms are characterized by symptoms such as general yellowing of the leaves, stunting, witches-broom growth of axillary shoots, and a change from floral to leaf-type structures in the flowers (phyllody). Protozoa cause a viruslike disease in oil palm.

2. *Toxins produced by arthropods*   Insects and other arthropods feeding on plants may secrete very potent toxins, which move systemically through the plant and may produce viruslike symptoms.

3. *Genetic abnormalities*   Numerous cultivated varieties of ornamental plants have been selected by horticulturists because they possess heritable leaf variegations or mosaics. These are often the result of maternally inherited plastid defects. The variegated patterns produced sometimes resemble virus-induced mosaics quite closely.

4. *Nutritional deficiencies*   Plants may suffer from a wide range of nutritional deficiencies that cause abnormal coloration, discoloration,

or death of leaf tissue. Some of these conditions can be fairly easily confused with symptoms due to virus infection.

5. *High temperatures*   Growing plants at substantially higher temperatures than normal may sometimes induce viruslike symptoms.

6. *Herbicide damage*   Herbicides such as 2,4-D, may produce viruslike symptoms of leaf distortion in some plants. Tomatoes and grapes are particularly susceptible. Others induce yellowing.

7. *Air pollutants*   Many air pollutants inhibit plant growth, and give rise to symptoms that could be confused with a virus disease.

2.2 HISTOLOGICAL CHANGES

The histological changes induced in susceptible host plants are of three main types: (1) necrosis or death of cells, tissues, or organs. For example, potato virus X (PVX) strain N may cause death of phloem tissue; (2) hypoplasia; for example, the mesophyll cells in yellow areas in leaves showing mosaic disease are frequently less differentiated than normal; and (3) hyperplasia, in which cell division may commence in fully differentiated cells, or abnormal divisions may occur in cambial tissue.

2.3 CYTOLOGICAL EFFECTS

There are two kinds of cytological effects to be considered—the effects of infection on normal cell structures, and virus-induced structures in the cell.

2.3.1 Effects on Cell Structures

Nuclei

Many viruses have no detectable cytological effects on nuclei. Others, such as some potyviruses, may give rise to intranuclear inclusions of various sorts, and may affect the nucleolus or the size and shape of the nucleus, even though they appear not to replicate in this organelle. The replication of some viruses, such as the caulimoviruses, the geminiviruses, and some rhabdoviruses, takes place at least in part in the nucleus, where cytological effects may be observed.

Mitochondria

The development of abnormal membrane systems within mitochondria has been described for several virus infections. They have no established relation to virus replication and are probably degenerative effects.

Chloroplasts

The small peripheral vesicles and other changes in and near the chloroplasts closely related to turnip yellow mosaic virus (TYMV) replication were discussed in Chapter 7, Section 4.5. TYMV infection can cause many other cytological changes in the chloroplasts, most of which appear to constitute a structural and biochemical degeneration of the organelles. The exact course of events in any mesophyll cell depends on (1) the developmental stage at which it was infected, (2) the strain of virus infecting, (3) the time after infection, and (4) the environmental conditions. Members of some other virus groups also cause modifications to chloroplast structure.

Cell Walls

The plant cell wall tends to be regarded mainly as a physical supporting and barrier structure. In fact, it is a distinct biochemical and physiological compartment containing a substantial proportion of the total activity of certain enzymes in the leaf. Three kinds of abnormalities have been observed in or near the walls of virus-diseased cells:

1. *Abnormal thickening,* due to the deposition of callose, may occur in cells near the edge of virus-induced lesions.
2. *Cell wall protrusions* involving the plasmodesmata have been reported for several unrelated viruses.
3. *Depositions of electron-dense material* between the cell wall and the plasma membrane may extend over substantial areas of the cell wall.

Cell Death

Drastic cytological changes occur in cells as they approach death. These changes have been studied by both light and electron microscopy, but the observations do not tell us how virus infection actually kills the cell.

2.3.2 Virus-Induced Structures in the Cytoplasm

The specialized virus-induced regions in the cytoplasm that are, or appear likely to be, the sites of virus synthesis and assembly (viroplasms) were discussed in Chapter 6 through 8. Here other types of inclusions will be described. These are usually either crystalline inclusions consisting mainly of virus, or the pinwheel inclusions characteristic of the potyviruses.

Crystalline and Paracrystalline Inclusions

Virus particles may accumulate in an infected cell in sufficient numbers, and exist under suitable conditions, to form three-dimensional crystalline arrays. These may grow into crystals large enough to be seen with the light microscope, or they may remain as small arrays that can be detected only by electron microscopy. The ability to form crystals within the host cell depends on properties of the virus itself, and is not related to the overall concentration reached in the tissue or to the ability of the purified virus to form crystals.

In tobacco leaves showing typical mosaic symptoms caused by TMV, leaf hair and epidermal cells lying over yellow-green areas may almost all contain crystalline inclusions. These can readily be observed under the light microscope in fresh tissue pieces from a leaf margin. The crystals contain about 60% water, and otherwise consist mainly of successive layers of closely packed parallel rods oriented not quite perpendicularly to the plane of the layers. Rods in successive layers are tilted with respect to one another.

The inclusions found in plants infected with beet yellows *Closterovirus* occur in phloem cells, and also appear in other tissues, for example, the mesophyll. By light microscopy, the inclusions are frequently spindle-shaped and may show banding (Fig. 11.6A). Electron microscopy reveals layers of flexuous virus rods (Fig. 11.6B). Most of the viral inclusions occur in the cytoplasm.

Many small icosahedral viruses form crystalline arrays in infected cells. Sometimes these are large enough to be seen by light microscopy. Icosahedral viruses that do not normally form regular arrays may be

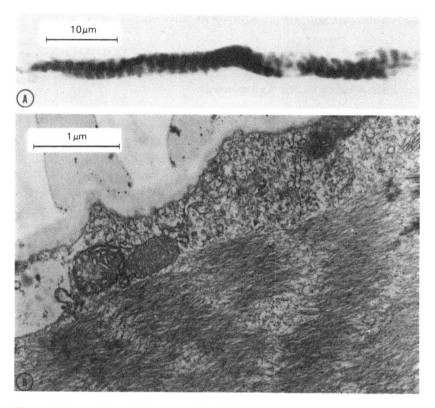

Figure 11.6 Inclusion bodies caused by BYV in parenchyma cells of small veins of *Beta vulgaris*. The banded forms of these inclusions is shown (A) by light microscopy and (B) by electron microscopy. Bands are made up of flexuous virus particles in more or less orderly array. From Esau *et al.* (1966).

induced to do so by wilting or plasmolyzing the tissue to remove some of the water.

Pinwheel Inclusions

Potyviruses induce the formation of characteristic cylindrical inclusions in the cytoplasm of infected cells. The most striking feature of these inclusions, viewed in thin cross-section, is the presence of a central tubule from which radiate curved *arms* to give a pinwheel effect. Reconstruction from serial sections shows that the inclusions consist of a series of plates and curved scrolls with a finely striated substructure. A viral-coded protein is found in these inclusions (Chapter 7, Section 1).

2.4 EFFECTS ON PLANT METABOLISM

Many variables interact to change the biochemical and physiological processes of a virus-infected plant. These include the virus and host plant; the tissue sampled; the age of the tissue; the time after inoculation; diurnal and seasonal variations; and variability between and within leaves, especially in leaves showing mosaic disease. Thus, although there have been many studies on the effects of virus infection on metabolic processes, it is difficult to make firm generalizations, or to relate the effects in a meaningful way to the processes of virus replication.

The physiological and biochemical changes most commonly found in virus-infected plants are (1) a decrease in rate of photosynthesis, often associated with a decrease in photosynthetic pigments, chloroplast ribosomes, and ribulose bisphosphate carboxylase; (2) an increase in respiratory rate; (3) an increase in the activity of certain enzymes, particularly polyphenoloxidases; and (4) decreased or increased activity of plant growth regulators. In many virus diseases, the general pattern of metabolic change appears to resemble an accelerated ageing process. Metabolic changes induced by virus infection are often nonspecific. Similar changes may occur in disease caused by cellular pathogens or following mechanical or chemical injury.

A single gene change in the host, or a single point mutation in the virus may change an almost symptomless infection into a severe disease. Thus, there is no reason to suppose that major disturbances of host-plant metabolism are necessarily determined by the major processes directly concerned with virus replication. Some minor initial effect of virus invasion and replication may lead to profound secondary changes in the host cell. Such changes may obscure important primary effects, even at an early stage after infection.

3 THE RESPONSES OF RESISTANT HOSTS

Two forms of resistant response to inoculation with a virus were defined in Table 11.1. In extreme hypersensitivity, virus multiplication is limited to the initially infected cells because of an ineffective viral-coded movement protein. In this response, no effect of inoculation could be seen with the naked eye. In the second kind of resistance, infection is limited by a host response to a zone of cells around the first infected cell, usually giving rise to a necrotic local lesion with no subsequent systemic movement of the virus. This second type of resistance is discussed here.

3.1 LOCAL AND SYSTEMIC ACQUIRED RESISTANCE

The effect of TMV inoculation on various tobacco varieties has been the most-studied example of local and systemic acquired resistance. Some varieties of tobacco respond to infection with certain strains of TMV at normal temperatures by producing necrotic local lesions and no systemic spread, instead of the usual chlorotic local lesions followed by mosaic disease. In several *Nicotiana* varieties, the reaction is under the control of a single dominant gene, the *N* gene, found naturally in *N. glutinosa*. This has been incorporated into *N. tabacum* cultivar Samsun NN; *N. tabacum* cultivar Burley NN; and *N. tabacum* cultivar Xanthi nc. Although other genes with similar effects are known, most experimental work has been carried out with tobacco varieties containing the *N* gene.

In Samsun NN tobacco grown at 21 to 24°C, a high degree of resistance to TMV develops in a 1- to 2-mm zone surrounding each necrotic local lesion (Fig. 11.7). The zone increases in size and resistance up to about 6 days after inoculation. Genes such as NN that induce a hypersensitive necrotic reaction in intact plants or excised leaf pieces fail to do so when isolated protoplasts are infected. The reason for this effect has not been established.

Following the appearance of localized acquired resistance in the inoculated tissue, systemic acquired resistance develops in other leaves. Systemic acquired resistance is measured by the reduction in diameter of the lesions. Lesions are about one fifth to one third the size found in control leaves, but lesion number is not reduced. Resistance is detectable in 2 to 3 days, rises to a maximum in about 7 days, and persists for about 20 days. Leaves that develop resistance are free of virus before the challenge inoculation. No conditions have been found that yield complete resistance. In plants held at 30°C, resistance fails to develop. Mechanical or chemical injury that kills cells does not lead to resistance, nor does infection with viruses that do not cause necrotic local lesions.

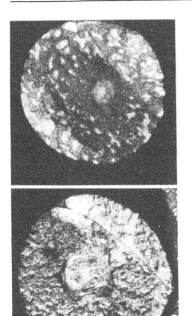

Figure 11.7 Acquired resistance to infection. (Upper) A disk cut from a Samsum NN tobacco leaf inoculated first with TMV (large lesion in center) and 7 days later challenge-inoculated with a concentrated TMV inoculum. Note absence of lesions from the second inoculation in a zone around the original lesion. (Lower) A similar experiment in which PVX was the challenge virus. No zone free of lesions is present. From Ross (1961).

3.2 HOST PROTEINS INDUCED BY THE HYPERSENSITIVE RESPONSE

3.2.1 The Pathogenesis-Related Proteins

Following inoculation of tobacco leaves responding in a hypersensitive way to TMV, a set of soluble host proteins appears. These have been called pathogenesis-related or PR proteins. In Samsun NN leaves infected with TMV, at least 14 acidic PR proteins appear. These can be placed in five groups based on MW, serology, and amino acid composition (Table 11.2). Some of the acidic proteins have basic homologs.

Important aspects of these PR proteins are as follows:

1. *Induction of messenger RNAs (mRNAs)* The appearance of PR proteins follows induction of a series of corresponding mRNAs in the inoculated leaf.

2. *Lack of specificity* PR proteins can be induced by various chemical treatments (Fig. 11.8) and by bacterial and fungal pathogens, as well as by virus infections that cause necrosis in hosts other than those with the NN genotype. However, these various PR proteins are not

Table 11.2 Extracellular Proteins Induced in Samsun NN Tobacco by TMV Infection[a]

Group	Acidic PR proteins			Basic homologs		Function
	Name	MW × 10⁻³	Estimated no. of genes	MW × 10⁻³	Estimated no. of genes	
1	1a, 1b, 1c	15	8	19	8	Unknown
2	2, N, O	40	?	33	?	β-1, 3-Glucanases
3	P, Q	29–30	4	32–34	4	Chitinases
4	R, S	23–24	?	—	—	Unknown
5	r_1, r_2, s_1, s_2	13–14.5	?	—	—	Unknown

[a] From Bol and van Kan (1988) and Kauffmann *et al.* (1989).

induced following infection with a virus that does not cause necrotic local lesions in the host used (Fig. 11.8, lane A).

3. *Functions of PR proteins* Group 2 PR proteins are β-1,3-glucanases, and those of group 3 are chitinases (Table 11.2). PR proteins of groups 1 and 4 are probably enzymes concerned with the production of various metabolites such as ethylene and phytoalexins.

4. *Widespread occurrence* Proteins homologous to the tobacco PR proteins have been found in at least 20 plant species in both monocotyledons and dicotyledons. Amino acid sequences are substantially conserved between different species.

From the above properties, we can conclude that the induction of PR proteins is a generalized defense reaction in plants, rather than a specific response to virus infections that cause necrosis. This view is strongly supported by the fact that chitin and β-1,3 glucan, the substrates of group 2 and 3 PR proteins, are found in fungal cell walls and not in plants.

3.2.2 Other Host Proteins

Increases in the activity of a variety of enzymes besides those among the PR proteins have been observed during the hypersensitive response to viruses. Peroxidase, polyphenoloxidase, and ribonuclease activities are increased. Peroxidases are not usually considered to be PR proteins, but they have similar properties. They are soluble at pH 3.0, are protease resistant, and acidic and basic isozymes accumulate in the intercellular spaces. The metabolism of phenylpropanoid compounds is

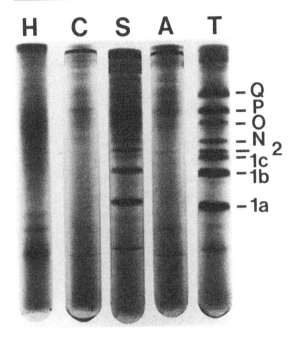

Figure 11.8 Accumulation of PR proteins in the intercellular fluid of tobacco after various treatments. Plants were sprayed with water (H), p-coumaric acid (C), salicylic acid (S), or inoculated with AMV (A) or TMV (T). Samples of the intercellular fluid were electrophoresed in nondenaturing polyacrylamide gels. The position of the major PR proteins is indicated in the margin. From Bol and van Kan (1988).

strongly activated by infection with various pathogens, including viruses that induce a hypersensitive response. This activation involves *de novo* enzyme synthesis and leads to the accumulation of compounds derived from phenylalanine, such as flavonoids and lignin.

3.2.3 Possible Role of Ethylene

One of the earliest detectable events in the interaction between a plant host and a pathogen that induces necrosis is a rapid increase in the production of ethylene, which is a gaseous plant-stress hormone. In the hypersensitive response to viruses, there is an increased release of ethylene from leaves. The fact that ethephon (a substance releasing ethylene), introduced into leaves with a needle, can mimic the changes associated with the response of Samsun NN to TMV is good evidence that ethylene is involved in the initiation of this hypersensitive response. An early burst of ethylene production is associated with the virus-localizing reaction. The increase in ethylene production is determined not by the onset of necrosis, but by a much earlier event.

3.2.4 Roles of Host-Coded and Viral-Coded Proteins

It is not yet established whether any of the PR proteins plays a role in the limitation of virus spread and acquired resistance. The PR proteins with known enzymatic activities appear to have a defense role against fungi, bacteria, or insects rather than viruses. If any of the PR proteins are, in fact active in the antiviral response, they may be found among those proteins for which a function has not yet been established. However, when expressed to high levels in transgenic tobacco plants with the NN constitution, individual PR proteins of Group 1 did not affect the necrotic response to inoculation with TMV.

Necrosis is not essential for the limitation of virus spread. For example, no necrosis follows inoculation of cucumber cotyledons with TMV, but virus is limited to local infections. By contrast, PR proteins of Group 1 are produced in *Nicotiana rustica* leaves developing necrotic local lesions after inoculation with TMV, but the virus moves systemically. On balance, present evidence suggests that the PR proteins are of little if any significance in limiting virus movement in the hypersensitive response.

Other experiments suggest that viral-coded proteins may be involved in the local-lesion response. For example, evidence for involvement of the coat protein comes from experiments with recombinant viruses. A common strain of TMV produces systemic infection in *Nicotiana sylvestris*, which contains the N' gene. Strain L produces a hypersensitive response. By switching the coat-protein genes between the two strains, it was shown that the hypersensitive response was a property of the coat-protein gene of strain L.

3.2.5 Systemic Acquired Resistance

The action at a distance involved in systemic acquired resistance presumably involves the translocation of some substance or substances. Various experiments, in which the conducting tissues were blocked in some way, support this view. The nature of the material that is translocated is unknown, but ethylene is a good candidate. When PR proteins are induced chemically, resistance to virus infection is usually confined to an area near the site of treatment. Ethylene is the exception in that it induces both PR proteins and resistance in leaves other than the treated leaf.

The mechanism of systemic acquired resistance in uninfected leaves is unknown. This mechanism may or may not be the same as that in the zone of tissue around necrotic lesions. PR proteins appear in resistant uninfected leaves after resistance can be detected. Thus, these proteins appear to have no significant role in eliciting the resistance response. Transgenic plants expressing PR proteins showed no resistance to inoculation with AMV.

4 THE ROLE OF VIRAL GENES IN THE INDUCTION OF SYSTEMIC DISEASE

4.1 GENERAL CONSIDERATIONS

Various lines of evidence show that viral genes must be involved in quite specific ways in the induction of disease. For example, closely related strains of TYMV produce two distinct, and mutually exclusive, pathways of change in the chloroplasts of TYMV-infected cells. In one, the chloroplasts first become rounded and clumped and then develop a large vacuole. In the other, they become angular before clumping, and then fragment to yield many small pieces. These different pathways must be activated by a viral gene or genes.

The existence of viruses with genomes consisting of two or more pieces of nucleic acid has allowed the production of pseudo-recombinants in the laboratory. Reassorting genomic segments has allowed determinants of specific disease symptoms to be located on particular genome segments. Many such experiments have been reported, and while they have given useful information, they do not always pinpoint the gene or genes involved, since genome segments often code for more than one protein product. Furthermore, experiments with only a few virus strains, tested on only a few hosts, may not reveal the full range of genes and genome segments involved.

In normal circumstances, all the gene products of a viral genome will have one or more functions in the complete virus life cycle. There is no evidence for the existence of viral genes that have a role only in inducing disease. In considering the induction of disease, it will be useful to make a distinction between the *functions* of viral genes in the virus life cycle, and the *effects* of the genes on the host.

If we consider a range of host species and environmental conditions (Section 6), a given virus can cause far more different kinds of symptoms than it has different gene products, or combinations of two or more gene products acting together. We must conclude that a particular viral gene may have a variety of effects on the kind of disease produced, depending on the host plant involved, the environmental conditions, and possible interactions with other viral genes. At present, we can distinguish three kinds of effects of viral genes: (1) those based on a specific requirement for an essential virus function. The small vesicles induced by TYMV in chloroplast membranes (see Fig. 7.10) may be an example of this sort; (2) those based on a defect in a viral gene, for example, a cell-to-cell transport function, that results in limitation of infection to the point of entry; and (3) those based on effects of virus infection that are quite inconsequential as far as virus replication or movement are concerned, unless they so damage host cells that further replication is inhibited. Most macroscopically observable disease symptoms may be this sort of effect.

Many virus infections show no observable macroscopic disease symptoms. However, apart from the cryptic viruses, most produce observable and often characteristic cytological effects. Thus, many cytological effects may represent essential requirements for virus replication and movement. Perhaps macroscopic disease symptoms should be regarded as effects of a virus on its cellular environment, rather than as part of the viral phenotype.

4.2 SPECIFIC VIRAL GENE PRODUCTS

The possibilities opened up by recombinant DNA technology have allowed new approaches to be made in investigating the role of viral genes in disease induction. Some examples are given in this section.

4.2.1 Cauliflower Mosaic Virus (CaMV)

Besides containing a host-range determinant, the product of Gene VI of CaMV is involved in determining the kind of disease that develops. Recombinant viruses have been constructed from strains of CaMV differing markedly in the symptoms they induce. The experiments showed that typical disease expression, consisting of leaf chlorosis and mottling, mapped to a genome segment containing Gene VI, but other regions of the genome also influenced the disease pattern.

In another experimental approach, when a segment of CaMV DNA containing Gene VI was transferred to tobacco plants using the *A. tumefaciens* Ti plasmid, the resulting transgenic plants showed viruslike symptoms. Gene VI from two different virus isolates produced different symptoms—either mosaiclike or a bleaching of the leaves. Symptom production was blocked by deletions or frameshift mutations in Gene VI. Production of symptoms was closely correlated with the appearance in the leaves of the 66-K gene product, as shown by immunoblots.

Other experiments with CaMV show that the effects of Gene VI on disease symptoms can depend on the host species. In addition, several other viral loci have been identified that can affect disease development.

4.2.2 Tobacco Mosaic Virus

Coat Protein

Various studies using recombinant viruses have demonstrated that TMV coat protein plays a role in disease induction. These studies were discussed in Chapter 7, Section 3.4. A single amino acid change in the coat protein may be the only difference detected between strains causing different diseases. Tests with a variety of mutant constructs containing insertions, deletions, frameshifts, or single base changes at defined sites in the coat protein demonstrated that a number of sites in the gene affect the necrotic response to TMV in *Nicotiana* varieties containing the N′ genotype. Point mutations in the coat protein also influenced systemic symptoms in *N. tabacum* with the n genotype. Other experiments have shown that the coat protein itself rather than the mRNA is involved in disease development.

The 130-K and 180-K Proteins

Comparison of the nucleotide sequences of two almost symptomless mutants of TMV (LII and LIIA) with the parent type strain showed that a change from cysteine to tyrosine at amino acid position 348 of the 130-K and 180-K proteins was involved in loss of symptom production. Two other amino acid changes in these proteins may also have been involved.

4.2.3 Cucumber Mosaic Virus (CMV)

Strain M of CMV induces severe systemic chlorosis in tobacco. Pseudorecombinants with other strains producing milder symptoms located the gene responsible on M CMV RNA3. Further experiments with recombinant RNA3s transcribed from engineered cDNAs showed that the symptom in tobacco was controlled by the coat-protein gene. There were eight amino acid differences between the M CMV coat protein and that of the other strain used.

4.2.4 Conclusion

It has been assumed for some years that virus–host cell interactions must involve specific recognition or lack of recognition between host and viral macromolecules. Interactions might involve activities of viral nucleic acids, or specific viral-coded proteins, or host proteins that are induced or repressed by viral infection. From the evidence available, it is probable that for some viruses at least, several viral genes play a part in determining the disease that develops. The protein products of the gene appear to be involved, but we cannot rule out the possibility of direct effects due to genomic or subgenomic nucleic acids.

5 PROCESSES INVOLVED IN DISEASE INDUCTION

In the previous sections, we have seen that viral genes play a role in the induction of disease and that host-coded proteins are induced as part of the disease process. In this section, I discuss some of the processes involved in the induction of disease. At present, we are unable to implicate specific viral- or host-coded proteins with the initiation of any of these processes.

5.1 SEQUESTRATION OF RAW MATERIALS

The diversion of supplies of raw materials into virus production, thus making host cells deficient in some respect, is an obvious mechanism by which a virus could induce disease symptoms. This mechanism is very probably a factor when the host plant is under nutritional stress. For example, in mildly nitrogen-deficient chinese cabbage plants, the local lesions produced by TYMV have a purple halo, the purple coloration being characteristic of nitrogen starvation. However, except under conditions of specific preexisting nutritional stress, it is unlikely that the actual sequestration of amino acids and other materials into virus particles has any direct connection with the induction of symptoms. There are many examples where large amounts of virus are produced in plants showing no obvious signs of disease.

5.2 EFFECTS ON GROWTH

5.2.1 Stunting

There appear to be three biochemical mechanisms whereby virus infection could cause stunting of growth.

A Change in the Activity of Growth Hormones

There are many possible ways in which virus infection could influence plant growth by increasing or decreasing the synthesis, translocation, or effectiveness of various hormones in different organs and at different stages of growth. In most situations, virus infection decreases the concentration of auxin and giberellin and increases that of abscissic acid. Ethylene production is stimulated by virus-induced necrosis and by development of chlorotic local lesions, but not where virus moves systemically without necrosis.

Reduction in the Availability of Fixed Carbon

Apart from any effects of hormone balance, plants will become stunted (on a dry weight basis, at least) if the availability of carbon fixed in photosynthesis is limiting. A reduction in available fixed carbon could be brought about in several ways: (1) by direct effects on the photosynthetic apparatus. Many such effects, both biochemical and cytological, are known; (2) by effects on stomatal opening; and (3) by reducing translocation of fixed carbon in the phloem.

Reduced Leaf Initiation

Reduction in leaf number is a very small factor in the stunting of herbaceous plants following virus infection. In the systems that have been studied in detail, infected plants have on average one leaf less than healthy plants. The transient reduction in leaf initiation occurs soon after the plants are infected.

5.2.2 Epinasty and Leaf Abscission

Epinasty (unequal growth of two surfaces leading to downcurling of the whole leaf) and leaf abscission, which occur in some virus diseases of herbaceous plants, are probably brought about by a burst of ethylene production soon after infection begins.

5.2.3 Virus-Induced Tumors

All the plant viruses belonging to the Reoviridae, except rice dwarf virus, induce galls or tumors in their plant hosts, but not in the insect vectors in which they also multiply. There is a clear organ or tissue specificity for the different viruses. For example, tumors caused by wound tumor virus (WTV) predominate on roots, and to a lesser extent, stems. Fiji disease virus (FDV) causes neoplastic growths on veins. Thus, we can be reasonably certain that some function of the viral genome induces tumor formation, but we are quite ignorant as to how this is brought about.

5.3 MOSAIC DISEASES

5.3.1 Dark Green Tissue

In most mosaic diseases in dicotyledons, irregular dark green islands of tissue are present. The tissue in these islands contains little or no detectable virus, is cytologically and biochemically normal, and appears to be resistant to superinfection with the virus that is causing the mosaic. Dark green islands arise only in developing young leaves still undergoing cell division. The dark green islands of tissue may persist in an essentially virus-free state for the life of the leaf. More often, *break down* occurs, leading to virus replication. This usually takes place after a period of weeks, or after a sudden elevation in temperature.

There is no convincing evidence for any of the various theories that have been put forward to explain the nature of dark green islands. They certainly do not consist merely of tissue that has escaped infection. There is no evidence for the presence of very mild strains of virus in dark green islands. It is possible that the cells in dark green islands are occupied by defective strains, but such strains would have to produce little or no intact virus or viral antigen, and would have to replicate without detectable cytological effects.

5.3.2 Leaf Ontogeny and Mosaic Diseases

In some diseases the mosaic may consist of tissue infected predominantly with one strain of the virus interspersed with islands of dark green tissue, as with TMV. In others, several different strains causing more or less severe effects on the chloroplasts in different blocks of tissue may be present, as wtih TYMV (Fig. 11.2).

The following lines of evidence show that these mosaic patterns are laid down at a very early stage of leaf ontogeny: (1) the type of mosaic pattern developing in a leaf at a particular position on the plant depends not on its position but on its stage of development when infected by the virus. There is a critical leaf size at the time of infection greater than which mosaic disease does not develop. This critical size is about 1.5 cm (length) for tobacco leaves infected with TMV; (2) it has been shown for TYMV in chinese cabbage that the mosaic pattern is already laid in very small apical leaves that have not yet developed significant amounts of chlorophyll; (3) in chinese cabbage leaves infected with TYMV, as well as the macroscopic mosaic, there is a microscopic arrangement of infected and dark green layers of cells in some areas of the leaf. The patterns of these layers show a striking resemblance to the layers seen in homoplastidic periclinal chimeras. These microscopically observable layers and mosaic patterns develop only in leaves that are less than 1–2 mm long at the time they become invaded by the virus; (4) in contrast to the patterns seen in dicotyledons, the macroscopic islands of tissue in mosaic diseases in monocotyledons almost always consist of stripes, streaks, or elongated blocks of tissue lying parallel to the axis of the leaf (Fig. 11.3). This arrangement is almost certainly owing to the fact that these leaves elongate by means of a basal meristem, producing files of cells that become committed to being dark green or infected with a virus at a very early stage.

6 FACTORS INFLUENCING THE COURSE OF INFECTION AND DISEASE

In earlier sections of this chapter, the kinds of host response to inoculation with a virus have been described, and our knowledge of the roles of host and viral genes has been summarized. This section is concerned with inherent variables in the host plant itself, environmental factors, interactions between unrelated viruses, and between viruses and some other agents of disease.

6.1 PLANT FACTORS

6.1.1 Age

In fairly mature herbaceous plants inoculated with a virus in the younger leaves, the lower leaves may never become infected. With increasing age, there is a greater tendency for infection to remain localized. Physiological age rather than actual age is the significant factor.

6.1.2 Plant Genotype

The Kinds of Host Response

The genetic make-up of the host plant has a profound influence on the outcome following inoculation with a particular virus. The kinds of host response were defined in Table 11.1. Defining the response of a particular host species or cultivar to a particular virus must always be regarded as provisional. A new mutant of the virus may develop in the stock culture, or a new strain of the virus that causes a different response in the plant may be found in nature. This applies particularly to nonhost immunity. Nevertheless immunity with long-term durability must occur frequently in nature. On the other hand, it is probable that many of the plants described in the past as immune to a particular virus were in fact infectible, but resistant and showing extreme hypersensitivity, as defined in Table 11.1.

Genetics of Resistance to Viruses

The following points concerning the effects of host genes on the plant's response to infection emerge from many different studies: (1) both dominant and recessive Mendelian genes may have effects. However, whereas most genes that affect host responses are inherited in a Mendelian manner, cytoplasmically transferred factors may sometimes be involved; (2) there may or may not be a gene-dose effect; (3) genes at different loci may have similar effects; (4) the genetic background of the host may affect the activity of a resistance gene; (5) genes may have their effect with all strains of a virus, or with only some; (6) some genes influence the response to more than one virus; (7) plant age and environmental conditions may interact strongly with host genotype to produce the final response; (8) the route of infection may affect the host response. Systemic necrosis may develop following introduction of a virus by grafting into a highly resistant host that does not permit systemic spread of the same virus following mechanical inoculation; and (9) resistance to viruses in most crop plant–virus combinations is controlled by a single dominant gene. However, this may merely reflect the fact that most resistant cultivars were developed in breeding programs aimed at the introduction of a single resistance gene.

The Gene-for-Gene Hypothesis

Viruses and virus strains may be described in relation to the various host responses defined in Table 11.1. Thus, if a particular plant species or cultivar shows immunity or resistance to a virus, that virus is said to be *nonpathogenic* for that species or cultivar. A virus is *pathogenic* if it usually causes systemic disease in a species or cultivar. A gene for immunity or resistance introduced into such a species or cultivar may make the virus *avirulent;* however, a mutant strain may overcome the host gene resistance, and it would then be *virulent.*

Gene-for-gene relationships are well known between host plant and fungal or bacterial pathogens. They have been established primarily on the basis of genetic analyses of both plants and pathogens. With these parasites, each allele in the host that confers resistance may be reflected in a complementary virulence locus in the parasite that can overcome the resistance.

A well-studied example of the gene-for-gene hypothesis applied to a plant virus is the resistance of tomato to tomato mosaic *Tobamovirus* (ToMV) (Table 11.3). There are three resistance genes Tm-1, Tm-2, and

Table 11.3 The Gene-for-Gene Hypothesis Applied to Plant–Virus Interactions[a]

Host genotypes	Virus genotypes					
	0	1	2	2^2	1.2	1.2^2
Wild type	M	M	M	M	M	M
Tm-1	R	M	R	R	M	M
Tm-2[b]	R	R	M	R	M	R
Tm-2²[b]	R	R	R	M	R	M
Tm-1/Tm-2	R	R	R	R	M	R
Tm-1/Tm-2²	R	R	R	R	R	M
Tm-2/Tm-2²	R	R	R	R	R	R
Tm-1/Tm-2/Tm-2²	R	R	R	R	R	R

[a]Genetic interactions between ToMV-resistant tomato plants and strains of the virus. M, systemic mosaic, R, resistance.

[b]Plants with these genotypes may show local and variable systemic necrosis rather than mosaic when inoculated with virulent strains. Modified from Fraser (1985), with permission from Martinus Nijhoff/Dr. W. Junk, Dordrecht, The Netherlands.

Tm-2². The virus has evolved variants that can overcome all three host-resistance genes. The virulent virus strains are numbered according to the host genes they can overcome; for example, strain 0 is avirulent. Strain 2 overcomes gene Tm-2; strain 1.2² overcomes genes 1 and 2². No virus strains are yet known that overcome host genes 2 plus 2² or 1, 2 and 2.2². The host genotypes differ in their *durability* in the field. Thus, strain 1 isolates appeared within a year in commercial crops containing only the Tm-1 gene. By contrast, most commercial ToMV-resistant tomato cultivars now contain the genes *Tm-1/Tm-1:Tm-2/Tm-2²* or Tm-1/+:Tm-2/Tm-2² and these appear to be highly durable in their resistance to TMV.

The relationships summarized in Table 11.3 bear a superficial similarity to the gene-for-gene relationships seen between bacterial and fungal pathogens and their hosts. Such pathogens contain large numbers of genes, and could easily maintain or develop a suite of genes that either allow or overcome host resistance. By contrast, ToMV contains four genes, all with functions involved in virus replication or movement. The change to virulence by the virus must involve mutational events in one or more of the four genes or in the controlling elements of the viral genome. This has been shown to be true for some of the viral genotypes listed in Table 11.3. Resistance-breaking strains of ToMV have been found to contain single base substitutions that lead to a changed amino acid at a specific site in either the 130-K and 180-K gene products or in the 30-K movement proteins. Thus for viruses, virulence or avirulence can be controlled by a single point mutation in genes essential for virus replication or movement, rather than some change in genes dedicated to overcoming host resistance.

Molecular Mechanisms of Host Immunity and Resistance

Extracts of many plants contain substances that inhibit infection by viruses when mixed with the inoculum. Some may inhibit virus replication in experimental systems, and many are known to be proteins or glycoproteins. The relevance of any of these substances to plant resistance to viruses has not been established. However, recent studies are revealing specific mechanisms for particular host–virus combinations.

Resistance of Cowpea to Cowpea Mosaic Virus (CPMV) Seedlings of the cowpea cultivar Arlington are resistant to CPMV, and isolated protoplasts are resistant. This cultivar is therefore immune, as defined in Table 11.1. This immunity is governed by a single dominant gene as determined by crosses with the susceptible Black Eye variety. Various activities inhibitory for CPMV have been found in extracts of Arlington cowpea protoplasts. One of these, an inhibitor of polyprotein processing, was specific for CPMV and had the coinheritance expected for an agent mediating the immunity to CPMV. It can be reasonably concluded that the proteinase inhibitor is the host-coded gene product responsible for immunity to CPMV. This is the first such gene product identified for plant viruses and the first example of immunity for which a clear molecular mechanism has been established.

Ineffective Viral Movement Proteins As discussed in Chapter 10, Section 4.4, many viruses code for a gene product that is necessary for cell-to-cell movement. If a movement protein is ineffective in a particular host plant, then virus replication is confined to the initially infected cells. Consequently, the plant shows extreme hypersensitivity (Table 11.1) and is effectively resistant to the virus. A mismatch between viral-movement protein and plant species may be a quite common mechanism for plant resistance. A virus that does not normally move systemically in a particular plant species may do so in the presence of an unrelated virus that does infect systemically.

Tolerance

The classic example of genetically controlled tolerance is the Ambalema tobacco variety. TMV infects and multiplies through the plant, but in the field, infected plants remain almost normal in appearance. This tolerance is owing to a pair of independently segregating recessive genes r_{m1} and r_{m2}, and perhaps to others as well, with minor effects.

Kinds of Symptoms

The symptoms that develop in plants that are neither resistant nor tolerant will be influenced in many ways by the host genotype. For example, differences in the mosaic disease induced by strains of TYMV in *Brassica pekinensis* and *B. rapa* are under the control of a nuclear gene.

6.2   ENVIRONMENTAL FACTORS

The environmental conditions under which plants are grown before inoculation, at the time of inoculation, and during the development of disease can have profound effects on the course of infection. A plant that is highly susceptible to a given virus under one set of conditions may be completely resistant under another. If infection occurs, the plant may support a high or low concentration of virus and develop severe disease or remain almost symptomless, depending on the conditions.

6.2.1   Factors Affecting Susceptibility to Infection

Any factor that alters the ease with which the surface of the leaf is wounded will alter the probability of successful entry of virus introduced by mechanical inoculation, whereas physiological changes in the leaf may make the cell more or less suitable for virus establishment. As a broad generalization, greenhouse-grown plants will have the greatest susceptibility when they are grown and used under the following conditions: (1) mineral nutrition and water supply that do not limit growth; (2) moderate to low light intensities; (3) a temperature in the range of 18 to 30°C, depending on virus and host; and (4) inoculation carried out in the afternoon. This last point arises because it has been shown for several host–virus combinations that there is a regular diurnal cycle of susceptibility, with a maximum in the afternoon and a minimum just before dawn.

6.2.2 Factors Affecting Virus Multiplication and Disease Expression

Light

Light intensity and duration affect virus production and disease expression in different ways with different viruses, but generally speaking, high light intensities and long days favor multiplication.

Temperature

Over the range of temperatures at which plants are normally grown, increasing temperature usually increases the rate at which viruses multiply and move through the plant. Like other biological phenomena, however, increase in temperature above a certain point leads to a reduction in the rate of multiplication. The kind of disease that develops is often markedly affected by temperature. The species of host plant, strain of virus, and age of the leaf in which the virus is multiplying may have a major effect on the way the virus behaves with changes in temperature.

Nutrition

To obtain a "standard" expression of disease for any particular host–virus combination, test plants should not be suffering from any nutritional deficiency. Deficiencies in particular elements may either obscure virus symptoms or make them more severe.

Time of Year

The complex of factors including day length, light intensity, air and soil temperatures, and water supply that change during the seasonal cycle will affect plant growth and thus the disease produced by a given virus, and the extent to which a virus multiplies.

6.3 INTERACTION WITH OTHER AGENTS

The disease produced by a particular virus in a given host and the extent to which it multiplies are sometimes markedly influenced by the presence of a second independent and unrelated virus or by infection with cellular parasites. These effects are discussed next.

6.3.1 Interactions between Unrelated Viruses

The interaction between tobacco necrosis virus (TNV) and satellite TNV (STNV), discussed in Chapter 9, Section 2.1, involves a complete dependence of one unrelated virus on the presence of another for replication. Another kind of situation exists between two viruses in which both are normally associated together with a recognized disease in the field. For example, the important tungro disease of rice is caused by a mixture of rice tungro bacilliform virus, a DNA virus, and rice tungro spherical virus, an RNA virus, and both are transmitted by leafhopper vectors. The DNA virus causes severe disease. The speckles disease that is found in lettuce, sugar beet, and spinach in California is caused by a complex of two viruses (beet western yellows *Luteovirus* and lettuce speckles mottle virus) transmitted by an aphid vector. Apart from these kinds of relationship, there are many examples known in which unrelated viruses, both capable of infecting and multiplying independently, interact when by chance they infect the same plant.

There are also instances in which a virus may allow another virus, normally restricted to a particular tissue, to invade the plant more widely. For example, bean golden mosaic *Geminivirus* normally confined to the phloem, invades nonphloem tissue in double infections with TMV. Not uncommonly, a mixed infection with two viruses produces a more severe disease than either alone. The classic example here is the mixture of PVX and PVY in tobacco, which produces a severe veinal necrosis instead of the milder mottling or vein-banding diseases seen with either virus separately.

6.3.2 Interactions between Viruses and Fungi

Effect of Virus Infection on Fungal Diseases

Numerous examples are known in which virus infection reduces susceptibility to, or development of fungal and bacterial parasites. The development of resistance of this sort may well involve the non-specific host responses discussed earlier (Section 3.2). On the other hand, there are many instances of increased susceptibility to fungi following virus infection. For example, prior infections of wheat or barley with BYDV predispose the ears to infection by *Cladosporium* spp. and *Verticillium* spp.

Effects of Fungal Infection on Susceptibility to Viruses

Infection by rust fungi may greatly increase the susceptibility of leaves to viruses. Other fungi may induce resistance.

7 DISCUSSION AND SUMMARY

Depending on the response to inoculation with a virus, plants can be described as either immune or infectible. If a plant is immune, it is a nonhost for the virus, and the virus does not replicate in any cells of the intact plant or in isolated protoplasts. Immunity usually applies to all the members of a given species, but sometimes it is applicable to just a particular cultivar. The molecular mechanism for one example of such immunity has been established. Cowpeas of the cultivar Arlington are effectively immune to CPMV through the presence of a single dominant Mendelian gene. This gene codes for an inhibitor that blocks the action of the viral-coded proteinase needed for the specific processing of the CPMV polyprotein in the early stages of viral replication.

In infectible species or cultivars, the virus can infect and replicate in isolated protoplasts. The plant may be either resistant or susceptible to infection. We can distinguish two kinds of resistance. In resistance involving *extreme hypersensitivity,* virus multiplication is limited to the initially infected cells, because the viral-coded protein necessary for cell-to-cell movement of virus is nonfunctional in the particular host. This gives rise to a subliminal infection. In the past, many examples of this type of resistance were described as immune.

In the second kind of resistance, infection is limited by a host response to a zone of cells around the initially infected cell, which usually results in necrotic local lesions. Uninfected tissue surrounding these lesions becomes resistant to infection. This is called *acquired resistance.* Acquired resistance involves the induction of at least 14 host proteins known as pathogenesis-related (PR) proteins. The synthesis of these proteins following virus infection appears to be part of a generalized nonspecific defense reaction against bacterial and fungal parasites, insects, or chemical and mechanical injury, as well as viruses. In many host species, the necrotic hypersensitive response (as opposed to a non-necrotic response leading to systemic infection and disease) is governed by a single dominant Mendelian gene.

There are other kinds of *cultivar resistance* to virus infection. Most of these are determined by one or a few dominant genes. Some are controlled by incompletely dominant genes, and a few by apparently recessive genes.

A virus that does not cause systemic disease in a particular plant is *nonpathogenic* for that plant. If a virus or virus strain causes systemic disease in a particular species or cultivar, it is *pathogenic*. A gene for resistance introduced into such a susceptible species or cultivar may make the virus *avirulent*. However, the virus may mutate and overcome the host resistance to become a *virulent* strain. Thus, both host and viral genes interact to determine the outcome of inoculation. The change from an avirulent to a virulent virus strain may involve no more than a single amino acid change in a viral-coded protein.

In species or cultivars that are susceptible, the virus replicates and moves systemically. In a *sensitive* reaction, disease ensues. If the plant is *tolerant*, there is no obvious effect on the plant, and a *latent* infection results. The consequences of infection for a susceptible plant are determined by both host and viral genes. For example, a single base change in the TMV coat-protein gene may be sufficient to alter the nature of the resulting disease.

The actual processes involved in the induction of disease are not well understood. Many of the biochemical changes involved may not be directly connected with virus replication. Stunting probably involves changes in the balance of growth hormones. The formation of mosaic patterns in virus-infected leaves involves events that occur in the early stages of leaf ontogeny.

Many environmental factors influence the course of infection and disease. These include light, temperature, water supply, nutrition and the interactions between these factors during the growing season. Complex interactions may occur when plants are infected with two unrelated viruses or with a virus and a cellular pathogen.

FURTHER READING

Bos, L. (1978). "Symptoms of Virus Diseases in Plants," 3rd Ed. Pudoc, Wageningen, The Netherlands.

Fraser, R. S. S. (1987a). Resistance to plant viruses. *Oxf. Surv. Plant Mol. Cell Biol.* 4, 1–45.

Fraser, R. S. S. (1987b). "Biochemistry of Virus-Infected Plants." Research Studies Press, Letchworth, England.

Matthews, R. E. F. (1991). "Plant Virology," 3rd. Ed. Academic Press, New York.

van Loon, L. C. (1987). Disease induction by plant viruses. *Adv. Virus Res.* **33**, 205–255.

van Loon, L. C. (1989). Stress proteins in infected plants. *In* "Plant–Microbe Interactions: Molecular and Genetic Perspectives" (T. Kosuge and W. E. Nester, eds.), pp. 198–237. McGraw-Hill, New York.

VARIABILITY

Like other living entities, viruses remain substantially like the parent during their replication, but can change to give rise to new types or *strains*. Because RNA viruses lack an error-correcting mechanism during genome replication, they give rise to many mutants involving one or a few nucleotide changes. There are several other mechanisms whereby new variants arise. New strains provide the raw material for virus evolution. Our knowledge of the pathways of virus evolution (Chapter 16) is quite fragmentary, but there is no doubt that viruses have undergone and continue to undergo evolutionary change, and that this change is sometimes rapid.

1 ISOLATION OF STRAINS

The property of a new strain that first allows it to be distinguished from other known strains of a virus has been almost always biological—usually a difference in disease symptoms in some particular host. Where the virus is mechanically transmissible, it is usual to pass new isolates through several successive single, local-lesion cultures, if a suitable host is available. There is good evidence that a single virus particle can give rise to a local lesion. On the other hand there is ample evidence that new mutants soon appear.

From a consideration of the mutation rate of viruses, discussed later, it is highly improbable that any virus culture actually consists entirely of a single strain. Many mutants arise even during the development of a single local lesion. The various chemical and physical methods described in this chapter for the characterization of strains give useful information only because they are not sufficiently sensitive to detect the small proportion of any particular mutant or variant present in a culture.

1.1 STRAINS OCCURRING NATURALLY IN PARTICULAR HOSTS

Different strains of a virus frequently occur in nature, either in particular host species or varieties or in particular locations. These strains can be cultured in appropriate host plants in the greenhouse.

1.2 ISOLATION FROM SYSTEMICALLY INFECTED PLANTS

Plants systemically infected with a virus frequently show atypical areas of tissue that contain strains differing from the predominant strain in the culture. These areas of tissue may be different parts of a mosaic disease pattern or they may be merely small necrotic or yellow spots in systemically infected leaves. When such areas or spots are dissected out and inoculated into fresh plants, they may be shown to contain distinctive strains.

1.3 SELECTION BY PARTICULAR HOSTS
OR CONDITIONS OF GROWTH

A particular strain may multiply and move ahead of others in a certain plant. Such a host can be used to isolate the strain. Similarly, strains may differ in the rate at which they multiply and move at different temperatures in a given host.

1.4 ISOLATION BY MEANS OF VECTORS

Vectors may be used in three ways to isolate strains. First, by using short feeding periods on the plants to be infected, only one strain out of a mixture may be transmitted. Second, particular vectors may preferentially transmit certain strains of a virus. Third, inoculation of insect vectors with diluted inoculum followed by repeated selection of infected plants for a particular type of symptom can result in the isolation of a variant virus.

1.5 ISOLATION OF ARTIFICIALLY INDUCED MUTANTS

Experimentally induced mutants have been used for several important kinds of investigations in plant virology. Nitrous acid-induced mutations in the coat-protein gene of tobacco mosaic virus (TMV) were of considerable importance in determining the nature of the genetic code and in confirming the nature of mutation. Nitrous acid also has been used to induce temperature sensitive (*ts*) mutants, mainly in TMV, to aid in the delineation of the *in vivo* functions carried by the viral genome. Nitrous acid mutants have been used as a source of mild virus strains to give disease control by cross protection (Chapter 15, Section 3.3.2).

One convenient method for isolating mutants is to inoculate a necrotic local-lesion host under conditions in which most lesions will have arisen from infection with single virus particles. Sometimes mutants give a recognizably different necrotic local lesion, very frequently smaller than normal. To detect other symptom differences, single lesions are dissected out and inoculated into hosts giving systemic symptoms. Mutants may then be recognized by the different symptoms they produce.

2 THE MOLECULAR BASIS FOR VARIATION

Viruses with RNA genomes, as well as those with DNA, are subject to genetic variation by molecular mechanisms that are very similar to most of those that occur in cellular organisms.

2.1 POINT MUTATIONS

There is no doubt that a single base change giving rise to a single amino acid substitution in the protein concerned is a frequent source of virus variability under natural conditions *in vivo*. Similarly, in the laboratory, treatment with mutagenic agents such as nitrous acid can lead to a single base change giving rise to an amino acid substitution in the protein coded for. Deletions or additions of single amino acids in a coding region will give rise to a change in reading frame, with many amino acid substitutions downstream from the mutation. Such changes will frequently be lethal unless the original reading frame is restored by a second addition or deletion.

2.2 RECOMBINATION

For many years recombination was considered to be a genetic mechanism confined almost entirely to organisms with DNA as their genetic material. It is now known that recombination occurs in plant viruses with genomes consisting of either DNA or RNA.

Recombination *in vivo* can be demonstrated in various ways. For example, plants can be coinoculated with two strains or mutants of the virus, each of which is defective in a different function. The appearance of competent virus indicates that recombination may have occurred. To demonstrate this, it is necessary to characterize the parent mutants and the presumed recombinants by such methods as restriction enzyme mapping or nucleotide sequencing. This physical characterization is necessary because other mechanisms such as *trans*-acting gene products may allow a mixture of two defective mutants to replicate.

The mechanism of recombination in RNA viruses is not firmly established, but three diverse mechanisms have been suggested: (1) enzymatic cutting and religation; (2) a copy-choice model between two templates that by chance lie close together during replication (of these two, the latter mechanism is more widely favored); and (3) there are several lines of evidence showing that nonviral RNAs can be reverse-transcribed and incorporated into the eukaryotic genome. Thus it is possible that rearrangement of viral RNA sequences might occur when the RNA is reverse-transcribed into DNA form, for example, by a transcriptase belonging to another virus or a cellular retrotransposon. The rate of recombination in different RNA viruses varies from uncommon to very frequent, a feature that is not yet understood.

2.3 DELETIONS

Deletions of up to several hundred base pairs in nonessential genes have been found in naturally occurring isolates of DNA viruses such as cauliflower mosaic virus (CaMV). These have been demonstrated by nucleotide sequence analysis.

Substantial deletions of genomic nucleic acid give rise to the defective interfering (DI) RNAs discussed in Chapter 9, Section 3. Deletions can occur in some viruses with multiparticle genomes. For example, when wound tumor virus (WTV) was cultured continuously in plants only, for a long period, some cultures lost their ability to be transmitted by leafhoppers. They also lost up to 85% of genomic RNA segments numbers 2 and 5, giving rise to 5' and 3' terminally conserved remnants that are functional with respect to transcription, replication, and packaging.

With natural isolates of a virus, it is usually difficult to estimate how frequently deletions occur in a virus stock. However, infectious synthetic RNA transcripts from viral copy DNAs (cDNAs) offer two major advantages for studying the deletion process: (1) they provide completely defined starting materials; and (2) they provide a way of "setting the clock to zero" with respect to deletion events.

2.4 ADDITIONS

Many examples have been described of mutant viruses with additional base sequences that have been generated by *in vitro* modification of recombinant DNA plasmids. Naturally occurring examples of such additions are much less common. In a group of naturally occurring variants, it may sometimes be difficult to establish whether a difference in length is the result of an addition or a deletion of nucleotides.

Repetition of blocks of sequences is known in some RNA viruses. For example, sequences of 56 and 75 nucleotides are duplicated in the leaders of the RNAs 3 of strain S and L of alfalfa mosaic virus (AMV). The duplications are next to one another in the leader sequences, and they do not appear to be essential for replication. Examples are known where viral genomes have acquired host gene sequences.

2.5 NUCLEOTIDE SEQUENCE REARRANGEMENTS

Some satellite RNAs appear to differ from one another mainly through rearrangement of blocks of nucleotide sequences.

2.6 REASSORTMENT IN MULTIPARTITE GENOMES

Genetic reassortment experiments have been carried out with most of the known multipartite viruses. These experiments have demonstrated beyond doubt that new variants can arise by reassortment of the preexisting segments of two related viral genomes, both in the laboratory and in nature.

2.7 THE ORIGIN OF STRAINS IN NATURE

All the kinds of molecular change noted earlier in this section will contribute to the evolution of strains in nature. A single base change resulting in a single amino acid change in a protein is probably one of the most common events giving rise to natural variation. The primary structure of the coat proteins of naturally occurring strains of TMV supports this view.

We can envisage strains diverging further and further from the parent type, as changes brought about by various mechanisms accumulate in the various proteins specified by the virus. The survival and spread of such strains will usually depend on their competitive advantage within the host in which they happen to arise, and in others that they subsequently infect. Selection of strains by host species is discussed in Section 4.2.

3 CRITERIA FOR THE RECOGNITION OF STRAINS

A virus might be defined simply as a collection of strains with similar properties. Sometimes we wish to ask whether two similar virus isolates are identical or not; on other occasions, we will have to decide whether two isolates are different viruses or strains of the same virus. Two kinds of properties are available for the recognition and delineation of virus strains—structural criteria based on the properties of the virus particle itself and its components, and biological criteria based on various interactions between the virus and its host plant and its insect or other vectors. These criteria are further discussed in Chapter 16 in relation to the general problem of virus classification. Serological properties are based on the structure of the viral protein or proteins, but because of their practical importance, serological criteria are considered here in a separate section.

3.1 STRUCTURAL CRITERIA

3.1.1 Nucleic Acids

Heterogeneity in Viral RNA Genomes

All the RNA genomes that have been examined have been found to exist not as a single nucleotide sequence, but as a distribution of sequence variants around a consensus sequence. Thus, sequence microheterogeneity is always present in natural populations. This nonuniformity in genome sequence has led to the concept of a *quasi species* for many viruses. The high mutation rate can be attributed to the high error rate in RNA-dependent RNA polymerases. This error rate is thought to be the result of the lack of any error-correcting mechanism for these polymerases. They have error rates in the range of 10^{-3} to 10^{-5} misincorporations per site, which is three or four orders of magnitude higher than the rate for DNA polymerases. This fact, coupled with the very high rate of virus replication that is possible, can potentially lead to rapid change in virus populations. This potential for rapid change must be borne in mind when considering nucleic acid differences as criteria for recognizing virus strains. Most of the variants in a culture of a particular virus strain will normally consist of base substitutions at various sites, perhaps with some deletions or additions of nucleotides.

Methods for Assessing Nucleic Acid Relationships

Nucleotide Sequences   Nucleotide sequences give valuable information about the extent of relationships between viruses. Determination of even relatively short sequences can give useful information. However, in order to study effectively the relationships between strains of a virus, it is necessary to determine the full nucleotide sequence of at least one, or preferably several isolates. When cDNA clones of RNA viruses are used for sequencing, the resulting sequence may, by chance, contain sequences that differ from the consensus sequence. This possibility is checked by the sequencing of several clones covering the same section of the genome. To ensure that a functional genome has been sequenced, it is necessary to use a cloned DNA that has been shown to be infectious or a cloned cDNA that can be transcribed into infectious RNA. In considering nucleotide sequences as a measure of relatedness between virus strains, it must be remembered that the extent and distribution of such differences may vary quite widely depending on the virus and the parts of the genome examined.

Nucleotide sequence information may sometimes reveal surprisingly large differences between strains of a single virus. For example, the RNA2 of the TCM strain of tobacco rattle virus (TRV) is considerably larger than that of other strains that have been sequenced (Fig. 12.1). Angenent *et al.* (1989) showed that the greater length was owing partly to a repetition of 1099 3' nucleotides from RNA1, which includes a 16-K open-reading frame (ORF), and partly to a 29-K ORF that was unique to this RNA2 (Fig. 12.1). The 29-K ORF has no significant similarity with

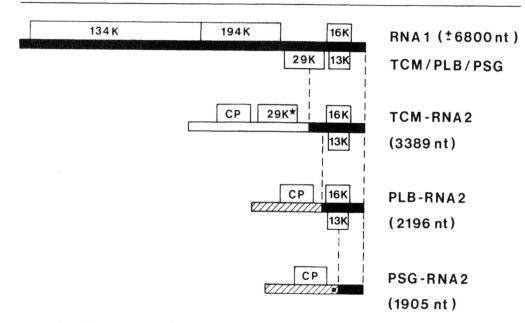

Figure 12.1 Variation in length of RNA2 in strains of TRV. Schematic representation of the genome structure of strains TCM, PLB, and PSG. RNA1 sequences in the RNA2s are indicated by solid bars connected by dotted lines. Sequences that are identical in PLB RNA2 and PSG RNA2 are indicated by hatched boxes. The asterisk in TCM RNA2 indicates the sequence that is unique to TCM RNA2. CP, coat protein. The M_r of other gene products is indicated. From Angenent *et al.* (1989).

the 28.8-K ORF of PSG RNA1, and its origin, and function if any, are not established.

Hybridization   The basis for nucleic acid hybridization was discussed in Chapter 2, Section 2.2.2. Nucleic acid hybridization experiments can give valuable information concerning the degree of base sequence homology between the nucleic acids of different virus isolates, but interpretation of the data may not be straightforward. A significant advantage of hybridization procedures is that a comparison can be made between total genome RNAs, or RNA segments. Results obtained will depend on various conditions in the tests, especially the stringency of hybridization (Fig. 12.2). Figure 12.2 also illustrates the use of the dot-blot technique in a semiquantitative manner.

Heterogeneity Mapping   Heterogeneity mapping is a method based on RNA hybridization, which allows the detection of single point

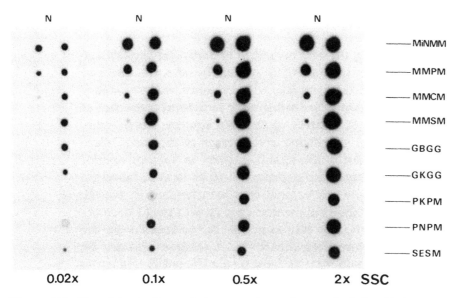

Figure 12.2 Use of the dot-blot technique to estimate degree of relationship between strains of a virus. Autoradiograph showing the extent of sequence homology between maize streak *Geminivirus* (MSV), strain MNM, and other MSV strains. Four identical filters were each spotted, on the right with 2 ng of DNA from the different MSV strains under test, and on the left, with doubling dilutions (2 ng → 7.8 pg) of MSV-MNM (N) as controls. Filters were hybridized with a full-length, nick-translated clone of MSV-MNM (N) before washing under conditions of different stringency (0.02, 0.1, 0.5 and 2 standard saline citrate [SSC] at 65°C). From Boulton and Markham (1986).

mutations, provided they occur in a significant proportion of the molecules. The method takes advantage of the ability of RNase A to recognize and excise single base mismatches in RNA heteroduplexes. Labeled minus-sense RNA probes are transcribed from cDNA clones of the RNAs. Following hybridization with the test RNA, and digestion by RNase A and T, fragments are separated and sized by polyacrylamide gel electropheresis (PAGE). The method makes possible the assessment of entire populations of RNA molecules for major sites of heterogeneity. By contrast, the sequencing of individual clones gives precise and detailed information on a few molecules that may make up a tiny fraction of the virus population.

Restriction Endonuclease Mapping Restriction endonuclease mapping has been used to characterize the double-stranded DNA (dsDNA) genomes of some caulimoviruses.

3.1.2 Proteins

For the small RNA viruses, the coat protein is of particular importance for the delineation of viruses and virus strains. Besides the intrinsic properties of this protein (size, amino acid sequence, and secondary and tertiary structure), many other measurable structural properties of the virus depend largely or entirely on the coat protein. These properties include serological specificity, architecture of the virus, electrophoretic mobility, cation binding, and stability to various agents. Thus, ideas on relationships within groups of virus strains, based on properties dependent on the coat protein, may be rather heavily biased. On the other hand, if mutations in the non-coat-protein genes have occurred more or less at the same rate as in the coat protein during the evolution of strains in nature, then such views on relationships may be reasonably well based. Another factor must be considered in relation to coat proteins. There is little or no significant amino acid sequence similarity between coat proteins of different groups and families of plant viruses. By contrast, nonstructural proteins frequently show some sequence similarities.

The *Potyvirus* group will serve to illustrate the use of the properties of coat proteins in the delineation of virus strains. Potyviruses have been one of the most difficult virus groups to study taxonomically. The group contains about 22% of the 711 known plant viruses (Table 16.1). The viruses infect a wide range of host plants, and exist in nature as many strains or pathotypes, differing in biological properties such as host range and disease severity. It has been considered by some workers that strains of potyviruses may form a continuous spectrum between two or more otherwise distinct viruses, making delineation of viruses and groups of strains difficult or impossible. However, recent comparisons between the amino acid sequences of the coat proteins of several viruses, and many strains indicate that this approach may provide a useful basis for taxonomy with the group. The data give no support for the *continuous spectrum* idea among the potyviruses. Distinct viruses showed major differences in length of their coat proteins. Major differences in amino acid sequence were near the N-termini, which are located near the surface of the virus, while homology was high in the C-terminal half of the proteins. On the other hand, strains had very similar N-termini. Analysis of the 136 possible pairings between a set of viruses and strains revealed a clear-cut bimodal distribution, with distinct viruses having an average sequence homology of 54% overall, whereas strains averaged 95% (Fig. 12.3).

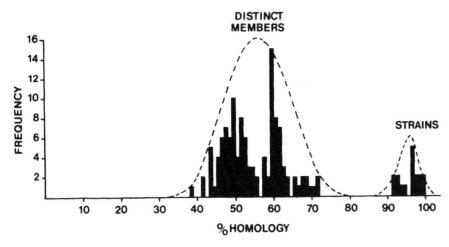

Figure 12.3 Demarcation between the extent of amino acid sequence homologies in coat proteins among distinct individual potyviruses (left-hand distribution) and between strains of the same virus (right-hand peak). The 136 possible pairings between 17 strains of 8 distinct viruses were analyzed. The homologies between distinct viruses had a mean value of 54.1% and a standard deviation of 7.29%, whereas the homologies between strains of individual viruses showed a mean of 95.4% and standard deviation of 2.56%. The dotted curves show that all values for distinct viruses and strains fall within ±3 standard deviations from their respective mean values. From Shukla and Ward (1988).

3.2 SEROLOGICAL CRITERIA

The nature of antigens and antibodies, the basis for serological tests, and their advantages and limitations were discussed in Chapter 3. This section considers the use of serological criteria to delineate viruses and virus strains.

For the small RNA viruses, antibodies are formed that react with the viral coat protein. The N-terminal region of this protein is often at the virus surface and is thus important in eliciting antibodies to intact virus. When the coat protein used for immunization is in a disaggregated state, the exposed sides of the subunits will present new antigenic sites. For those virus groups that have been studied sufficiently, the N-terminal surface region is the most variable, and therefore the most useful in delineating virus strains.

There is no good evidence that single-stranded (ss) plant virus RNAs can elicit RNA-specific antibodies. However, dsRNAs do elicit antibodies that react nonspecifically with other dsRNAs.

3.2.1 Experimental Variables

A number of important experimental variables can affect the estimated degree of serological relationship between viruses and strains. These include the following: (1) variability in antisera, both in successive bleedings from the same animal and in sera from different individuals. The proportion of cross-reacting antibody present in a series of bleedings taken over a period of months from a single animal may vary widely; (2) the extent to which antisera to two virus strains cross-react is usually correlated with the antibody content of the serum. Sera of low titer show lower cross-reactivity, and those with high titers show greater cross-reactivity. Thus, to detect serological differences between closely related strains using polyclonal antisera, it is preferable to use antisera of fairly low titer. To demonstrate distant serological relationships, it may be necessary to use high-titer antisera; (3) many virus preparations used for immunization and for antibody assay may contain varying amounts of free coat protein or coat protein in various intermediate states of aggregation or in a denatured state. Coat protein in the intact virus may lose amino acids through proteolysis. Antibodies reactive with coat protein in these various forms may or may not indicate the same sorts of relationships as antibodies against intact virus; and (4) the method used to detect and assay cross-reacting and strain-specific antibodies may affect the apparent degree of relationship.

3.2.2 The Serological Differentiation Index

In spite of all these variables, useful assessment of degrees of serological relationship can be obtained by testing successive bleedings from many animals and pooling the results. Most quantitative measurements of degrees of serological relationship have been carried out using precipitation titers, but enzyme-linked immunosorbent assay (ELISA) tests can also be used. The extent of serological cross-reactivity can be expressed by a serological differentiation index (SDI). The SDI is the number of twofold dilution steps separating homologous and heterologous titers. The SDI values are equal to the difference in those titers expressed as negative log 2.

3.2.3 Procedures Used for Delineating Viruses and Strains

Assay Methods

The various serological methods that are used in the detection and assay of viruses were discussed in Chapter 3, Section 2. Most of these procedures have been used for delineating viruses and virus strains, but ELISA procedures have become the most popular.

Monoclonal Antibodies (MAbs)

The advantages and disadvantages of using MAbs for assay, detection and diagnosis of viruses were summarized in Chapter 3, Section 3. The outstanding value of MAbs in the delineation of virus strains is that their molecular homogeneity ensures that only one antigenic determinant is involved in a particular reaction. The high specificity of this single interaction is not swamped in a large number of other interactions as with a polyclonal antiserum. Provided a MAb can be found that recognizes a small antigenic change between two virus strains, then very fine distinctions can be made in a reproducible manner.

However, there are several limitations in the use of MAbs: (1) there is usually no immunoprecipitation between MAbs and viral protein monomers; (2) MAbs are often sensitive to minor conformational changes in the antigen, such as may be caused by detergent or by binding of antigen to an ELISA plate; (3) among a set of virus strains, the relative reactivity of different MAbs may vary considerably; (4) MAbs may be heterospecific; that is, they may frequently react more strongly with other antigens than with the virus used for immunization; and (5) another potential limitation of MAbs in the delineation of strains can occur if two strains have an identical antigenic determinant in common. If, by chance, the MAb specificity is directed against this determinant, the strains will appear identical, even though they may have substantially different determinants elsewhere in the molecule.

3.3 BIOLOGICAL CRITERIA

3.3.1 Symptoms

Macroscopic Symptoms

As noted earlier, symptom differences are of prime importance in the recognition of mutant strains. However, the extent of differences in disease symptoms may be a quite unreliable measure of the degree of relatedness between different members of a group of strains. Symptoms produced by different virus strains in the same species and variety of host plant may range from the symptomless *carrier* state, through mosaic diseases of varying degrees of severity, to lethal necrotic disease. Figure 12.4 illustrates the range of systemic symptom types produced by four strains of tobacco streak virus (TSV) in tobacco.

The diseases produced by a given set of strains in one host plant

Figure 12.4 Control of disease expression by the viral genome. Variation in chronic disease symptom type caused by four tobacco streak *Ilarvirus* isolates in tobacco. (a) The *standard* North American strain. Tobaccos became more or less symptomless. (b) A strain causing toothed margins on the leaves. (c) A strain in which tobaccos continue to show mosaic and necrotic symptoms. (d) A strain causing severe chronic stunting. These symptom types can be artificially reassorted by making crosses between top, middle, and bottom components of the various strains (see Fulton, 1975). Courtesy of R. W. Fulton.

may not be correlated at all with the kinds of disease produced in another host species. A set of defined cultivars that give differential local-lesion responses may provide a particularly useful and rapid method for delineating strains among field isolates of a virus. However, the important influence of environmental conditions on local-lesion response must be controlled.

A named variety of host plant, especially a long-established one, may come to vary considerably in its reaction to a given strain of virus, owing for example to the fact that seed merchants in different localities may make different selections for propagation. This may add a further complication to the identification of strains by means of symptoms produced on named cultivars. Nevertheless, a systematic study of symptoms produced on several host species or varieties under standard conditions may help considerably to delineate strains among large numbers of field isolates of a virus.

Cytological Effects

The cytological changes induced by different strains of a virus are often readily distinguished. Differences are of three kinds: (1) in the effects on cell organelles; (2) in the virus-induced structures within the cell; or (3) in the distribution or aggregation state of virus particles within the cell.

3.3.2 Host Range and Host-Plant Genotype

Host ranges of viruses generally are discussed in Chapter 10, Section 5. Many strains of a virus may have very similar host ranges, but others may differ considerably. Similar responses of a set of host-plant genotypes to two viruses may provide good evidence that they are related strains. On the other hand, a loss in ability to infect a particular host may be brought about by a single mutation.

3.3.3 Methods of Transmission

Different arthropod vector species or different races of a single species may differ in their transmission of various strains of the same virus. Differences may be of the following kinds: (1) in the percentage of successful transmissions; (2) in minimal acquisition time by the vector; (3) in the length of the latent period; (4) in the time the vectors remain infective; and (5) some strains may not be transmitted at all by particular vectors.

Patterns of transmissibility by three aphid species have allowed large numbers of field variants of barley yellow dwarf (BYDV) found in North America to be placed into groups that have facilitated studies on the distribution of virus variants both geographically and in successive seasons.

3.3.4 Cross-Protection

When plants systemically infected with a mild strain of a virus are given a second inoculation with a severe strain, they may be protected from infection with the second strain. This phenomenon, known as cross-protection, has been used for many years as an indication that the two viruses are related strains. Different strains may vary in the extent to which they confer protection against a severe strain (Fig. 12.5).

Figure 12.5 Cross-protection by strains of PVX in *Datura tatula*. (A) Healthy leaf inoculated with a strain giving necrotic local lesions. (B) Leaf previously systematically infected with a very mild strain of the virus, and showing complete protection against inoculation with the necrotic strain. (C) Leaf previously systemically infected with a mottling strain, and showing only partial protection. From Matthews (1949).

Within a set of isolates that are undoubtedly related strains, all possibilities may exist—reciprocal cross-protection of varying degrees of completeness, unilateral cross-protection, and no cross-protection at all. The other factor that may make cross-protection tests ambiguous is that there can be quite strong interference between some unrelated viruses. The mechanism of cross-protection is discussed in Section 4.1.

3.3.5 Productivity

Different strains of a virus may vary widely in the amount of virus produced in a given host under standard conditions.

3.3.6 Proportion of Particle Classes

The proportion of particles with differing sedimentation rates found in purified virus preparations or in crude extracts may vary quite widely with different strains of a virus or members of a virus group. Variation of three kinds can be distinguished: (1) in relative amounts of top component (empty protein shells) for viruses where these are produced; (2) the proportion of nucleic acid components encapsidated may vary in different strains of viruses with multipartite genomes; and (3) abnormal particle classes may be produced by particular strains.

It should be remembered that the proportion of particle classes can be affected by factors other than the strain of virus. These include (1) time after infection; (2) host species; (3) environmental conditions; (4) system of culture; and (5) isolation procedure.

3.3.7 Genome Compatibility

The possibility of carrying out viability tests with mixtures of components from different isolates of viruses with multipartite genomes provides a functional biological test of relationship. Genome compatibility can be tested in a more direct fashion when the gene products can be isolated and their function is known. For example, the protease coded for by cowpea mosaic virus (CPMV) does not process the primary translation products of other comoviruses.

3.3.8 Activation of Satellites

Particular isolates of tobacco necrosis virus (TNV) will support the replication of some satellite TNVs (STNVs), but not others. Similarly, among the cucumoviruses and the small satellite RNAs found in association with them, some viruses support the replication of particular satellites, whereas others do not (Chapter 9, Section 2.2).

In considering use of the various possible criteria for the delineation of virus strains, we must bear in mind that, from a strictly genetic point of view, complete nucleotide sequence data would be sufficient to establish relationships between strains. Nevertheless, small changes in nucleotide sequence could have very different phenotypic effects. At one extreme, a single base change in the coat-protein gene could give rise to changes in several of the phenotypic properties noted earlier. On the other hand, several base changes might give rise to no phenotypic effects at all. For practical purposes, phenotypic characters such as host range, disease symptoms, and insect vectors must usually be given some weight in delineating and grouping virus strains (see Chapter 16).

4 VIRUS STRAINS IN THE PLANT

In the previous sections we have considered ways of isolating virus strains, the molecular mechanisms by which they originate, and the criteria that can be used for distinguishing them. Here certain activities of strains in the infected plant are discussed.

4.1 THE MECHANISM OF CROSS-PROTECTION

There is increasing evidence that more than one mechanism must be involved in cross-protection. Since it was shown that transgenic plants expressing a coat-protein gene may be resistant to infection with intact virus, there has been greatly renewed interest in the phenomenon, especially regarding the possibility of using appropriate transgenic plants for the control of virus diseases (Chapter 15, Section 3.3.2).

4.1.1 General Hypotheses

Competition for Replication Sites

If all available virus-specific replication sites were occupied by the first virus, a second strain might not be able to establish itself. A specific example of this theory might involve the viral replicases. Certain replicases may involve a combination of host-coded and viral-coded subunits. Replication of a second strain might be prevented because all or most of the host component had been preempted by the first strain.

Shortage of Essential Metabolites

In this mechanism, the first strain uses the essential metabolites required by the second strain. There are various reasons for thinking that this is most unlikely. All viruses are made from essentially the same selection of amino acids and nucleotides; thus, sequestration of metabolites should not account for the observed specificity. In well-grown plants, the amount of virus produced is probably not limited by the availability of amino acids and nucleotides.

The Development of Protective Substances

Although transgenic plants expressing appropriate mammalian genes for antibody subunits can produce functional antibodies, there is no equivalent in plants to the immune system in vertebrates. Virus-inhibitory substances are found in extracts from many plant species. However, there is no evidence to show that they prevent virus replication *in vivo*.

4.1.2 Hypotheses Depending on Activities of the Coat Protein

Adsorptive Properties of Infected Cells

An incoming virus particle of the same or of a sufficiently closely related strain might become adsorbed in one of the aggregates of virus already present. This theory has some appeal, but there is no good evidence to support it.

Encapsidation of the Superinfecting RNA

An incoming RNA molecule might become coated with the viral subunits already present in the cell before it could begin replication.

Prevention of Initial Partial Uncoating

It has been proposed for tobamoviruses that the infecting particle loses some protein subunits from the 5' end, allowing translation to begin on the exposed RNA (Chapter 7, Section 3.7). The presence of coat protein of the first infecting strain might prevent partial uncoating of the superinfecting strain. This idea would be unlikely to apply where uncoating of infectious RNA is a rapid all-or-nothing event, as it is for a virus like turnip yellow mosaic virus (TYMV) (Chapter 7, Section 4.7).

Evidence for Factors Other than Coat Protein Being Involved

The results of experiments using transgenic plants discussed in the following Section 4.1.4 strongly favor a role for coat protein. Nevertheless, there is good evidence that other factors must be involved: (1) cross-protection can occur between viroids; (2) a TMV mutant with a protein defective for assembly can cross-protect; (3) assortment experiments with multiparticle viruses have indicated that the ability of a strain to cross-protect can map on a genome segment other than that coding for the coat protein; (4) a mutant of TMV called DT-1G produces no coat protein that can be detected by highly sensitive ELISA tests. However, provided highly infectious inoculum of this strain is used, it protects against superinfection of inoculated leaves of Samsun tobacco with the UI strain of TMV; and (5) plants transgenic for the 54-K putative protein of TMV are resistant to infection with TMV (see Chapter 15, Section 3.3.1).

4.1.3 Hypotheses Depending on Activities of the Viral Nucleic Acid

Gene Recombination

In the gene recombination theory, the incoming strain gets lost, in effect, through genetic recombination with the strain already present. This mechanism does not account for all the known facts of the cross-protection phenomenon, although it might be a significant factor with tomato spotted wilt virus (TSWV) and some other viruses.

Negative-Strand Capture

Annealing of the first produced minus-strand copies of the superinfecting virus with plus strands of the virus already present in the cell might block further replication of the incoming virus. It is known that fully base-paired dsRNAs of ssRNA viruses are noninfectious. This idea would explain how RNA viruses with defective coat proteins (or none) can prevent superinfection. It receives support from the fact that antisense RNAs produced in transgenic plants can block the mRNAs for which they have the complementary sequences. However, it must be remembered that most of the plus-strand RNA of the protecting strain will be in the form of virus particles. We may also ask, if negative strand capture is effective, why the negative strands of the first infecting virus are not rendered ineffective by the virus' own plus strands. Furthermore, RNA–RNA interaction cannot explain why transgenic plants expressing a coat-protein gene can show resistance to superinfection with the same virus or related strains (see the following Section 4.1.4).

Translation Competition

Two strains of TMV that cross-protect in the plant have been shown to exhibit translational competition in the reticulocyte lysate and wheat germ systems if the second RNA was added after translation of the first had commenced. The phenomenon was virus specific. In principle, translation competition could occur in the plant, but again this mechanism would not explain the resistance shown by transgenic plants expressing coat protein.

4.1.4 Transgenic Plants

Resistance in Plants Expressing a Viral Coat-Protein Gene

Over the past few years, it has been shown for several viruses in various host species that transgenic plants expressing the coat-protein gene are resistant to superinfection. Such plants either escaped infection following inoculation or develop symptoms significantly later than in plants not expressing coat protein. Various experiments have demonstrated that the coat protein itself rather than the mRNA is responsible for resistance. Some early event following inoculation is probably involved.

Relation between Natural Cross-Protection and Resistance in Transgenic Plants

The question arises as to whether the resistance generated in transgenic plants is in fact related to the natural phenomenon of cross-protection. There are several similarities that would support the idea: (1) in both situations, the degree of resistance depends on the inoculum concentration, with high concentrations reducing the observed resistance; (2) both are effective against closely related strains of a virus, less against distantly related strains, and not against unrelated viruses; (3) in some circumstances, cross-protection can be substantially overcome when RNA is used as inoculum rather than whole virus. Similarly, for several viruses, the resistance of transgenic plants expressing the coat protein is substantially but not completely overcome when RNA is used as inoculum; and (4) when cross-protection between related strains of a virus is incomplete, the local lesions produced may be much smaller than in control leaves. This indicates reduced movement and/or replication of the superinfecting strain. Local lesions that form in transgenic tobacco plants expressing the potato virus X (PVX) coat are smaller than those of the controls, in line with the result for PVX shown in Fig. 12.5

Nevertheless, coat protein may not be the only viral gene product involved in cross-protection. Plants transgenic for the 54-K protein of TMV are also resistant to superinfection; plants transgenic for the 30-K TMV movement protein are not.

4.2 SELECTIVE SURVIVAL IN SPECIFIC HOSTS

When a virus culture that has been maintained in an apparently stable state in one host species is transferred to another species and then inoculated back to the original host, it is sometimes found that the dominant strain in the culture has been changed. The various host-selection phenomena that have been observed may well involve differences in the cell-to-cell movement protein coded for by strains of the virus. Certain mutations in the movement protein may favor movement in particular hosts. Satellite RNAs such as those associated with cucumber mosaic virus (CMV) may undergo differential replication in particular hosts. This may provide another basis for variation in symptoms following culture of a virus in a given species.

5 DISCUSSION AND SUMMARY

The study of variability is one of the most important aspects of plant virology. It is important from the practical point of view, because strains vary in the severity of disease they cause in the field, and because strains can mutate to break crop plant resistance to a virus. It is important also for developing our understanding of how viruses have evolved in the past, and how they are evolving at present.

A range of procedures is available for isolating virus variants either from nature or following some form of mutagenesis or other manipulation outside the plant. Mutants of the *ts* type have been particularly useful in studying various aspects of virus structure and replication. Because of the very high mutation rate for RNA viruses, all cultures of RNA plant viruses consist of a mixture of numerous strains even after single lesion passage. However, a *consensus* genome sequence will usually dominate in the culture, and many variants will not be detected.

The molecular mechanisms by which variation within a virus population is produced are like many of those found in cellular organisms, except that for many plant virus groups, the material on which variation operates is RNA rather than DNA. Mechanisms include mutations involving single nucleotide changes or the addition or deletion of one or a few nucleotides; recombination; deletions or additions of blocks of nucleotide sequences; rearrangement of nucleotide sequences; reassortment among multipartite genomes; duplication of gene sequences; and acquisition of genes from another virus or the host genome.

A range of structural, serological, and biological criteria is available for delineating viruses and virus strains within a group or family of viruses. The kinds of criteria to be used will depend on the purpose of the study. If we are studying evolutionary relationships within a virus group or family, or among the variants of a single virus, then the full or partial nucleotide sequences of the viruses concerned will be of prime importance, but a knowledge of the functional products of the viral genome will often be needed as well. Amino acid sequences of viral coat proteins are proving particularly useful for delineating viruses and virus strains in some groups, as is illustrated for the *Potyvirus* group in Fig. 12.6.

If the full nucleotide sequences are known for representative viruses, then other methods, such as various forms of nucleic acid hybridization, can be usefully interpreted for additional viruses and strains. If we are interested in developing methods for reliably and rapidly diagnosing viruses and virus strains from the field, then other methods may be appropriate. Dot-blot serological assays using some form of ELISA are emerging as an important type of test. Polyclonal antisera of wide specificity or MAbs of very narrow specificity can be used in such tests as appropriate. Biological criteria such as disease

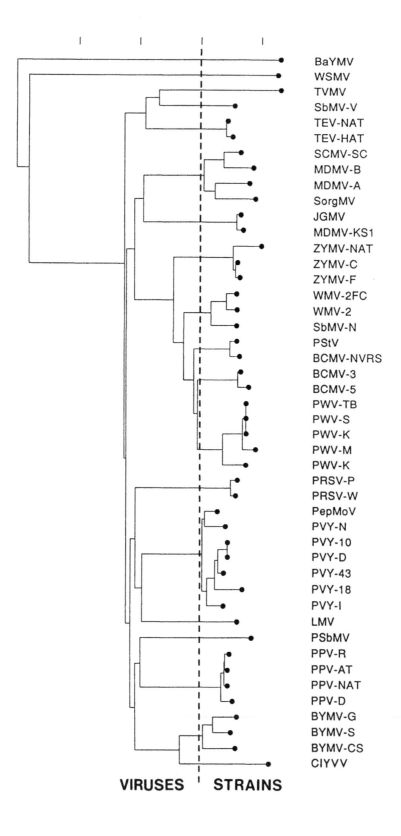

BaYMV
WSMV
TVMV
SbMV-V
TEV-NAT
TEV-HAT
SCMV-SC
MDMV-B
MDMV-A
SorgMV
JGMV
MDMV-KS1
ZYMV-NAT
ZYMV-C
ZYMV-F
WMV-2FC
WMV-2
SbMV-N
PStV
BCMV-NVRS
BCMV-3
BCMV-5
PWV-TB
PWV-S
PWV-K
PWV-M
PWV-K
PRSV-P
PRSV-W
PepMoV
PVY-N
PVY-10
PVY-D
PVY-43
PVY-18
PVY-I
LMV
PSbMV
PPV-R
PPV-AT
PPV-NAT
PPV-D
BYMV-G
BYMV-S
BYMV-CS
CIYVV

VIRUSES | **STRAINS**

symptoms, host range, methods of transmission, and cross-protection may be important in defining viruses and virus strains.

There is renewed interest in the subject of cross-protection between virus strains, because it has been shown for several viruses that transgenic plants expressing the coat-protein gene are resistant to superinfection, and that this phenomenon resembles natural cross-protection in several respects. The mechanism of resistance in plants transgenic for coat protein is not fully understood, but it certainly involves the coat protein in some way. In natural cross-protection, it is likely that more than one mechanism is involved in the protective response.

The extent to which virus species have been clearly delineated varies widely among the different groups and families of viruses. There are dangers in formalizing virus species or virus groups before a sufficient number and diversity of strains have been investigated. When various criteria fit together to suggest the same relationships between sets of virus strains, the proposed taxonomy is greatly strengthened. For some groups, such as the potyviruses, *a common set or pattern of correlating stable properties* is emerging that can allow the grouping of virus strains into species with some degree of confidence.

The relative importance, or weight, to be placed on different properties of a virus for purposes of classification remains a difficult problem. An adequate understanding of the significance to be placed on the various properties may come only when we have a detailed knowledge of the structure of the viral genome, the polypeptides it codes for and their functions, and the regulatory or other roles of any translated or untranslated regions in the genome. Particular gene functions may be

---

Figure 12.6 Relationships between some potyviruses and their strains. The dendrogram is based on the amino acid sequence of the core of the viral coat proteins, omitting the more variable terminal sequences. The core regions were aligned, and the percentages of different amino acids in each sequence (that is, their difference) were calculated. Gaps were counted as a twenty-first amino acid. The distances were then converted into the dendrogram using the neighbor-joining method of Saitou and Nei (1987). The divisions in the horizontal scale represent 10% differences. The broken vertical line indicates a possible boundary between viral species and strains. Some of the viruses listed do not appear in Appendix 1 (JGMV, Johnson grass mosaic virus; SorgMV, sorghum mosaic virus; PStV, peanut stripe virus). The classification suggested here differs from some of the conclusions put forward in the text. The evolutionary distances indicated here fit with known vector relationships. BaYMV is transmitted by fungi; WSMV, by mites; whereas all the others listed here have aphid vectors. Diagram courtesy of A. J. Gibbs.

of particular ecological, and therefore practical significance, for example, mutations in a viral gene that affects insect vector specificity.

Even with such knowledge, considerable difficulties will remain.

1. Disease induction, which is a complex process, has been shown for some viruses to depend on the functions of two or more viral genes.
2. Various possible mechanisms are now known whereby a single mutation could have effects on two or more functions.
3. A single gene product may have two or more functions, differing in importance for the virus life cycle.

Thus, from a practical point of view, it may sometimes be an oversimplification to establish relationships between viruses and strains within a family or group solely on the basis of nucleotide sequences.

FURTHER READING

Beachy, R. N. (1988). Virus cross-protection in transgenic plants. *In* "Temporal and Spatial Regulation of Plant Genes" (D. P. S. Varma and R. B. Goldberg, eds.), pp. 313–331. Springer-Verlag, Wien, New York.

Matthews, R. E. F. (1991). "Plant Virology," 3rd Ed. Academic Press, New York.

Van Vloten-Doting, L., and Bol, J. F. (1988). Variability, mutant selection, and mutant stability in plant RNA viruses. *In* "RNA Genetics" Vol. III (E. Domingo, J. J. Holland, and P. Ahlquist, eds.), pp. 37–51. CRC Press, Boca Raton, Florida.

Ward, C. W., and Shukla, D. D. (1991). Taxonomy of potyviruses: Current problems and some solutions. *Intervirology* 32, 269–296.

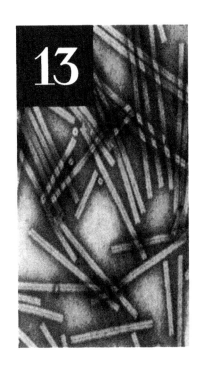

13

RELATIONSHIPS BETWEEN PLANT VIRUSES AND INVERTEBRATES

The transmission of viruses from plant to plant by invertebrate animals is of considerable interest from two points of view. First, such vectors provide the main method of spread in the field for many viruses that cause severe economic loss. Second, there is considerable general biological interest in the relationships between vectors and viruses, especially as some viruses have been shown to multiply in the vector. Such viruses can be regarded as both plant and animal viruses. Even with those that do not multiply in the animal vector, the relationship is usually more than just a simple one involving passive transport of virus on some external surface of the animal. Transmission by invertebrate vectors is usually a complex phenomenon involving the virus, the vector, the host plant, and the environmental conditions.

1 VECTOR GROUPS

Of about 13 orders of invertebrates that contain at least some members that feed regularly on living green land plants, 10 have been shown to contain vectors of plant viruses. These are the Dorylamida in the Nematoda; the Homoptera, Hemiptera, Thysanoptera, Orthoptera, Dermoptera, Coleoptera, Lepidoptera, and Diptera in the Insecta (Arthropoda); and the Eriophyidae in Acari (Arachnida; Arthropoda). Of these orders, the Homoptera in the Insecta, which contains a very diverse group of plant-feeding insects, is by far the most important, with about 300 vector species among the aphids, whiteflies, leafhoppers, planthoppers, and mealybugs. Almost all viruses that have invertebrate vectors are transmitted by only one of the known vector groups.

2 NEMATODES (NEMATODA)

Several widespread and important viruses in two virus groups are transmitted through the soil by nematodes, which feed on root epidermal cells (Fig. 13.1). Members of the *Nepovirus* group are transmitted by species in the genera *Xiphinema* and *Longidorus*. Members of the *Tobravirus* group are transmitted by species of *Trichodorus*.

Nematodes are difficult vectors to study experimentally because of their small size and their rather critical requirements with respect to soil moisture content, type of soil, and, to a lesser extent, temperature. A common method for detecting nematode transmission has been to set out suitable "bait" plants (such as cucumber) in a sample of the test soil. These plants are allowed to grow for a time to allow any viruliferous nematodes to feed on the roots and transmit the virus, and for any transmitted virus to replicate. Extracts from the roots and leaves of

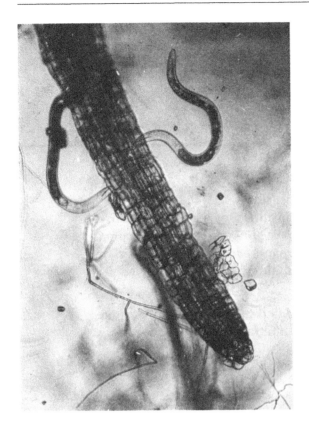

Figure 13.1 Nematodes of the species *Trichodorus christei* feeding on a root of blueberry (*Vaccinium corymbosum*). Courtesy of B. M. Zuckermann.

the bait plants are then inoculated mechanically into a suitable range of indicator species.

Once acquired, viruses may persist in transmissible form in starved nematodes for up to a year or more depending on the species. Viruses do not appear to replicate in nematode vectors, and virus particles have never been observed inside nematode cells. Electron microscope studies show that in different vector genera, and for different viruses, particles adhere to different portions of the lining of the alimentary tract. This adsorption is involved in the specificity of transmission. Genetic reassortment experiments and other observations have shown that the coat protein of the virus determines the specificity of adsorption. Presumably, when a nematode carrying virus begins feeding on a fresh plant, and the stylets penetrate a cell, some virus particles are washed from the gut lining into the plant cell by the passage of salivary secretions. Whereas the nematode–virus interaction determines the specificity of transmission, the efficiency of transmission may be influenced by the plant host (as well as by nematode species and virus strain).

3 APHIDS (APHIDIDAE)

Among insects, the aphids have evolved to be the most successful exploiters of higher plants as a food source, particularly in flora of temperate regions. It is, therefore, not surprising that they have also developed into the most important group of virus vectors. About 66% of approximately 370 viruses with invertebrate vectors are transmitted by aphids. There are three kinds of variability in aphids that may affect their ability to transmit a virus.

1. An aphid species may contain different clones or races, with or without obvious morphological differences, but which are associated with different efficiencies of virus transmission.
2. Aphids can exist in different forms. In particular, there are apterous (wingless) and alate (winged) forms.
3. Each form develops through a series of larval stages, the larvae being known as nymphs.

Successive molts by the developing insect define the number of stages or instars. Under favorable conditions, female aphids breed very rapidly by an asexual process (Fig. 13.2).

3.1 FEEDING HABITS

The mouth parts of aphids consist of two pairs of flexible stylets, held within a groove of the labium. They are extended from the labium during feeding. The maxillary stylets have a series of toothlike projections near their tips (Fig. 13.3). At the beginning of feeding, a drop of gelling saliva is secreted. The stylets then rapidly penetrate the epidermis, and in exploratory probes, the aphid may feed there temporarily. Penetration usually continues into the deeper layers, with a sheath of gelled saliva forming during penetration. The stylets usually move between cells until they reach a phloem sieve tube (a process that may take minutes or hours). Only the maxillary stylets enter the sieve tube. Compression by the cell

Figure 13.2 A group of apterous female adults of the cereal aphid *Rhopalosiphum padi* L., together with various nymphal stages, feeding on a leaf of wheat. This species transmits barley yellow dwarf *Luteovirus*. From Lowe (1964).

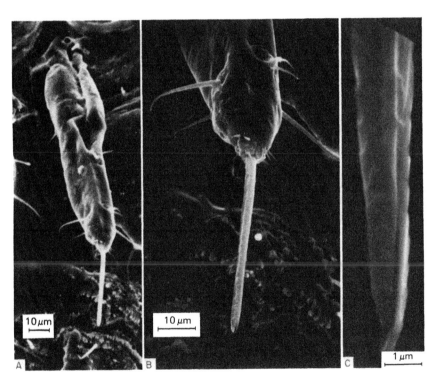

Figure 13.3 Mouth parts of *Myzus persicae* revealed by scanning electron microscopy. (A) Labium with joint area and bristles. Mandibular stylets protrude from the labium (the aphid was frozen in liquid nitrogen immediately after it had withdrawn its stylets from a leaf). (B) Tip of labium and mandibular stylets at higher magnification. (C) Tip of mandibular stylets showing ridges on both stylets, and the overlap of the tip of one stylet. From De Zoeten (1968).

wall causes the tips to open, exposing the end of the food and salivary canals. While they feed in the phloem, aphids appear to secrete a watery, enzyme-bearing, nongelling saliva. No gelling-sheath saliva is secreted during feeding in a sieve tube, but on withdrawal, such saliva is used to seal the lumen in the salivary sheath that had been occupied by the stylets. This feeding process causes minimal damage to the sieve tube and to surrounding cells in the stylet path.

Both physical and chemical features of the plant may markedly affect aphid feeding behavior. Such features include the physical state of the wax on the leaf surface; the density of trichomes; and specific chemicals that may attract or inhibit feeding by particular aphid species. Virus infection itself may make a plant more suitable for aphid growth and reproduction.

Environmental conditions, particularly temperature, humidity, and wind, may have marked effects on aphid movement and feeding. These various effects are discussed in relation to virus ecology and control of virus diseases in Chapters 14 and 15. Environmental factors may also influence transmission through effects on plant susceptibility, and on the concentration of virus in source plants.

3.2 TYPES OF APHID–VIRUS RELATIONSHIPS

Various terms have been used to describe the different ways in which aphids transmit viruses, and the viruses that are aphid-transmitted. I shall mainly follow the usage of Harris (1983). Table 13.1 summarizes the main properties of the different kinds of relationships. Some definitions are needed.

1. *Inoculativity* is the ability of an aphid or other insect to deliver infectious virus into a healthy plant.
2. The *acquisition feed* is the feeding process by which the insect acquires virus from an infected plant.
3. The *inoculative feed* is the feed during which virus is delivered into a healthy plant.
4. The *latent period* is the time after an acquisition feed for which the aphid is unable to transmit a virus.

Some workers have used the term *stylet-borne* for nonpersistent viruses, but it is by no means certain that inoculum virus is in fact carried on the stylets. The weight of evidence favors the food canal in the maxillae and the foregut as sites of virus retention. The term *foregut-*

Table 13.1 Categories Describing Aphid Transmission or the Transmitted Viruses

Circulative (persistent)

Main features: Virus acquired through the food canal and translocated
A latent period follows acquisition feeding
Infection feeding involves ejection of virus in saliva from the maxillary saliva canal

Subcategories for circulative

Propagative	Nonpropagative
Virus multiplies in insect	Virus not known to replicate in insect

Foregut borne (noncirculative)

Main features: No detectable latent period
Loss of vector inoculativity following a moult
No evidence for virus in hemocoel or salivary system

Subcategories for foregut borne

Property	Nonpersistent	Semipersistent
Mean time for retention of inoculativity following inoculative feeding, without further access to virus	A few minutes	Hours
Effect of starving before virus acquisition feeding	Severalfold increase in inoculativity	No effect
Time for acquisition and inoculation thresholds	Seconds	Several minutes
Effect of continued acquisition feeding	Acquisition probes >1 min lead to a marked drop in transmission	Inoculative capacity increases for several hours
Effect following a long inoculative feed	Rarely inoculative after a long feed	No marked effect
Effects of starving compared with feeding	Starved aphids remain inoculative longer than those allowed to feed	Inoculativity not affected by starving or feeding

borne has been proposed for viruses that do not circulate within the body of the vector (noncirculative). It is not known how foregut-borne viruses are released from their site of retention to reenter a plant cell, but vector saliva may play a role. Most aphid-transmitted viruses fall clearly into one of the categories defined in Table 13.1. However, cauliflower mosaic virus (CaMV) can be transmitted both nonpersistently and semipersistently. This is known as *bimodal* transmission (Section 3.2.3).

3.2.1 Nonpersistent Transmission

When they alight on a leaf, aphids may make brief probes into the leaf (usually less than 30 sec). These probes are to test the suitability of the leaf as a food source. The weight of evidence indicates, at least for aphids like *Myzus persicae,* that the stylets enter the epidermal cells during probing. Thus, the initial behavior of such aphids on reaching a leaf is ideally suited to rapid acquisition of a nonpersistent virus. Many such viruses are of substantial economic importance. Of the approximately 250 known aphid-borne viruses, most are nonpersistent. The virus groups with definite members transmitted in a nonpersistent manner are *Potyvirus, Carlavirus, Caulimovirus* (by *M. persicae*), *Cucumovirus,* alfalfa mosaic virus, and *Fabavirus.* These groups include helical and isometric viruses; DNA and RNA viruses; and viruses with mono-, bi-, and tripartite genomes.

Different strains of the same virus may vary in the efficiency with which they are transmitted by a particular aphid species. Some strains may not be transmitted by aphids at all. Examples are also known in which a given virus strain either gained or lost the ability to be transmitted by a particular aphid. Properties of the viral coat protein appear to determine the specificity of transmission. Aphid species vary widely in the number of different viruses they can transmit. At one extreme, *M. persicae* is known to be able to transmit a large number of nonpersistent viruses. Other aphids have been found to transmit only one virus. Certain viruses can be transmitted by aphids in a nonpersistent fashion only when another virus is also present in the source plant. The helper factor is a specific gene product of the helper virus. Such factors have been found in two kinds of virus, the *Potyvirus* and *Caulimovirus* groups.

Potyvirus helper proteins have M_rs in the range of 50 to 60 K (e.g., Fig. 7.1). Antisera against the proteins specifically block aphid transmission in *in vitro* uptake tests. The way in which the helper protein makes aphid transmission possible has not been established. The most likely kind of effect is that the protein makes it possible for the virus to attach to sites within the aphid in a way that allows it to be transmitted.

CaMV, and presumably other caulimoviruses, requires a helper component (or aphid-transmission factor) when being transmitted in a nonpersistent manner by *M. persicae*. The helper component is the product of open reading frame II (ORFII) in the CaMV genome (Fig. 8.1). A viable ORFII product is not necessary for virus replication in the plant. Natural isolates are known that are not aphid transmissible. These have a defective ORFII protein product that may be expressed in the plant. Non-aphid-transmissible isolates of CaMV are transmitted if the aphids are first fed on plants infected with some other caulimovirus containing a viable aphid-transmission protein.

3.2.2 Semipersistent Transmission

About 15 aphid-transmitted viruses have transmission properties that are intermediate between nonpersistent and circulative viruses, although as a group, the viruses do not show very uniform transmission properties.

The best studied of the semipersistent viruses are beet yellows virus (BYV) and citrus tristeza virus. Both of these closteroviruses are found particularly in the phloem, which may account for some of their transmission characteristics. The various features of semipersistent transmission are best accounted for by assuming that the virus particles taken up by the vector aphids are selectively adsorbed to the surfaces of the foregut, resulting in an ingestion–egestion mechanism for transmission. Semipersistent viruses such as BYV are confined to the phloem elements and related cells. It takes several minutes, or longer, for the aphid stylets to penetrate to the phloem. This accounts in part for the longer acquisition feeding time compared with that of nonpersistent viruses, where the aphid can pick up virus in seconds from an epidermal cell. Aphids must also feed on the phloem for a period of several minutes to become inoculative.

Helper viruses may be involved in semipersistent transmission. The aphid *Cavariella aegopodii* transmits both anthricis yellows virus (AYV) and parsnip yellow fleck virus (PYFV) in a semipersistent manner. PYFV is a small isometric RNA virus with a diameter of ≈30nm, as is AYV. The aphid transmits PYFV only when it is carrying AYV, a virus that is not sap transmitted. Unlike the helper for some nonpersistent viruses, the AYV itself is the helper agent for PYFV. However, PYFV can be acquired from plants infected singly with this virus only if the aphids are already carrying the helper AYV. This suggests that a helper protein may be involved, as with some nonpersistent viruses.

3.2.3 Bimodal Transmission

A few viruses, for example CaMV and pea seed borne mosaic *Potyvirus,* can be transmitted by the same aphid species both nonpersistently and semipersistently. When the percentage of infected plants is plotted against the acquisition feeding time, a bimodal curve is obtained. The first peak represents nonpersistent acquisition and inoculation by individual aphids. These aphids retain the ability to transmit for only short times. The second peak is from semipersistent transmission following longer feeds.

3.2.4 Circulative Transmission

The main features of circulative transmission are summarized in Table 13.1. Viruses transmitted in a circulative manner are usually transmitted by one or a few species of aphid. Yellowing and leafrolling symptoms are commonly produced by infection with circulative viruses.

Rhabdoviruses such as sonchus yellow net virus (SYNV) transmitted by aphids replicate in their vectors. The latent period of SYNV in the vector is long, and depends strongly on temperature. Characteristic bacilliform particles have been observed in the nucleus and cytoplasm of cells in the brain, subesophageal ganglion, salivary glands, ovaries, fat body, mycetome, and muscle. Virus particles appear to be assembled in the nucleus. The virus can be serially transmitted from aphid to aphid by injection of hemolymph, and infection is associated with increased mortality of the aphids. Decreased life span varied with different virus isolates. However, since infected aphids lived through the period of maximal larviposition, the intrinsic rate of population growth is little affected. The virus is transmitted through the egg of *Hyperomyzus lactucae,* about 1% of larvae produced being able to infect plants. The virus has been shown to multiply in primary cultures of aphid cells.

Some other viruses transmitted in a persistent manner by aphids do not replicate in their vectors (nonpropagative). The most important viruses with this kind of transmission belong to the *Luteovirus* group, in which virus is confined to the phloem. Barley yellow dwarf virus (BYDV) has been the most studied. Penetration of a sieve element appears to be necessary for both acquisition and transmission of the virus. The current hypothesis for nonpropagative circulative transmission of luteoviruses by aphids involves specific interactions between the viral coat protein and receptor sites in the accessory salivary glands of the aphid.

As with certain nonpersistent viruses, some persistent viruses require a helper virus to be present in the plant before aphid transmission can occur. For persistent viruses dependent on another virus, the presence of the virus itself in a mixed infection provides the assistance. BYDV is the best studied of the persistent viruses for which dependent transmission can occur.

There is a high degree of vector specificity among the luteoviruses, including BYDV. For example, one form of BYDV, RPV, is transmitted efficiently by *Rhopalosiphum padi* but not by *Sitobion avenae*. For another form, MAV, the reverse is true. Virus can enter the hemolymph of inefficient vectors. Thus, the block in transmission appears to lie in the inability of a virus to move from the hemolymph, via the saliva to the plant. Experiments in which strain-specific antisera were used to block aphid transmission showed that dependent transmission is the result of phenotypic mixing during virus replication in the plant. Some RNA genomes of the dependent strain become encapsidated in the coat protein of the helper strain.

Certain luteoviruses are required as helper viruses for the transmission of other unrelated viruses. About six of these complexes are known. Perhaps the most interesting is the groundnut rosette virus, which requires the groundnut rosette assistor *Luteovirus* for aphid transmission. Again the assistance is brought about by the packaging of the RNA of the rosette virus in the assistor *Luteovirus* coat protein. This phenotypic mixing can take place only in plants infected with both viruses. Recent work has revealed a further complication. Groundnut rosette virus depends on its satellite RNA as well as on the assistor virus for transmission by *Aphis craccivora*.

4 LEAFHOPPERS AND PLANTHOPPERS (AUCHENORRHYNCHA)

In the Cicadellidae (leafhopper family), there are about 15,000 described species in about 2000 genera. Of these, only 49 species from 21 genera have been reported as being virus vectors. Two out of 60 subfamilies contain species that are virus vectors. The Agallinae have herbaceous dicotyledonous hosts, whereas most Deltocephalinae feed on monocotyledons.

There are about 20 families of planthoppers (Fulgoroidea), but only the Delphacidae have definite virus vector species. Members of this family feed on monocotyledons, primarily members of the Poaceae. Thus, all the viruses known to be transmitted by members of this family have hosts in the Poaceae. These cause important diseases of cereal crops, including rice, wheat, and maize. Unlike aphids, leafhoppers have a simple life cycle in which the egg hatches to a nymph, which feeds by sucking, and passes through a number of molts before becoming an adult (Fig. 13.4). There may be one or several generations per

Figure 13.4 *(Top to bottom)* Eggs, nymph, long- and short-winged adults of the plant hopper *Laodelphax striatellus,* vector of several reolike plant viruses. From Conti (1984).

year. Different species overwinter as the egg, as the adult, or as imma-
ture forms.

4.1 SEMIPERSISTENT TRANSMISSION

No viruses have been found to be transmitted in a nonpersistent manner by hoppers, or by purely mechanical means. Some have the characteristics of semipersistent transmission. Because virus-like particles
have been found attached to the cuticular linings of the anterior alimentary canal, the term *foregut borne* has been used to describe this
kind of transmission. Hoppers egest material from the foregut from
time to time during feeding. Thus, transmission of semipersistent viruses may involve an ingestion–egestion mechanism.

The virus complex causing the important rice tungro disease consists of two viruses, rice tungro spherical virus and rice tungro
bacilliform virus. Both are hopper transmitted in a semipersistent manner. The bacilliform virus depends on the spherical virus for its transmission. Several experiments using virus-specific neutralizing antisera
suggest that a factor from plants infected with the spherical virus, but
not the virus itself, is required by the bacilliform virus for transmission.
The mechanism for vector specificity is not known, but may involve the
properties of the cuticular lining of the foregut and/or the nature of salivary secretions.

4.2 CIRCULATIVE TRANSMISSION

Circulative transmission by hoppers involves movement of ingested
virus to the salivary glands. As with the aphid vectors, some circulative
viruses replicate in the hopper vector (propagative) and some do not
(nonpropagative).

4.2.1 Nonpropagative Transmission

Some geminiviruses are transmitted by leafhoppers, and all of these are
transmitted in a nonpropagative manner. The properties of the gut wall
appear to regulate vector specificity. Among geminiviruses, the coat

protein determines whether the virus is transmitted by leafhoppers or whiteflies.

4.2.2 Propagative Transmission

Four families and groups have members that replicate in their hopper vectors. These are the Reoviridae and Rhabdoviridae, and the *Tenuivirus* and *Marafivirus* groups.

The classic experiments demonstrating that some plant viruses could also replicate in their animal vectors were carried out with plant reoviruses and their leafhopper vectors. In the most direct experiments, growth curves of virus and viral antigen in the hopper vector have demonstrated that replication occurs. The latent, or incubation period in propagative transmission by hoppers is usually 1–3 weeks. During this period, the virus replicates and invades most tissues. Ability to transmit appears to coincide with arrival of the virus in the salivary glands. Hoppers then usually retain infectivity for the rest of their life span. In propagative transmission, the virus may be transmitted through a proportion of the eggs of infected females.

When wound tumor *Phytoreovirus* was maintained over a period of years in cuttings made from clover shoots, and without access to the hopper vector, a variety of deletion mutants were obtained. In certain selected isolates, RNA2 or RNA5 was missing. In spite of this, the virus replicated normally in sweet clover and had a full capacity to induce tumors. On the other hand, these isolates were entirely unable to replicate in the leafhopper vector or in vector cell monolayers. Thus, segments 2 and 5 contain genes required for replication in the insect, but not in the plant.

Eighteen members of the plant Rhabdoviridae have hopper vectors. Each virus has Cicadellid or Delphacid vectors, but not both. Various kinds of experiments have demonstrated that the viruses replicate in their hopper vectors. Minimal acquisition times range from less than 1 min to about 15 min for different viruses. The longer times are associated with rhabdoviruses confined to the phloem and nearby cells. The latent period may be days or months. There may be a high degree of vector specificity, even between strains of the same virus. It is probable that the G (glycosylated) protein, which is exposed at the virus surface, is involved in this specificity.

4.2.3 Factors Affecting Propagative Transmission

Virus can multiply in hoppers feeding on an immune host. Eggs may overwinter and provide a source of virus for spring crops, in the absence of diseased plants. Thus, persistence of virus in the insect and transovarial transmission, and the factors that affect its efficiency, may be of considerable economic importance. The following may be important factors: (1) *age of the vector when infected.* Nymphs tend to be more efficient vectors than adults, and adults decrease in efficiency as they age; (2) *time after infection.* This factor has variable effects; (3) *temperature* affects both the ability of hoppers to infect plants and to transmit virus through the egg. High temperatures ($\geq30°C$) decrease the efficiency of both processes; temperature can also affect the duration of the latent period; and (4) *genetic variation in the leafhopper.* Different lines or races within a vector species may vary widely in their efficiency as vectors.

5 INSECTS WITH BITING MOUTHPARTS

Within the order Coleoptera, interest centers on the family Chrysomelidae, which consists of 55,000 species of plant-eating beetles. About 30 of these are known to transmit plant viruses, and each species feeds on a limited range of host plants. Most vector species are found in the subfamilies Galerucinae and Halticinae (flea-beetles).

Leaf-feeding beetles do not have salivary glands. The Chrysomelid beetles tend to eat the parenchyma tissues between vascular bundles, thus leaving holes in the leaf, but with heavy infestation, damage may be more severe. The beetles regurgitate during feeding, which bathes the mouth parts with plant sap, as well as with viruses, if the plant fed on is infected. It was once thought that transmission by beetles involved simply a mechanical process of wounding in the presence of virus. This is not so because (1) some very stable sap-transmissible viruses such as TMV are not transmitted by beetles; (2) some transmitted viruses may be retained by beetle vectors for long periods; and (3) there is a substantial degree of specificity between viruses and vector beetles.

The viruses transmitted by beetles belong to the *Tymovirus, Comovirus, Bromovirus,* and *Sobemovirus* groups. Most of the viruses in these groups are not transmitted by members of other arthropod groups. They are usually quite stable and reach high concentrations in infected tissues. They have small isometric particles (25–30 nm diameter) and are readily transmitted by mechanical inoculation. The viruses tend to have relatively narrow host ranges, as do their beetle vectors.

Beetles can acquire virus very quickly—even after a single bite—but efficiency of transmission increases with longer feeding, as does retention time. Viruses appear quickly in the hemolymph after beetles feed on an infected plant. Insects become viruliferous after injection of virus into the hemocoel. Retention time varies between about 1 and 10 days with different beetles. However, under dormant, overwintering conditions, beetles may stay viruliferous for periods of months. Beetles can transmit the virus with their first bite on a susceptible plant. There is no good evidence for the existence of a latent period following virus acquisition, and no evidence for virus replication in beetle vectors. Trans-stadial transmission has not been demonstrated, probably because of the processes involved in pupation.

The regurgitant fluid is a key factor in determining whether a virus will or will not have beetle vectors. This discovery was made possible by a gross wounding technique, which involved cutting disks from a leaf with a glass cylinder contaminated with virus-regurgitant mixture, thus mimicking the kinds of wounds made by feeding beetles. When virus was mixed with regurgitant, only viruses normally transmitted by beetles were transmitted by the gross wounding technique. Non-beetle-transmitted viruses are reversibly inactivated by beetle regurgitant. The inactivating agent is a protein with RNase activity. Ability to be translocated in the xylem and thus to infect nonwounded tissues may be another important feature of beetle-transmitted viruses.

These experiments shed little light on the problem of specificity among beetle species—why some species are highly efficient vectors of a particular virus, and others are not. Thus the apparently simple transmission of viruses by beetles is in fact quite a complex process.

6 OTHER VECTOR GROUPS

6.1 MEALYBUGS (COCCOIDEA AND PSEUDOCOCCOIDEA)

Mealybugs are much less mobile on the plant than are other groups of vectors such as aphids and leafhoppers, a feature that makes them relatively inefficient as virus vectors. They spread from one plant to another in contact with it, and the crawling nymphs move more readily than adults. Ants that tend the mealybugs may move them from one plant to another. Occasional long-distance dispersal by wind may occur. Mealybugs feed on the phloem. They have been established as the vectors of several viruses affecting tropical plants. The most important economically is the cocoa swollen shoot group of viruses. The relationship between the cocoa swollen shoot virus and mealybugs has some similarities to that of the nonpersistent aphid-transmitted viruses.

6.2 WHITEFLIES (ALEYRODIDAE)

Whiteflies are known to transmit about 70 disease agents, mainly of tropical and subtropical plants. Many of the agents transmitted by whiteflies cause mosaic disease of a bright yellow or golden nature. Less commonly, the diseases involve marked curling of the leaves or generalized yellowing. The diseases are of substantial importance in tropical regions, but are not confined to these areas. The agents causing many of the diseases have not been characterized. They are transmitted by nonpersistent, semipersistent, and persistent mechanisms. At least three species of whitefly are involved, and it is probable that members of at least seven groups of viruses are transmitted by whiteflies. These are geminiviruses, certain viruses with particles like closteroviruses, carlaviruses, potyviruses, nepoviruses, luteoviruses, and a DNA-containing rod-shaped virus. With respect to whitefly vectors, the best-studied group are the geminiviruses.

The most-studied vector is *Bemisia tabaci* (Gennadius), which is known to transmit a number of virus diseases. Only the first instar of the larvae is mobile, and it does not move far. Adults are winged, and many generations may be produced in a year. The nymphs of *B. tabaci* are phloem feeders.

There is no evidence that the viruses are transmitted through the egg of the whitefly or that they multiply in the vector. The virus–vector relationship is probably closest to the circulative nonpropagative situation found with some aphid-transmitted viruses.

6.3 BUGS (MIRIDAE AND PIESMATIDAE)

The mirid bugs feed by means of stylets, but their biology and taxonomy are not well understood. *Cyrtopeltis nicotianae* has been shown to be a vector of velvet tobacco mottle virus, southern bean mosaic virus (SBMV), and several other, but not all, sobemoviruses. There is no evidence for virus replication in the vector. The transmission was like that of beetles in several respects.

6.4 THRIPS (THYSANOPTERA)

The known vector species of thrips are all in the family Thripidae. The very small size of thrips compared to leafhoppers, or even aphids, makes them difficult to handle experimentally. *Thrips tabaci* (Lindeman) is cosmopolitan, feeding on at least 140 species from over 40 families of plants. It reproduces mainly parthenogenetically. The larvae are rather inactive, but the adults are winged and very active. *Thrips tabaci* feeds by sucking the contents of the subepidermal cells of the host plant. Adults live up to about 20 days, and several generations can develop in a year. At least nine species of thrips can transmit tomato spotted wilt virus (TSWV), which is an important virus in many countries, with a wide host range. Only larvae can acquire TSWV; thus, trans-stadial transmission through the pupa to the adult must occur. TSWV is not passed through the egg, and there is no evidence for its replication in the insect. Thrips can also transmit a virus such as tobacco streak *Ilarvirus,* by carrying pollen from an infected to a healthy plant and introducing the virus while feeding on leaves (Greber *et al.,* 1991).

6.5 MITES (ARACHNIDA)

Several members of the mite family Eriophyidae transmit three potyviruses and several other viruses. They have limited powers for independent movement, but can sometimes be wind-blown several kilometers downstream from infested fields. Their very small size (≈ 0.2 mm) makes virus–vector relationships difficult to study experimentally. There is no good evidence for virus replication in these arthropods.

7 POLLINATING INSECTS

Viruses transmitted through infected pollen, and which also infect insect-pollinated host plants, probably have the infecting pollen distributed by pollinating insects. For example, field experiments showed that blueberry leaf mottle *Nepovirus* is transmitted via pollen carried by foraging honey bees.

FURTHER READING

Fulton, J. P., Gergerich, R. C., and Scott, H. A. (1987). Beetle transmission of plant viruses. *Annu. Rev. Plant Pathol.* **25**, 111–123.

Harris, K. F. (1983). Sternorrhynchous vectors of plant viruses: Virus–vector interactions and transmission mechanisms. *Adv. Virus Res.* **28**, 113–140.

Matthews, R. E. F. (1991). "Plant Virology," 3rd Ed. Academic Press, New York.

Nault, L. R., and Ammar, E. D. (1989). Leafhopper and planthopper transmission of plant viruses. *Annu. Rev. Entomol.* **34**, 503–529.

14

ECOLOGY

In order to survive, a plant virus must have (1) one or more host plant species in which it can multiply; (2) an effective means of spreading to, and infecting fresh individual host plants; and (3) a supply of suitable healthy host plants to which it can spread. The actual situation that exists for any given virus in a particular locality, or on the global scale, will be the result of complex interactions between many physical and biological factors. An understanding of the ecology of a virus in a particular crop and locality is essential for the development of appropriate methods for the control of the disease it causes. As with most other obligate parasites, the dominant ecological factors to be considered are usually the way viruses spread from plant to plant and the ways that other factors influence such spread.

1 BIOLOGICAL FACTORS

1.1 PROPERTIES OF VIRUSES AND THEIR HOST PLANTS

1.1.1 Physical Stability of Viruses and Concentrations Reached

For viruses depending on mechanical transmission, a virus that is stable, both inside and outside the plant, and that reaches a high concentration in the tissues, is more likely to survive and spread than one that is highly unstable. The survival and spread of certain viruses appears to depend mainly on a high degree of stability and the large amounts produced in infected tissues. For example, TMV may survive for long periods in dead plant material in the soil, which is then a source of infection for subsequent crops. TMV also survives tobacco processing and is found in cigarettes.

1.1.2 Rate of Movement and Distribution within Host Plants

Viruses or virus strains that move slowly through the host plant from the point of infection are less likely to survive and spread efficiently than those that move rapidly. Speed of movement is important as measured relative to the lifetime of individual host plants. Viruses that infect long-lived shrubs or trees can afford to move much more slowly through their hosts than those affecting annual plants. Viruses that can move into the seed and survive there have an important advantage in spread and survival.

1.1.3 Severity of the Disease

A virus that kills its host plant with a rapidly developing systemic disease is much less likely to survive than one that causes only a mild or moderate disease that allows the host plant to survive and reproduce effectively. There is probably a natural selection in the field against strains that cause rapid death of the host plant.

1.1.4 Mutability and Strain Selection

The extent to which a virus is able to mutate to produce strains that can cope effectively with changes in the environment may well affect survival and dispersal of the virus. The following are known to provide important selection pressures in certain circumstances: (1) the species and variety of host plant; (2) the species and race of invertebrate vectors; (3) seasonal variation in climate, particularly in temperature. Different strains of a virus may dominate in a crop in different seasons; and (4) agricultural practices may result in selection of certain strains of a virus, particularly in crops reproduced vegetatively.

1.1.5 Host Range

Viruses vary greatly in the range of species they are able to infect. Some viruses affecting strawberries appear to be confined to the genus *Fragaria*. Other viruses may be able to infect a wide range of plants; for example, cucumber mosaic virus (CMV) can infect species in over 100 families. Viruses with very narrow host ranges presumably survive either because their host is perennial, because it is vegetatively propagated, or because the virus is transmitted efficiently through the seed.

A diversity of hosts gives a virus much greater opportunities to maintain itself and spread widely. Viruses that have perennial ornamental plants such as dahlia and lilies as hosts, as well as other agricultural and horticultural species, have become widespread around the world. Weeds, wild plants, hedgerows, and ornamental trees and shrubs may also act as virus reservoirs. The actual importance of these various sorts of hosts for neighboring crops will depend on circumstances, particularly on the presence of active invertebrate vectors.

1.2 DISPERSAL

Dispersal of viruses, whether by airborne or soil-inhabiting vectors, by seed and pollen, or over long distances by human activities, plays a key role in the ecology of viruses.

1.2.1 Airborne Vectors

Taking plant viruses as a whole, the flying, sap-sucking groups of insect vectors, particularly the aphids, are by far the most important agents of spread and survival. The pattern of spread in the crop and the rate and extent of spread will depend on many factors, including (1) the source of the inoculum—whether it comes from outside the crop, from diseased individuals within the crop arising from seed transmission or through vegetative propagation, from weeds or other plants within the crop, or from crop debris; (2) the amount of potential inoculum available; (3) the nature and habits of the vector; for example, with aphids, whether they are winged transients or colonizers; (4) whether virus is nonpersistent, semipersistent, or persistent in the vector; (5) the time at which vectors become active in relation to the lifetime of the crop; and (6) weather conditions.

In the absence of seed transmission, scattered single infections within a field would suggest that infection is being brought in from an outside source by flying vectors. Clumping of infected plants in groups indicates spread of infection from sources within the crop. Transient winged forms of aphids that do not colonize the crop at all, but move from plant to plant, may be especially important with nonpersistent viruses. Though they may form a small proportion of the total population, these winged aphids may introduce virus from outside, or acquire it from infected plants within the crop and spread it rapidly as they move about seeking suitable food plants. Colonizers will also be important if they move about within the crop.

Various kinds of traps have been used to try to determine the extent of aphid movement and potential for virus infection in a particular crop and location. Ideally, from the point of view of virus transmission, we want to know the numbers of infective aphids that are flying, and which of these will land on the crop of interest. To determine this may be a very tedious and labor-intensive task, especially when nonpersistent viruses are involved. Trapped aphids have to be placed on healthy test plants as soon as possible after capture.

The *infection pressure* to which a crop is subjected can also be assessed by placing sets of bait plants in pots within the field for successive periods of a few days. The sets of plants are then maintained in the glasshouse and observed for infection. Figure 14.1 illustrates such a trial in relation to the prevalence of potential vector species. The frequency with which aphids were trapped clearly implicates the early

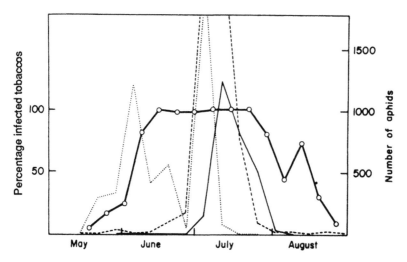

Figure 14.1 Infection pressure of PVY in a field of potatoes. Batches of 100 tobacco plants in pots were set out in the field for 7 days. The left axis and the solid line show the percentage of these plants that subsequently developed infection with PVY in the glasshouse. The number of aphids of three species trapped each week is given by the axis on the right. o ——— o, PVY[N] in tobacco in %; ———, *Myzus persicae;* ---, *Rhopalosiphum padi;* ...., *Cavariella aegopodii.* From van Hoof (1977).

flights of *Cavariella aegopodii* as being of prime importance in the spread of potato virus Y (PVY) in this particular field and season.

The fact that the majority of epidemiological studies have been carried out on aphid-borne viruses reflects the importance of this group of vectors. Other groups of airborne vectors such as hoppers and beetles are, of course, also of crucial importance in the ecology of the viruses they transmit.

1.2.2 Soilborne Viruses

Viruses with No Known Vectors

TMV is one of the few viruses transmitted through the soil to any significant extent without the aid of any known vector. The stability of this virus allows it to survive from season to season in plant remains, provided conditions are suitable. Viruses with no known vectors, such as members of the *Potexvirus* and *Tobamovirus* groups are widespread in soils of forest ecosystems in Germany.

Viruses with Fungal Vectors

The ecological implications of transmission by fungal vectors depend on the way in which the fungus carries the virus. Viruses such as soilborne wheat mosaic *Furovirus* are carried in the resting spores of the plasmodiophoromycete vectors. In these spores, the viruses may survive in air-dried soil for many years. Build-up of infection may take several seasons, but once established in a field, viruses with this type of vector may persist for many years, even in the absence of suitable plant hosts. Localized spread of these viruses by zoospores and resting spores through soil water is a major factor influencing movement. Viruses with this type of transmission tend to have rather narrow host ranges.

By contrast, viruses transmitted by *Olpidium brassicae* such as tobacco necrosis virus (TNV), and satellite TNV (STNV), are carried on the surface of the spores. Zoospores carrying virus probably survive only a few hours; however, virus that is free in the soil may be picked up and transmitted by newly released zoospores. Viruses transmitted by *Olpidium* do not survive in air-dried soil. In general, viruses of this type have wide host ranges, and probably survive in the soil by frequent transmission to successive hosts. Drainage water and movement of soil and root fragments are probably important in the spread of these viruses from one site to another.

Viruses with Nematode Vectors

The ecology of viruses with nematode vectors (tobraviruses and nepoviruses) differs considerably from that of viruses with airborne vectors. Nematodes are long-lived, may have wide host ranges, and are capable of surviving in adverse conditions and in the absence of host plants for considerable periods. Vector nematodes do not have a resistant resting stage, but can survive adverse soil conditions by movement through the soil profile. As soils become dry in summer or cold in winter, they move to the subsoil and return when conditions are favorable. Some of the viruses (e.g., arabis mosaic virus) may persist through the winter in the nematode vector.

Nematode-transmitted viruses usually have two other characteristics in common—a wide host range especially among annual weed species, and seed transmission in many of their hosts. Thus nematode vectors that lose infectivity during the winter may regain it in the spring from germinating infected weeds.

The spread of nematodes in undisturbed soil may be rather slow. Agricultural practices will increase distribution of the vectors during cultivation and probably in drainage or flood water. Farmworkers' footwear and machinery contaminated with infested soil may transmit nematodes and their associated viruses over both short and long distances. However, the pattern of infection observed in a crop may often depend largely on the vector and virus situation in the soil before the crop was planted. In a field already infected with nematodes and planted with a biennial or perennial crop, it may take 1 or 2 yr before the initial pattern of infection becomes apparent, since leaf symptoms may not show until about a year after infection. Subsequent localized spread may give rise to slowly expanding patches of infected plants within the field.

1.2.3 Seed and Pollen Transmission

Transmission by seed and pollen may be of crucial importance in the ecology of certain viruses. Survival in the seed can be particularly important for viruses that have only annual plants as hosts, and for those that have invertebrate vectors, such as nematodes, which normally move only slowly. Some viruses can persist in buried seeds for at least a year. Natural dispersal of virus-infected seeds by wind or water may be a factor in the transport of a virus.

1.2.4 Dispersal over Long Distances

Dispersal by Humans

There is little doubt that over the past few centuries, humans have been mainly responsible for the wide distribution around the world of many viruses that were previously localized in one or a few geographical areas. Viruses have been transported in plants or plant parts and perhaps occasionally in invertebrate vectors. Many of the virus diseases of potato and some of their vectors must have been brought to Europe with the potato from America and have since been spread to many other countries in tubers. Lettuce mosaic virus, because it is seed transmitted, has probably been distributed wherever this crop is grown. The fact that TMV can survive in infectious form in prepared smoking tobacco is probably sufficient to account for its presence wherever tobacco is grown commercially.

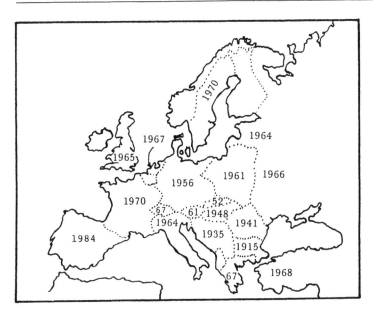

Figure 14.2 First reports of plum pox *Potyvirus* in Europe and Western Asia. Dates signify the year the virus was first recorded. Courtesy J. M. Thresh.

Other effects of human activities can be more precisely dated. Plum pox *Potyvirus* infects *Prunus* spp. and is transmitted in a nonpersistent manner by aphids. The disease was reported first in Bulgaria in 1915. It has since spread through Europe, as indicated in Fig. 14.2. Aphids spread the virus to nearby orchards. Long-distance spread has most probably been by means of vegetative plant material distributed by humans.

Airborne Vectors

Aphid vectors are important in the long-distance as well as local spread of viruses. This may be true for nonpersistent as well as persistent viruses. For example, detailed studies of aphid movements in the United Kingdom have shown that many important aphid species recolonize the whole island each year, as long as appropriate host plants are available. This movement involves distances of up to 1000 km, although not usually in one flight. Several successive colonizations may be involved.

On the other hand, under appropriate climatic conditions, a continuous long-distance journey may not be uncommon. High-altitude airstreams have almost certainly led to the mass transport of winged aphids from Australia to New Zealand, a distance over sea of some 2000 km. Lettuce necrotic yellows rhabdovirus (LNYV) and several vector species of aphids may have been introduced in this way. Leafhopper vectors may also travel long distances when blown by wind.

Waterborne Dispersal

Infectious plant viruses isolated from rivers and lakes include tobamoviruses; a potexvirus; TNV and STNV; tombusviruses; carnation mottle *Carmovirus;* CMV; and carnation ringspot *Dianthovirus.* Many of these viruses are very stable, and lack airborne vectors that would allow spread over long distances. They occur in high concentrations in infected plants and are released from infected roots, and can infect roots without a vector. Many have a wide host range. Most of the infectious virus probably moves in water while adsorbed onto organic and inorganic colloidal particles, especially clays. In this state, they would be substantially more resistant to inactivation than as free virus. It is known that some of these viruses can pass in an infectious state through the human alimentary tract, and there is circumstantial evidence that sewage may be a source of some waterborne viruses. Waterborne viruses may be a factor in the forest decline occurring in Europe, but further research is needed to resolve this question.

1.3 AGRICULTURAL AND HORTICULTURAL PRACTICES

Agricultural and horticultural practices have had, and continue to have, many and diverse effects on the incidence of virus diseases. Many practices have local effects on virus incidence. Others have effects on a country-wide or more global scale.

1.3.1 Practices Having Local Effects

Planting Date

Planting date may be critically important in relation to migrations of airborne vectors. Changes in planting practices that provide an overwintering crop may increase virus incidence.

Crop Rotation

The kind of crop rotation practiced may have a marked effect on the incidence of viruses that can survive the winter in weeds or volunteer plants. With certain crops, such as potatoes, volunteer plants that can carry viruses may survive in high numbers in a field for several years.

Soil Cultivation

Soil-cultivation practices may affect the spread and survival of viruses in the soil or in plant remains. Nematode and fungal vectors may be spread by movement of soil during cultivation, and their population sizes may be affected markedly.

Field Size

The influence of the size of field on the spread of a virus will depend to a great extent on the source of initial infection. If virus is coming from outside the crop, aggregation of the crop into large fields of compact shape will reduce infection from outside the crop to a minimum.

Plant Size and Planting Density

Airborne vectors bringing a virus into a crop from outside will infect a greater proportion of the plants in a given area when they are widely spaced than when they are close together. Large plants in a crop might be expected to become infected more readily by viruses with airborne vectors than small ones, since they are more likely to be visited by a vector.

Effects of Glasshouses

The use of glasshouses and polythene tunnels favors the survival of a stable virus such as TMV, since the structures remain at one site and are in use for intensive cultivation. On the other hand, they provide some protection against aphid-borne viruses. Glasshouses are normally used in regions with cool winters, and they will therefore favor the introduction into such regions of viruses such as tomato spotted wilt virus (TSWV), which are more adapted to tropical and subtropical climates.

Pollination Practices

The horticultural practice of planting mixtures of varieties as pollinators in orchards may favor the spread of pollen-borne viruses.

Nurseries as Sources of Infection

Nurseries, especially where they have been used for some years, may act themselves as important sources of virus infection.

1.3.2 Practices Having Large-Scale Effects

Plant Selection and Breeding

Plant selection and later plant breeding have given rise to new host genotypes with differing reactions to the existing viruses.

Grafting Procedures

Grafting procedures, apart from their role in spreading viruses, may have allowed the expression of new diseases by selection of virus strains, and by the introduction of viruses into previously uninfected species and varieties of plants.

Cultivation of Undisturbed Areas

Cultivation of areas previously undisturbed by humans brings new communities of plants together. These communities contain both the useful crop species and the weeds associated with cultivation. These agricultural communities may be neighbors of the indigenous flora. In this situation, many opportunities exist for the movement of viruses and vectors into previously uninfected host species and for the emergence of new strains of a virus, and of new virus vectors.

Movement of Crop Plants to Distant Countries

As well as distributing viruses around the world, humans have moved crop species to distant countries, often with disastrous consequences, as far as virus infection is concerned. Plant species that were relatively virus-free in their native land may become infected with viruses that have long been present in the countries to which they were moved for commercial purposes. It is often difficult to prove a sequence of events of this sort, especially if the movement took place before virologists could investigate and record events. Cacao swollen shoot virus is a significant example because of the importance of the cacao crop in several West African economies. Cacao was transferred from the Amazonian jungle to West Africa late last century, and since then, major commercial production has developed there. The swollen shoot disease, first reported in cacao in 1936, was very probably transmitted from natural West African tree hosts of the virus by the mealybug vectors that are indigenous to the region.

Movement of plant species between countries and continents has been carried out with increasing frequency during the last two centuries. Agriculture in India, North America, and Australasia is almost totally dependent on introduced crop plants. Thus, there has been ample opportunity for events such as that outlined for cacao to occur.

Monocropping

Cultivation of a single crop, or at least a very dominant crop, over a wide area continuously for many years may lead to major epidemics of virus disease, especially if an airborne vector is involved. Soilborne vectors may also be important from this point of view, for example, with grape vine fanleaf virus in vineyards, where the vines are cropped for many years. Monocropping may also lead to a build-up of crop debris, and the proliferation of weeds that become associated with the particular crop.

An outstanding example of the effects of monocropping is the rise in importance of certain virus diseases of rice and of their hopper vectors as a consequence of the *green revolution* that began about 25 yr ago. New rice varieties introduced in tropical and subtropical areas are a major factor in the greatly improved yields that have been achieved. However, the gains have been seriously impaired by the increased prevalence of several serious virus diseases of rice, and also of their hopper vectors. An important example is rice grassy stunt *Tenuivirus* (RGSV) and rice ragged stunt plant reovirus and their plant hopper vector (*Nilaparvata lugens*), known as rice brown planthopper (Thresh, 1989). As well as transmitting these viruses, the hopper can itself cause serious crop damage. The first new rice cultivars released between 1966 and 1971 were all susceptible to both vector and RGSV. Effective sources of resistance to both were found, and incorporated into new varieties that were released in 1974 and 1975. These were at first grown widely and successfully in the Philippines, Indonesia, and elsewhere. However, severe infestations with the planthopper were reported within 2 to 3 yr, and in 1982 and 1983, a resistance-breaking strain of RGSV was reported. No new source of virus resistance has been found. A new source of hopper resistance was identified and incorporated into new varieties, and it was successful for a few years until it, too, broke down with the emergence of a new hopper biotype.

1.3.3 Conclusion

Agricultural practices may have played a role in unconscious selection of mild strains of a virus but, generally speaking, the foregoing factors combine in various ways continually to produce new and unstable ecological situations in which outbreaks of disease can occur in agricultural and horticultural crops. On the other hand, in regions where a crop has been grown continually for long periods under conditions favoring infection with a virus, resistant or tolerant cultivars may have developed without a conscious selection procedure. For example, many cultivars of barley from the Ethiopian plateau have been found to have a high degree of resistance to barley yellow dwarf virus (BYDV) and barley stripe mosaic virus (BSMV). It is probable that barley has been grown in this region for millenia.

2 PHYSICAL FACTORS

2.1 RAINFALL

Rainfall may influence both airborne and soilborne virus vectors. The timing and extent of rainfall may alter the influence it has on vector populations. For example, some rainfall or high humidity is necessary for the build-up of whitefly populations, whereas continuous heavy rainfall may be a factor in reducing the size of such populations. Similarly, heavy rainfall just after airborne aphids have arrived in a crop may kill many potential vectors, thus reducing subsequent virus incidence.

2.2 WIND

Wind may be an important factor, not only in assisting or inhibiting spread of viruses by airborne vectors, but also in determining the predominant direction of spread. Windbreaks may affect the local incidence of vectors and viruses in complex ways. Winged aphids tend not to fly when wind speed is too great, although their direction of flight can be influenced by the prevailing wind. At low wind speeds some species may fly with the wind, and others against it. The direction of movement of leafhoppers and whiteflies may also be markedly influenced by wind speed and direction. Mealybugs and mites can also be carried some distance by wind.

2.3 AIR TEMPERATURE

Air temperature may have marked effects on the rate of multiplication and movement of airborne virus vectors. For example, winged aphids tend to fly only when conditions are reasonably warm. However, very high temperatures may be particularly effective in reducing certain aphid populations.

2.4 SOIL

Conditions in the soil can influence the incidence of virus disease in various ways. Highly fertile soils tend to increase the incidence of virus disease. Soil conditions can have a marked effect on the survival of TMV in plant debris. Moist well-aerated soils favor inactivation of the virus, compared with dry, compacted, or waterlogged soils. Soil temperature may have a marked effect on the transmission of viruses by nematodes. The optimal temperature and temperature range may vary with different viruses, hosts, and nematode species. However, in spite of seasonal fluctuations in temperature, populations of nematodes in the field may remain quite stable for years.

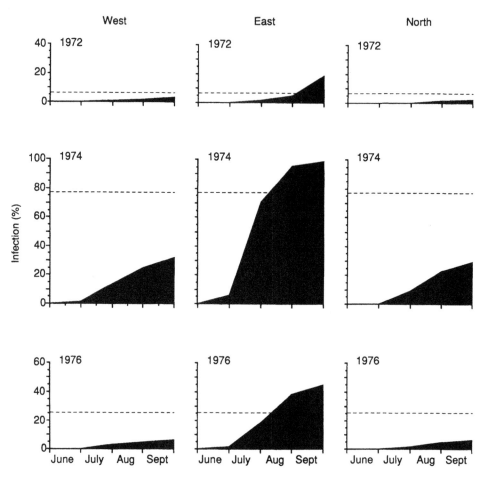

Figure 14.3 Variation in virus incidence with year and location. Mean monthly incidence of sugar beet yellows disease due to BYV and BMYV in different regions of England in years with greatly differing amounts of spread. Data supplied by G. D. Heathcote (Brooms Barn Experimental Station) for representative crops sown annually in March or April in eastern England (Suffolk, Essex, and Cambridgeshire), in wetter areas to the west (Shropshire and Worcestershire) and in cooler areas to the north (Yorkshire). The dashed lines indicate the mean incidence of yellows at the end of September as calculated for the entire English crop. From Thresh (1983).

2.5 SEASONAL VARIATION IN WEATHER AND THE DEVELOPMENT OF EPIDEMICS

The wide annual variation that can occur in the incidence of virus disease in an annual crop is illustrated in Fig. 14.3 for beet yellowing viruses in sugar beet in England. The size and time of migration of aphids into the crop may account for some of the variation shown in Fig. 14.3, but other factors must have been involved.

3 SURVIVAL THROUGH THE SEASONAL CYCLE

Viruses can survive a cold winter period or a dry summer season in various ways. Some have more than one means of survival.

1.   Many viruses readily survive from season to season in the same host plant or propagating material from it. These include viruses in perennial hosts, in tubers, runners, and so on, and viruses that are transmitted through the seed. Similarly, crops stored over winter in the field, such as mangolds, may act as a source of both viruses and vectors.

2.   Viruses with a wide host range are well adapted to survive, provided their hosts include perennial species, a group of annual species with overlapping growing seasons, or species in which seed transmission occurs.

3.   Biennial or perennial wild plants, ornamental trees and shrubs, or weed hosts such as *Plantago* spp. may be important sources for overwintering or oversummering of many viruses.

4.   Leafhopper-transmitted viruses that pass through the egg of the vector may overwinter in the egg, or in young nymphs.

5.   TMV and a few other viruses can survive through the winter under appropriate conditions in plant refuse, and perhaps free in the soil. TMV can also survive in plant litter left in curing sheds and so on.

6.   Viruses carried within the resting spores of their fungal vectors may survive for long periods in soil.

7.   Agricultural practices may allow the virus to survive in successive crops of the same plant grown in the same locality throughout the year. This may happen where the climate is suitable for production cropping throughout the year, where seed crops of the species are grown through the winter and overlap with production crops, or with wheat, where spring and winter crops are grown in the same area.

4 CONCLUSION

We can see from this brief account that many factors may affect virus spread and survival in the field. Each virus has some combination of properties that allows it to persist more or less effectively. TMV has no significant invertebrate vector, but survives because of its high resistance to inactivation, the high concentration it generally reaches in its host, the ease with which it can be mechanically transmitted, and its wide host range. By contrast, viruses that multiply in the insect vectors and are transmitted through the egg have a complete alternative to the plant host for survival.

TSWV is an example of a very unstable virus that is extremely successful in the field. It can retain infectivity only for periods of hours under normal conditions outside the plant. It has a specialized type of vector (species of thrips that can acquire virus only as nymphs), and yet it is widespread around the world and common within individual countries. Its success probably depends on several factors: (1) its very wide host range, including many perennial symptomless carriers that provide year-round reservoirs of the virus; (2) the widespread distribution of species of thrips capable of transmitting the virus; and (3) the existence of a wide range of strains in strain mixtures, and the probable occurrence of recombination or gene reassortment, allowing rapid adaptation of the virus to new hosts to give relatively mild disease.

FURTHER READING

Duffus, J. E. (1971). Role of weeds in the incidence of virus diseases. *Annu. Rev. Phytopathol.* 9, 319–340.

Maramorosch, K., and Harris, K. F. (eds.) (1981). "Plant Diseases and Vectors: Ecology and Epidemiology." Academic Press, New York.

Matthews, R. E. F. (1991). "Plant Virology," 3rd Ed. Academic Press, New York.

McLean, G. D., Garrett, R. G., and Ruesink, W. G. (1986). "Plant Virus Epidemics. Monitoring, Modeling, and Forecasting Outbreaks." Academic Press, New York.

Thresh, J. M. (1989). Insect-borne viruses of rice and the green revolution. *Trop. Pest Manag.* 35, 264–272.

ECONOMIC IMPORTANCE AND CONTROL

1 ECONOMIC IMPORTANCE

Accurate global figures for crop losses due to viruses are not available.
However, some idea of the scale can be obtained by considering that

plant disease losses worldwide have been assessed as high as (United States) $60 billion per year. It is generally accepted that, of the various plant pathogens, viruses rank second only to fungi with respect to the disease losses they cause. Losses differ greatly between and within different countries and regions. Not infrequently, it is in developing countries, that are dependent on one or a few major crops, that control measures are most inadequate and crops are affected most seriously.

Viroids as a group of agents are limited in the range of crop species they affect compared with viruses. Nevertheless, they can cause severe problems in specific crops, for example, cadang-cadange disease of coconuts, potato spindle tuber disease, and chrysanthemum stunt disease. The cadang-cadang viroid has killed over 30 million coconut trees over about 40 yr, and losses of about half a million trees each year continue.

1.1 MEASUREMENT OF LOSSES

It is often difficult to obtain reliable data either on direct crop losses due to virus disease or on the cost of control. There is wide variation in the extent of direct losses with different crops in different seasons and in various regions, so that average figures are not particularly useful. We can distinguish three general kinds of situations in which losses occur.

1. Perennial crops, often trees, where a lethal or crippling disease can have very serious consequences because of the time and land invested in such crops. Examples are the tristeza disease of citrus in most continents, and the swollen shoot disease of cocoa in West Africa.
2. Annual crops sown from seed, for example, many vegetables and grains, where severity of a particular disease may fluctuate very greatly from season to season (see Fig. 14.3).
3. Vegetatively reproduced plants, for example, potatoes and many fruit and ornamental species. In such crops, mild virus infection may be very widespread, often occurring in every plant, and reducing performance every year by a relatively small amount, perhaps about 10%.

Field trials to assess crop losses need careful design if the results are to reflect actual losses in commercial plantings. For example, yield reductions in individual cassava plants due to African cassava mosaic

Geminivirus were found to be much greater in plots where diseased plants were intermixed with healthy ones and subjected to interplant competition, than when they were assessed in separate healthy and infected plots.

Additional complexities in assessing effects on yield are involved when the species concerned is grown as a member of an ecological community containing other species not affected by the virus in question. Losses may be particularly severe in certain regions of the world. For example, it has been estimated that although 11% of the world area devoted to maize is in tropical Africa, that region produces only 3% of the total crop. Maize streak *Geminivirus* is often a major contributor to this lower productivity.

1.2 BIOLOGICAL AND PHYSICAL FACTORS
   AFFECTING LOSSES

The extent to which yield is reduced will depend in any particular year and locality on many factors, including the varieties of host plants and the strains of the virus present, the incidence and activity of any vectors, the time at which infection occurs, the nutritional state of the crop, the weather, and the presence of other parasites, including another virus, or a satellite RNA. When new varieties are introduced into a region, they may suffer severe losses compared with varieties grown previously. This is because viruses may be present in the region for which the new variety has no resistance or tolerance.

Virus infection often reduces the quality of agricultural or horticultural product as well as the quantity. In an open market, the relationship between weight loss and financial loss for individual growers is further complicated by the question as to whether most growers supplying a market had similar losses. If losses due to a disease are fairly evenly distributed among the growers, price rises will tend to compensate for the losses sustained. Where only a few growers supplying a market are affected, the financial loss to these growers may be much more severe for the same loss of crop.

The cost of any control program must be measured in relation to the yield gain. Losses due to virus disease must also be viewed in relation to the economy of the country concerned, and the importance of the product in international trade.

2 DIAGNOSIS

Correct diagnosis of the virus or viruses causing a particular disease is essential if effective control measures are to be developed. Many of the procedures described in Chapters 3 and 4 can assist in diagnosis. The most important procedures are as follows.

2.1 DISEASE SYMPTOMS AND HOST RANGE

Disease symptoms in the crop plant in the field may be very misleading. For example, virus disease in lettuce is caused by about 14 viruses with aphid, leafhopper, thrip, nematode, or fungus vectors. Many of these viruses produce brown necrotic spots or bronzing on the leaves, followed by chlorosis and stunting. However, symptoms on a set of experimental indicator hosts may be very useful, and often can be essential for diagnosis. Host range on a set of indicator species or cultivars may help to distinguish viruses and strains within a group. Cytological effects studied by light or electron microscopy may also be useful.

2.2 CROSS-PROTECTION

As discussed in Chapter 12, Section 3.3.4, the phenomenon of cross-protection can be useful in diagnosis.

2.3 METHODS OF TRANSMISSION

The ways in which a virus is transmitted may give a useful guide to diagnosis, but on their own, they are not definitive.

2.4 STRUCTURE OF THE VIRUS

The structure of the virus particle revealed by electron microscopy of purified virus particles or in thin sections of infected tissue may indicate the group or family to which a virus belongs, especially for those viruses with a characteristic morphology, such as the geminiviruses and the rhabdoviruses.

2.5 SEROLOGICAL PROCEDURES

The serological tests described in Chapter 3 provide rapid and convenient methods for the identification of plant viruses, the main advantages being as follows: (1) the specificity of the reaction allows virus to be identified in the presence of host material or other impurities; (2) results are obtained in a few hours or overnight, compared with days for infectivity tests; (3) some serological detection procedures are more sensitive than infectivity measurements; (4) serological tests are particularly useful with viruses that are not sap transmissible; and (5) antisera can be stored, and comparable tests made over periods of years, and in different laboratories.

Currently the most useful forms of serological tests used in diagnosis are (1) enzyme-linked immunosorbent assay (ELISA) tests; (2) dot-blot immunobinding assays; (3) electrophoresis followed by electroblot immunoassay; and (4) serologically specific electron microscopy, as discussed in Chapters 3 and 12.

2.6 NUCLEIC ACID HYBRIDIZATION

Various techniques using nucleic acid hybridization, discussed in Chapters 4 and 12, are becoming increasingly useful in the diagnosis of virus diseases.

3 CONTROL MEASURES

Most of the serious virus disease problems around the world are the direct or indirect result of human activities. Important activities leading to epidemics of virus disease include (1) introduction of viruses into new areas through transport of infected seed or vegetative material; (2) introduction of virus vectors into new areas; (3) introduction of a new variety of a crop into an area when that variety is especially susceptible to a virus already present; (4) use of monocultures, that is, the planting of genetically uniform crops over large areas, replacing traditional polycultures; (5) use of irrigation to prolong the cropping season; (6) repeated use of the same fields for the same crop; and (7) increased use of fertilizer and herbicides or other forms of weed control.

In recent years, the most active areas of research into the control of virus diseases have been (1) the breeding of resistant or immune cultivars by classical genetic procedures; (2) the control of vectors by various strategies; (3) the production of virus-free stocks of seed and vegetative propagules; and (4) the production of transgenic plants containing viral genes that confer resistance to the virus.

Data for many of the control measures discussed in this chapter are derived from laboratory and field trials. Because of the many variables, and the large number of countries involved, it is often difficult to assess the extent to which any particular procedure or set of procedures has actually been adopted into commerical practice on a regular basis.

More and more attention is being given to the possibilities of integrated control involving several strategies. Several of these are noted in this chapter. Again, it is often difficult to assess whether these are actually being effectively used in the field, or whether they remain optimistic dreams.

When adequate facilities and expertise are available, a multidisciplinary approach may be useful. Cassava is the third largest plant source of calories in the world, being used for human and animal food and for the production of industrial ethanol. Recently an international project has been set up to attempt to control the virus diseases in this crop on several continents (Fauquet and Beachy, 1989).

3.1 REMOVAL OR AVOIDANCE OF SOURCES OF INFECTION

3.1.1 Removal of Sources of Infection in or near the Crop

It is obvious that there will be no virus problem if the crop is free of virus when planted and there is no source of infection in the field, and none near enough to allow virus spread into the crop. The extent to which it will be worthwhile to attempt to eliminate sources of infection in the field can be decided only on the basis of a detailed knowledge of such sources and of the ways in which the virus is spreading from them into a crop.

Sources of infection include

1. *Living hosts of the virus*   These may include weed or ornamental species or crop plants of the same or another species. The extent to which removal of other hosts of a virus from an area may succeed will depend largely on the host range of the virus. It may be practicable to control alternative hosts where the virus has a narrow host range, but

with others, such as cauliflower mosaic virus (CMV) and tomato spotted wilt virus (TSWV), the task is usually impossible.

2. *Plant remains* Plant remains in the soil, or attached to structures and implements used for propagation and cultivation may harbor a stable mechanically transmitted virus such as tomato mosaic virus (ToMV) and act as a source of infection for the next crop.

3. *Roguing* If the spread is occurring rapidly from sources outside the crop, roguing the crop will have no beneficial effect. However, if virus spread is slow relative to the lifetime of the crop and occurring mainly from within the crop, then roguing may be worthwhile, especially early in the season for annual crops.

4. *Contaminated workers and equipment* For some mechanically transmitted viruses, and particularly for tobacco mosaic virus (TMV) or ToMV, human activities during cultivation and tending of a crop are a major means by which the virus is spread. Once TMV or ToMV enters a crop like tobacco or tomato, it is very difficult to prevent its spread during cultivation, and particularly during such processes as lateraling and tying of plants. Control measures consist of treatment of implements and washing of the hands with a solution such as 3% trisodium orthophosphate. Workers' clothing may become heavily contaminated with TMV and thus spread the virus by contact.

While TMV is the most stable of the mechanically transmitted viruses, others can also be transmitted more or less readily on cutting knives, and other tools. These include tulip-breaking *Potyvirus*, cymbidium mosaic *Potexvirus*, potato virus X, and potato spindle tuber viroid. These mechanically transmitted agents may be a particular problem in glasshouse crops, where lush growth, close contact between plants, high temperatures, and frequent handling are important factors facilitating virus transmission.

3.1.2 Virus-Free Seed

Where a virus is transmitted through the seed, such transmission may be an important source of infection, since it introduces the virus into the crop at a very early stage, allowing infection to be spread to other plants while they are still young. In addition, seed transmission introduces scattered foci of infection throughout the crop. Where seed infection is the main or only source of virus, and where the crop can be grown in reasonable isolation from outside sources of infection, virus-free seed may provide a very effective means for control of a disease. The most successful example is the control of lettuc mosaic *Potyvirus* through the use of virus-free seed.

Tomato seed from ToMV-infected tomatoes carries the virus on the surface of the seed coat. As the seed germinates, virus contaminates the cotyledons and is inoculated into the plant by handling during pricking out. Seed can be cleaned by extraction in HCl, or treatment with tri-sodium orthophosphate or sodium hypochlorite.

3.1.3 Virus-Free Vegetative Stocks

For many vegetatively propagated plants, the main source of virus is chronic infection in the plant itself. With such crops, one of the most successful forms of control has involved the development of virus-free clones—that is, clones free of the particular virus under consideration. Two problems are involved. First, a virus-free line of the desired variety with good horticultural characteristics must be found. When the variety is 100% infected, attempts must be made to free a plant or part of a plant, from the virus. Second, having obtained a virus-free clone, a foundation stock or *mother* line must be maintained virus-free, while other material is grown up on a sufficiently large scale under conditions where reinfection with the virus is minimal or does not take place. These stocks are then used for commercial planting.

Methods for Identification of Virus-Free Material

Visual inspection for symptoms of virus disease is usually quite inadequate when selecting virus-free plants. Appropriate indexing methods are essential. A variety of methods is available, and the most suitable will depend on the host plant and virus. For many viruses, especially those of woody plants, the rather laborious process of graft-indexing to one or more indicator hosts is essential. Mechanical inoculation to indicator hosts can be used with some viruses, but other methods of diagnosis now rival infectivity tests in their sensitivity.

Methods for Obtaining Virus-Free Plants

Naturally-Occurring Virus-Free Plants   Occasionally, individual plants of a variety, or plants in a particular location may be found to be free of the virus.

Heat Therapy   Heat treatment has been a most useful method for freeing plant material from viruses. Many viruses have been eliminated from at least one host plant by heat treatment.

Two kinds of plant material have been used. Dormant plant parts, such as tubers or budwood, can generally stand higher temperatures than growing tissues, and the method probably depends on direct heat inactivation of the virus. Growing plants are much more widely used, and hot air rather than hot water is applied. Temperatures in the range of 35° to 49° C for periods of weeks are commonly employed. This form of treatment gives a better survival rate for growing plant material. Details of the treatment vary widely and have to be worked out empirically for each host–virus combination. Very frequently, small cuttings are taken from the shoot tips immediately after the heat treatment, as these may be free of virus when the rest of the plant is not.

Meristem Tip Culture   Culture of meristem tips has proved an effective way of obtaining vegetatively propagated plants free from certain viruses. Meristem tip culture has been defined as aseptic culture of the apical meristem dome plus the first pair of leaf primordia. This piece of tissue is about 0.1- to 0.5-mm long in different plants. The minimum size of tip that will survive varies with different species. The smaller the excised tips are at the time of removal, the better the chance that they will give rise to virus-free plants, although they are more difficult to regenerate.

Only a proportion of meristem tip cultures yield virus-free plants. It is not always clear to what extent the success of the method depends on (1) the regular absence of virus from meristem tissue, some tips being accidentally contaminated; (2) some meristem regions in the plant containing virus and others containing none; or (3) virus present in the meristem being inactivated during culture on the synthetic medium.

Tissue Culture   Culture of single cells or small clumps of cells from virus-infected plants may sometimes give rise to virus-free plants.

Chemotherapy   Attempts to free infected plant material of a virus solely by the application of antiviral chemicals have been disappointing. There have been several reports of such cures, but they have often been based on very small numbers of plants, or the results have been open to other interpretations.

Importance of Using Selected Clones

The selection of horticulturally desirable clonal material free of virus infection is an important aspect of any program. Selection may need to be carried out both before and after the material has been rendered virus-free by any of the preceding procedures.

The Importance of Adequate Virus Testing

Plants found to be apparently free of the virus at an early stage of growth may develop infection after quite a long incubation, so that in practice, it is very important that apparently virus-free plants be tested over a period before release.

3.1.4 Propagation and Maintenance of Virus-Free Stocks

Once suitable virus-free material of a variety is obtained, it has to be multiplied under conditions that preclude reinfection of the nucleus stock, and that allow the horticultural value of the material to be checked with respect to quality of type. Nucleus stock is then further multiplied for commercial use. This multiplication and distribution phase requires a continuing organization for checking on all aspects of the growth and sale of the material. A classic example is the potato-certification scheme in Great Britain, which, over 30 years or so, led to a two- to threefold increase in yield, much of which was owing to decreased incidence of virus infection. Tested Foundation Stocks, which are virtually free of virus, are grown in isolation in parts of Scotland that are unfavorable to early aphid migration and colonization. High-grade stocks are grown from this seed elsewhere in Britain, in areas selected because of the low incidence of aphid vectors. Health of the stocks is regularly checked.

Many such schemes are now in operation around the world for a variety of agricultural and horticultural plants, including stone and pome fruits, grapevines and berry fruits, as well as potatoes.

3.1.5 Modified Planting and Harvesting Procedures

Breaking an Infection Cycle

Where one major susceptible annual crop or group of related crops is grown in an area, and where these are the main hosts for a virus in that area, it may be possible to reduce infection very greatly by ensuring that there is a period when none of the crop is grown. This type of control will be difficult in areas where a major food plant is traditionally grown in an overlapping succession. Rice is the major example. The increased use of irrigation in the tropics and protected cropping in cold temperate regions both limit the options available to growers for breaking the infection cycle.

Changed Planting Dates

The effect of infection on yield is usually much greater when young plants are infected. Furthermore, older plants may be more resistant to infection, and virus moves more slowly through them. Thus, with viruses that have an airborne vector, the choice of sowing or planting date may influence the time and amount of infection.

Plant Spacing

Closer plant spacing may sometimes be a practical measure to reduce the percentage of virus-infected plants and thus increase yield, provided that increased plant competition does not negate the effect.

Isolation of Plantings

Where land availability and other factors permit, isolation of plantings from a large source of aphid-borne infection might give a useful reduction in disease incidence.

3.1.6 Prevention of Long-Distance Spread

Most agriculturally advanced countries have regulations controlling the entry of plant material to prevent the entry of diseases and pests not already present. Many countries have regulations aimed at excluding specific viruses and their vectors, sometimes from specific countries or areas. The setting up of quarantine regulations and providing effective means for administering them is a complex problem. Economic and political factors frequently have to be considered. Quarantine measures may be well worthwhile with certain viruses, such as those transmitted through seed, or in dormant vegetative parts such as fruit trees and bud wood.

The value of quarantine regulations will depend to a significant degree on the previous history of plant movements in a region. For example, active exchanges of ornamental plants between the countries of Europe has been going on for a long period, leading to an already fairly uniform geographical distribution of viruses infecting this type of plant.

An effective quarantine system requires an effective technological infrastructure that is capable of detecting viruses under a variety of circumstances. Furthermore, quarantine must be concerned not only with important crop species but also with species, unimportant in themselves, that may harbor viruses infecting major crops. For some virus diseases, natural spread over very long distances by invertebrate vectors may negate the effects of quarantine measures.

3.2 CONTROL OR AVOIDANCE OF VECTORS

Control or avoidance of invertebrate or fungal vectors is of prime importance for the limitation of crop damage by viruses that have such vectors. For this reason, more than one method for control may be used simultaneously for a particular crop, location, virus, and vector.

3.2.1 Airborne Vectors

Insecticides

A wide range of insecticides is available for the control of insect pests on plants. To prevent an insect from causing direct damage to a crop, it is necessary to reduce the population only below a damaging level. Control of insect vectors to prevent infection by viruses is a much more difficult problem, as relatively few winged individuals may cause substantial spread of virus. Contact insecticides would be expected to be of little use unless they were applied very frequently. Persistent insecticides, especially those that move systemically through the plant, offer more hope for virus control. Viruses are often brought into crops by winged aphids, and these may infect a plant during their first feeding, before any insecticide can kill them. When the virus is nonpersistent, the incoming aphid when feeding rapidly loses infectivity anyway, so that killing it with insecticide will not make much difference to infection of the crop from outside. On the other hand, an aphid bringing in a persistent virus is normally able to infect many plants, so that killing it on the first plant will reduce spread by this individual. It will also limit colonization and build-up of infected alate vectors within the crop.

With leafhopper vectors, the speed at which they are killed after feeding on a treated plant may be an important factor for virus control. From the point of view of the economics of insecticide use, it may be important to forecast whether applications are necessary, and, if so, to define optimal times for treatment.

Oil Sprays

In certain circumstances, spraying plants with oil may reduce the incidence of nonpersistent viruses. Such sprays do not prevent the spread of persistent viruses. The use of oil sprays has not become widespread.

Nonchemical Barriers against Infection

Various nonchemical barriers against infection have given positive results in field experiments. These include (1) a tall cover crop, such as maize over cucurbits; (2) reflective aluminum polythene mulches; (3) sticky yellow polythene strips around the crop; and (4) coarse white nets above the crop. However, none of these appears to have moved into commercial practice on a significant scale.

Plant Resistance to Vectors

In recent years there has been substantially increased interest in breeding crops for resistance to insect pests, as an alternative to the use of pesticide chemicals. This has been the result of various factors, including emergence of resistance to insecticides in insects, the costs of developing new pesticides, and increasing concern regarding environmental hazards and the effects on natural enemies. Along with these developments, there has been increased activity in breeding for resistance to invertebrates that are virus vectors. Some virus vectors are not pests in their own right, but others, especially leaf- and planthoppers, may be severe pests. In this situation, there is a double benefit in achieving a resistant cultivar, sometimes with striking improvements in performance.

The basis for plant resistance to virus vectors is not always clearly understood, but some factors have been defined, for example, heavy leaf pubesence. In general terms, two kinds of resistance are relevant to the control of vectors. First, *nonpreference,* which involves an adverse effect on vector behavior resulting in decreased colonization; and second, *antibiosis,* which involves an adverse effect on vector growth, reproduction, and survival after colonization has occurred. These two kinds of factors may not always be readily distinguished.

There may be various limitations on the use of vector-resistant cultivars: (1) sometimes such resistance provides no protection against viruses; (2) if a particular virus has several vector species, or if the crop is subject to infection with several viruses, breeding effective resistance against all the possibilities may not be practicable, unless a nonspecific mechanism is used (e.g., tomentose leaves); (3) perhaps the most serious problem is the potential for new vector biotypes to emerge following widespread cultivation of a resistant cultivar, as can also happen follow the use of insecticides.

This difficulty is well illustrated by the recent history of the rice brown planthopper. With the advent of high-yielding rice varieties in Southeast Asia in the 1960s and 1970s, the rice brown planthopper (*Nilaparvata lugens*) and rice grassy stunt *Tenuivirus,* which it transmits, became serious problems. A succession of resistance genes incorporated into rice varieties have been overcome by the emergence of new biotypes of the planthopper.

In spite of these difficulties, and the problems associated with the identification of plants with resistance to vectors, it seems certain that substantial efforts will continue to be made to improve and extend the range of crop cultivars with resistance to virus vectors. A combination of resistance to the vector and to the virus will frequently be the goal.

Control by Predators or Parasites

Under some circumstances, predators or parasites might play a part in limiting spread of a virus, but with aphids in particular, they will usually have little effect if they arrive after the early migratory individuals that are so important for virus spread.

3.2.2 Soilborne Vectors

Most work on the control of viruses transmitted by nematodes and fungi has centered on the use of soil sterilization with chemicals. However, several factors make general and long-term success unlikely: (1) huge volumes of soil may have to be treated; (2) a mortality of 99.99% still leaves many viable vectors; and (3) use of some of the chemicals involved has been banned in certain countries, and such bans are likely to be extended. In any event, chemical control will be justified economically only for high-return crops, or crops that can remain in the ground for many years.

Nematodes

In principle, the control of viruses transmitted by nematodes should be possible by treatment of infested soil with nematicides. Movement and dispersal of nematodes are generally slow, so that one treatment might be expected to remain effective for longer periods than with airborne vectors. On the other hand, infective nematodes may occur at depths of 100 cm or more.

At present, soil fumigation appears to be the only effective measure for the control of *Xiphinema index*, the vector of grapevine fanleaf virus. If proper care is taken with the fumigation process, vineyards may be grown successfully for 15 to 20 years. With vines, it is very important to kill all nematodes before replanting. The process may be assisted by treating old vines with a herbicide so that deep roots, and the nematodes they harbor, are killed before fumigation treatment of the upper soil.

Fungi

In general, attempts to control infection with viruses having fungal vectors by application of chemicals to the soil have not been successful.

3.3 PROTECTING THE PLANT FROM SYSTEMIC DISEASE

Even if sources of infection are available, and the vectors are active, there is a third kind of control measure available—protecting inoculated plants from developing systemic disease. Genetic protection is the most important means of protecting crops in this way.

3.3.1 Genetic Protection

Immunity

Although many searches have been made, true immunity against viruses and viroids, which can be incorporated into useful crop cultivars, is a rather uncommon phenomenon.

Resistance

Where genes for resistance can be introduced into agriculturally satisfactory cultivars, breeding for resistance to a virus provides one of the best solutions to the problem of virus disease. Attempts to achieve this objective have been made with many of the virus diseases of major importance. Genes for resistance have often been found, but it has frequently proved very difficult to incorporate such factors into useful cultivars.

Occasionally, resistance to a virus has been discovered even in a plant in which most known cultivars are highly susceptible. However, the designation of a variety or genotype as resistant to a particular virus must always be provisional, because it is always possible that a mutant of the virus will arise that can overcome the plant resistance. Thus, the major problem with resistance of any sort as a control measure is its durability. How long can it be deployed successfully before a resistance-breaking (virulent) strain of the virus emerges?

The costs of a breeding program must be weighed against the possible gains in crop yield. Many factors are involved, such as (1) the seriousness of the viral disease in relation to other yield-limiting factors; (2) the *quality* of the available resistance genes. For example, resistance genes against CMV are usually *weak* and short-lived, which may be due, at least in part, to the many strains of CMV that exist in the field; (3) the importance of the crop. Compare, for instance, a minor ornamental species with a staple food crop such as rice; and (4) crop quality. Good virus resistance, giving increased yields, may be accompanied by poorer quality in the product.

Sources of Resistant Genotypes   Quite frequently no resistance has been found for particular crops and viruses. Possible sources of resistance include an existing line or variety of the crop; individual plants showing good growth in a severely infected field; related species from other localities; artificially induced mutants; culture of cells as protoplasts to induce somaclonal variation; protoplast fusion of cells belonging to different genera; and intergeneric hybridization by classical methods. Most of these potential sources have not yet yielded commerically successful cultivars.

Low Seed Transmission   For those viruses affecting annual crops that are transmitted through the seed, resistance to seed transmission may be an important method for limiting infection in the field. For example, a single recessive gene for resistance to seed transmission of barley stripe mosaic (BSMV) in barley has been introduced into the variety Mobet.

Adequate Testing of Resistant Material   Varieties or lines showing resistance in preliminary trials must then be tested under a range of conditions. Important factors to be considered are strains of the virus, climatic conditions, and inoculum pressure.

The Need for Resistance to Multiple Pathogens   The difficulties in finding suitable breeding material are compounded when there are strains of not one, but several viruses to consider. Cowpeas in tropical Africa are infected to a significant extent by at least seven different viruses. In such circumstances, a breeding program may utilize any form of genetic protection that can be found. There is, of course, the further problem of combining these factors with multiple resistance to fungal and bacterial diseases.

Durability of Resistance   For some crop plants and viruses, resistance has proved to be remarkably durable. Thus, the resistance to bean common mosaic *Potyvirus* found in Corbett Refugee bean has been bred into most varieties of dry and snap beans in the United States, and resistance has not broken down after 45 yr. However, it is probable that most cultivars bred for resistance to a particular virus will become susceptible, as new strains of the virus emerge. For example the Tm-2^2 gene in tomato has been very useful for protection against ToMV for more than 10 yr. However, ToMV strains have now been found that can, in the laboratory, overcome the resistance due to the TM-2^2 gene. It is probably only a matter of time before these become prevalent in commercial glasshouses.

We must conclude that, on present knowledge, for most crops and most viruses, the search for new sources of resistance, and their incorporation into useful cultivars will be a continuing and very long-term process, as it is with many fungal and other parasitic agents.

Tolerance

Where no source of genetic resistance can be found in the host plant, a search for tolerant varieties or races is sometimes made. However, tolerance is not nearly so satisfactory a solution as genetic resistance for several reasons: (1) infected tolerant plants can act as a reservoir of virus for other hosts, and large populations of such plants may appear; (2) double infection with an unrelated virus may lead to severe disease; and (3) virus infection may increase susceptibility to cellular parasites.

However, tolerant varieties may yield very much better than standard varieties where virus infection causes severe crop losses and where large reservoirs of virus exist under conditions in which they cannot be eradicated. Thus tolerance has, in fact, been widely used. For example, cultivars of wheat and oats commonly grown in the midwestern United States have probably been selected for tolerance to barley yellow dwarf (BYDV) in an incidental manner, because of the prevalence of the virus.

Transgenic Plants

It is now possible to introduce almost any foreign gene into a plant and obtain expression of that gene. In principle, this should make it possible to transfer genes for resistance or immunity to a particular virus across species, genus, and family boundaries. At present the major problem with this approach is the difficulty in identifying the genes responsible for immunity or resistance. With the exception of those discussed in Chapter 11, Section 6.1, no such genes active against plant viruses have yet been identified. Nevertheless, several other approaches to producing transgenic plants resistant to virus infection are being actively explored. In particular, expression of viral coat-protein genes or other viral genes may provide transgenic plants with a useful degree of protection against the virus concerned.

Transgenic Plants Expressing a Viral Coat Protein   The development of transgenic plants expressing the coat-protein genes for several viruses was discussed in Chapter 12, Section 4.1.4. The mechanism for protection against superinfection with the virus concerned was compared with natural cross-protection. The extent of protection provided by such transgenic plants is correlated with the degree of expression of the coat-protein gene. The cover illustration shows the striking protection obtained with the transgenic expression of TMV coat protein in tobacco. It is not yet certain that these developments will lead to a practical and general method of control for virus diseases.

Some important questions remain to be answered, for example: (1) Will resistance be maintained to a useful degree when inoculation in the field is by means of invertebrate or fungal vectors? (2) Will seed transmission be reduced in transgenic plants that do become infected? (3) Will resistance be achieved to multiple viruses important in particular crops such as tomatoes and peppers? (4) Will a degree of resistance that protects a useful proportion of plants in an annual crop, be effective in protecting a perennial crop? and (5) How frequently can viruses mutate to overcome this kind of resistance? Answers to some of these questions will be forthcoming in the fairly near future.

Transgenic Plants Expressing Noncoat Viral Genes Encouraging results have recently been obtained with TMV. Golemboski *et al.* (1990) transformed tobacco plants with the nucleotide sequence coding for the 54-K putative protein of TMV (Fig. 7.3). The sequence lacked only the three 3′ terminal nucleotides. This sequence encodes a component of the putative replicase complex. The transgenic plants were resistant to infection with TMV U_1 or its RNA at high concentrations. It was also resistant to a U_1 mutant, but not to two other tobamoviruses or CMV. The transformed plants accumulated the expected sequence-specific transcripts, but no protein product could be detected. This situation parallels that found in natural TMV infections (Chapter 7, Sections 3.3). Experiments in protoplasts have shown that a very early event is involved. No 126-K or 183-K proteins could be detected in protoplasts derived from plants transgenic for the 54-K protein (M. Zaitlin, 1991, personal communication).

Transgenic Plants Expressing Satellite RNAs In two systems, plants transgenic for a satellite RNA have shown greatly attenuated disease systems, compared to control plants. These were CMV and transgenic tobacco plants containing DNA copies of a CMV satellite RNA, and tobacco ringspot (TRSV) and satellite TRSV (STRSV) in tobacco.

The use of satellite RNAs in transgenic plants to protect against the effect of virus infection has both advantages and disadvantages. The protection afforded is not affected by the inoculum concentration, as it is with viral coat-protein transformants. The losses that do occur in transgenic plants because of slight stunting will affect only the plants that become naturally infected in the field; whereas, if all plants are deliberately infected with a mild CMV–satellite combination, they will all suffer some loss (Section 3.3.2). Furthermore, the resistance may be stronger in transgenic plants than in plants inoculated with the satellite. Inoculation is not needed each season, and the mutation frequency is lower.

However, there are distinct risks and limitations with the satellite control strategy. The satellite RNA could cause virulent disease in another crop species or could mutate to a form that enhances disease rather than causing attenuation (Chapter 9, Section 2.2). Another risk is the reservoir of virus available to vectors in the protected plants. The satellite approach will be limited to those viruses for which satellite RNAs are known. However, CMV is a widespread and economically important virus.

Transgenic Plants Expressing Antisense RNAs One method of gene regulation in organisms is by complementary RNA molecules that are able to bind to the RNA transcripts of specific genes and thus prevent their translation. Such RNA has been called antisense or mic RNA (*m*essenger-RNA-*i*nterfering *c*omplementary RNA). Various laboratories have made transgenic plants containing genes for the production of antisense segments of plant viral RNA genomes. Present evidence suggests that the approach may not be particularly successful.

Transgenic Plants Expressing Other Genes Several other kinds of transgenic plants are currently being studied. These include plants expressing ribozymes designed to cleave specifically a viral RNA; plants expressing pathogenesis-related proteins constitutively; plants expressing a gene for an insect toxin that kills a virus vector, when the vector feeds on the plant; and plants expressing virus-specific antibodies. It has recently been shown that plants transgenic for the heavy and light chains of mammalian immunoglobulin can synthesize functional antibodies.

3.3.2 Cross-Protection

Infection of a plant with a strain of virus causing only mild disease symptoms may protect it from infection with severe strains (Chapter 12, Sections 3.3.4 and 4.1). Thus, plants might be purposely infected with a mild strain as a protective measure against severe disease. Whereas such a procedure could be worthwhile as an expedient in very difficult situations, it is not to be recommended as a general practice, for the following reasons: (1) so-called mild strains often reduce yield by about 5 to 10%; (2) the infected crop may act as a reservoir of virus from which other more sensitive species or varieties can become infected; (3) the dominant strain of virus may change to a more severe type in some plants; (4) serious disease may result from mixed infection, when an unrelated virus is introduced into the crop; and (5) for annual crops, introduction of a mild strain is a labor-intensive procedure. Nevertheless, the procedure has been used successfully with some crops.

Citrus tristeza virus provides the most successful example for the use of cross-protection. Worldwide, this is the most important virus in citrus orchards. In the 1920s, after its introduction to South America from South Africa, the virus virtually destroyed the citrus industry in many parts of Argentina, Brazil, and Uruguay. The method has been particularly successful with Pera oranges, with more than 8 million trees being planted in Brazil in 1980. Protection continues in most individual plants through successive clonal generations.

As another example, mild strains of CMV have been obtained by adding selected satellite RNAs to a CMV isolate. This procedure has been tested, with increased yields, in many areas of China for the control of CMV in peppers. The extent of the protective effect depended on such factors as inoculation time, percentage of plants inoculated, the nature of virulent CMV strains already in the field, and variety of pepper.

In other experiments, tomatoes were inoculated with CMV strain S carrying a non-necrogenic satellite called S-CARNA5. These were planted out at the seedling stage in spring on a farm in southern Italy where a severe epidemic of the tomato necrosis disease was expected. The epidemic occurred, with 100% of plants being destroyed in some fields. In the field containing the inoculated plants, protection against necrosis was almost 100%, whereas 40% of uninoculated plants developed lethal necrosis. Fruit yields were about doubled in the protected plants.

3.3.3 Antiviral Chemicals

Many substances isolated from plants and other organisms, as well as synthetic organic chemicals, have been tested for activity against plant viruses. Almost all of the substances showing some inhibition of virus infectivity do so only if applied to the leaves before inoculation or very shortly afterward. None has achieved any practical importance.

3.4 DISEASE FORECASTING

For annual crops, forecasting of disease epidemics can be of very great value. For example, appropriate use of organophosphorus insecticides can delay the spread of yellows viruses by aphids within the sugar beet crop. Whether such spraying will give an economic return depends on the timing and extent of virus spread in a given season. Figure 15.1 illustrates the extent to which infection of sugar beet by yellows disease can be predicted from weather factors known to be important for aphid vector and host-plant survival during the winter and early spring aphid migration. Forecasting methods are continually being refined, and these, in turn, are dependent on improved techniques for trapping and testing winged vectors.

3.5 DISCUSSION AND CONCLUSIONS

It is impossible to give precise measures of crop losses due to viruses. For a given virus, losses may vary widely with season, crop, country, and locality. Nevertheless, there are sufficient data to show that con-

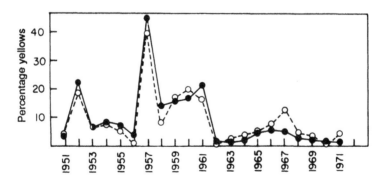

Figure 15.1 Disease forecasting using climatic factors important for aphid vector overwintering and early spring multiplication. ●—●, Observed values for percentage yellows disease in sugar beet crops in the United Kingdom; ○- - -○, values predicted by a regression using the number of frost days (ground temperature $< -0.3°$ C) in January, February, and March and mean April temperatures. From Watson *et al.* (1975).

tinuing effort is needed to prevent losses from becoming more and
more extensive. Three kinds of situations are of particular importance:
(1) annual crops of staple foods such as grains and sugar beet that are
grown on a large scale, and that under certain seasonal conditions may
be subject to epidemics of viral disease; (2) perennial crops, mainly

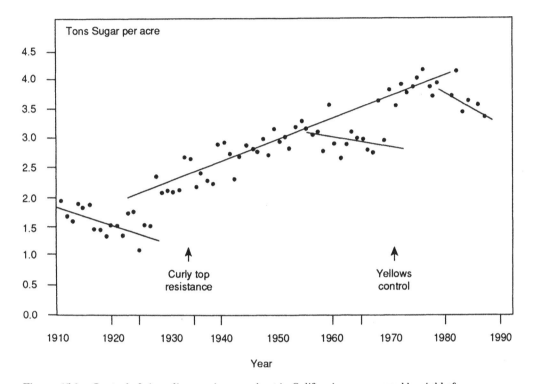

Figure 15.2. Control of virus diseases in sugar beet in California as measured by yield of
sugar per acre, for the period 1911 to 1985. Five periods can be delineated: (1) from 1911
to 1925 yields declined owing to curly top virus infection; (2) from about 1930 yields improved because of the removal of badly infected areas from production, and the introduction of curly top-resistant cultivars; (3) from about 1959 to 1969, yields declined again,
owing to more severe strains of curly top virus and the spread of yellows; (4) in 1967, a
beet-free period was widely maintained, together with a cooperative effort to clean up
weed beets. In 1968 varieties with moderate resistance to yellows were introduced, giving
a further period of increasing yield; (5) from about 1979, yields have again decreased because of a number of factors: (a) lettuce infectious yellows virus (Duffus *et al.*, 1986) first
occurred in the desert regions of California in 1981, and has since been a limiting factor
in sugar production in that region. In susceptible varieties, this virus causes a 30% reduction in yield; (b) in 1983, Rhizomania was first found in California and by 1989, affected
over 80,000 ha; (c) in recent years, growers in the Delta region have been careless in their
management of beet-free periods, resulting in renewed damage by beet yellows virus
(J. E. Duffus, 1991, personal communication). Courtesy of J. E. Duffus.

fruit trees with a big investment in time and land, where spread of a virus disease, such as citrus tristeza or plum pox, may be particularly damaging; and (3) high-value cash crops such as tobacco, tomato, cucurbits, peppers, and a number of ornamental plants that are subject to widespread virus infections.

Possible control measures can be classified under three headings: (1) removal or avoidance of sources of infection; (2) control or avoidance of vectors; and (3) protecting the plant from systemic disease. In principle, by far the best method for control would be the development of cultivars that are resistant to a particular virus on a permanent basis. Experience has shown that viruses continually mutate in the field, with respect to both virulence and the range of crops or cultivars they can infect. Thus, breeding for resistance, or the development of transgenic plants, is unlikely to give a permanent solution for any particular virus and crop. With almost all crops affected by viruses, an integrated and continuing program of control measures is necessary to reduce crop losses to acceptable levels. Such programs will usually need to include elements of all three kinds of control measure noted at the beginning of this paragraph. The long-term nature of control programs is well illustrated by the history of sugar beet and its virus diseases over a 76-year period in California (Fig. 15.2).

FURTHER READING

Bos, L. (1982). Crop losses caused by viruses. *Crop Protection* 1, 263–282.

Harris, K. F., and Maramorosch, K. (eds.) (1982). "Pathogens, Vectors, and Plant Disease. Approaches to Control." Academic Press, New York.

Jones, A. T. (1987). Control of virus infection in crop plants through vector resistance: A review of achievements, prospects, and problems. *Ann. Appl. Biol.* 111, 745–772.

Matthews, R. E. F. (1991). "Plant Virology," 3rd Ed. Academic Press, New York.

Matthews, R. E. F. (editor) (1992). "Diagnosis of Plant Virus Diseases." CRC Press, Boca Raton, Florida.

Shukla, D. D., Frenkel, M. J., McKern, N. M., and Ward, C. W. (1991). Immunological and molecular approaches to the diagnosis of viruses infecting horticultural crops. *In* "Horticulture—New Technologies and Applications" (J. Prakash and R. L. M. Pierik, eds.), pp. 311–319. Kluwer, Dordrecht, The Netherlands.

Thresh, J. M. (1982). Cropping practices and virus spread. *Annu. Rev. Phytophathol.* 20, 193–218.

NOMENCLATURE, CLASSIFICATION, ORIGINS, AND EVOLUTION

16

Attempts before 1966 to develop an effective system for the classification and nomenclature of viruses failed for two main reasons: (1) there was insufficient information about the viruses themselves, as opposed to the diseases they cause, and (2) most attempts at classification were made by individuals or small groups without any input from virologists at large. In 1966 at the International Congress for Microbiology held in Moscow, an organization was set up for developing an internationally agreed-upon taxonomy and nomenclature for all viruses. This organization is now known as the International Committee for Taxonomy of Viruses (ICTV). Meetings are held every 3 years in conjunction with the International Congress of Virology. At these meetings, decisions are made by the members of the ICTV, representing many different countries, concerning new developments in taxonomy and nomenclature. From time to time reports are published, summarizing the current state of viral taxonomy, the most recent being that of Francki *et al.* (1991).

As noted in Chapter 2, one of the most valuable uses of nucleotide sequence information is in the development of a meaningful classification of viruses. This is so at all levels in the classification. A firmly based classification is essential for the identification of viruses causing particular diseases and thus for their control. For theoretical biology, the origins and evolution of viruses are questions of general interest. Nucleotide sequence data are making possible the beginnings of a true systematics for viruses, that is, a classification of viruses based on evolutionary relationships.

1 NOMENCLATURE

1.1 SPECIES

No official names for individual plant viruses have yet been approved by the ICTV. Thus, viruses are still called by their vernacular, or common names, usually based on the English version. These names often include the name of the first-described or most common host plant, together with a major feature of the disease caused, e.g., *tobacco mosaic virus*. Such names can be confusing. For example, lily symptomless virus has major hosts other than lily in which it causes severe disease. For some years the ICTV hoped to introduce a Latin binomial system for viruses, but this has been dropped owing to lack of support from virologists at large. It is probable that most plant viruses will be called by their English vernacular names for many years to come.

1.2 FAMILIES AND GROUPS

The three families to which some plant viruses belong, the Reoviridae, Rhabdoviridae, and Bunyaviridae, were named by vertebrate virologists many years ago. Most plant virus group names are derived from sigla from the English vernacular name of the type member of the group followed by *virus,* for example *Tombusvirus* from *Tom*ato *Bus*hy stunt *virus.*

2 CLASSIFICATION

2.1 SPECIES

The species taxon is generally regarded as the most important in any taxonomy, but it has proved the most difficult to apply to viruses. The central property of viruses is that they consist of a set of genes that code for functional proteins. Through alterations in these genes, viruses undergo evolutionary change. These fundamental properties are like those of cells. Thus, the concept of species should be applicable. As recorded in the minutes of meeting of the Executive Committee of the ICTV in April, 1991, the following definition of a virus species was approved: *"A virus species is a polythetic class of viruses that constitutes a replicating lineage and occupies a particular ecological niche."* This definition will almost certainly receive final approval by a meeting of the full ICTV in 1993. A polythetic class differs from an ordinary class in that its members need not have any particular single property in common. Each member of a polythetic class is defined by more than one property, and no single property is necessary or sufficient for membership in the class. In day-to-day practice, virologists use the concept of a *virus* as being a group of fairly closed related strains, variants, or pathovars. A virus defined in this way is essentially a species, as defined by the ICTV.

Some plant virologists believe that the species concept should not be applied to viruses. Their main reason is that, because viruses reproduce asexually, the criterion of reproductive isolation cannot be used as a basis for defining virus species. In fact, the species concept has been usefully applied to many groups of organisms to which the criterion of reproductive isolation cannot be or has not been applied. Thus there is no theoretical or practical reason that the concept of species should not be applied to plant viruses.

The delineation of the kinds of viruses that exist in nature, that is to say virus species, is a practical necessity, especially for diagnostic purposes relating to the control of virus diseases. For some 20 years, B. D. Harrison and A. F. Murant have acted as editors for the Association of Applied Biologists (AAB) to produce a series of descriptions of plant viruses and plant virus groups. Each description is written by a recognized expert for the virus or group in question. The two editors use commonsense guidelines devised by themselves to decide whether a virus described in the literature is a new virus, or merely a strain of a virus that has already been described. When they publish a new virus description they are, in effect, delineating a new species of virus. The AAB descriptions of plant viruses are widely used by plant virologists, and are accepted as a practical and effective taxonomic contribution. The descriptions now cover a substantial proportion of the known plant viruses.

Properties of the nucleic acids can sometimes be used to delineate distinct viruses within a related group. Nucleotide sequence information, restriction endonuclease maps, and extent of cross-hybridization have been used. However, these criteria are not equally useful for all groups of viruses.

Properties of viral-coded proteins have been of particular importance in delineating virus species. From current knowledge, it appears that nonstructural viral proteins show more similarities within a related group or family of viruses than do the coat proteins. Thus, coat proteins have assumed particular importance in defining virus species. This is well illustrated in Figures 12.3 and 12.6 for the potyviruses. Amino acid composition, amino acid sequence data, and serological relationships based on the coat protein have all been used to delineate virus species.

Biological properties are usually not very reliable criteria, but may sometimes provide useful information. For example, the host range of most individual tymoviruses is confined to a single angiosperm family.

2.2 FAMILIES AND GROUPS

Early in the development of an internationally agreed-upon taxonomy of viruses, all virologists except those dealing with plant viruses agreed on a family and genus structure for their classification. Plant virologists retained the indeterminate concept of virus groups. However, by 1978, the very obvious and detailed similarities between reoviruses infecting vertebrates, invertebrates, and plants made it inescapable that the plant reoviruses should be placed as members of the already existing Reoviridae family. Similarly, plant rhabdoviruses became members of the Rhabdoviridae family. Two genera of plant reoviruses were delineated—*Phytoreovirus* and *Fijivirus*. More recently, tomato spotted wilt virus (TSWV) has been placed in the family Bunyaviridae.

Table 16.1 lists the groups and families of plant viruses currently approved by the ICTV. Most newly described viruses can be accommodated in these categories. However, nucleotide sequence data on viral genomes is emphasizing more than ever before the unity of virology. For this reason there is an urgent need to bring the taxonomy of plant viruses into line with that of other host groups of viruses. In fact, the Plant Virus Subcommittee of the ICTV has recently decided to recommend adoption of the Family–Genus–Species classification for all plant viruses (G. P. Martelli, personal communication, 1991). It is almost certain that these proposals will be adopted by the full ICTV at its next meeting in 1993.

A new family (the *Potyviridae*) to include all potyviruses has recently been proposed (Barnett, 1991). The Plant Virus Subcommittee has moved further, and is proposing six new plant virus families each with two or more genera (Table 16.2). Other possible genera, whose relationships are not yet clearly established, are also listed in Table 16.2.

Table 16.1 The 35 Families and Groups of Plant Viruses Approved by the ICTV[a]

Characterization	Family or group	Number of members	Number of probable or possible members	Total
dsDNA nonenveloped	*Caulimovirus*	11	6	17
	Commelina yellow mottle virus group	3	11	14
ssDNA nonenveloped	*Geminivirus*	35	13	48
dsRNA nonenveloped	*Cryptovirus*	20	10	30
	Reoviridae	6	2	8
ssRNA enveloped	Rhabdoviridae	14	76	90
	Bunyaviridae	1	0	1
ssRNA nonenveloped, *Monopartite genomes*				
Isometric particles	*Carmovirus*	8	9	17
	Luteovirus	12	12	24
	Marafivirus	3	0	3
	MCDV group	1	2	3
	Necrovirus	2	2	4
	PYFV group	2	1	3
	Sobemovirus	10	6	16
	Tombusvirus	12	0	12
	Tymovirus	18	1	19
Rod-shaped particles	*Capillovirus*	2	2	4
	Closterovirus	10	12	22
	Carlavirus	27	29	56
	Potexvirus	18	21	39
	Potyvirus	74	84	158
	Tobamovirus	12	2	14
Bipartite genomes				
Isometric particles	*Comovirus*	14	0	14
	Dianthovirus	3	0	3
	Fabavirus	3	0	3
	Nepovirus	28	8	36
	PEMV	1	0	1
Rod-shaped particles	*Furovirus*	5	6	11
	Tobravirus	3	0	3
Tripartite genomes				
Isometric particles	*Bromovirus*	6	0	6
	Cucumovirus	3	1	4
	Ilarvirus	20	0	20
Isometric and bacilliform particles	AMV	1	0	1
Rod-shaped particles	*Hordeivirus*	4	0	4
Quadripartite genome				
Rod-shaped particles	*Tenuivirus*	3	4	7
		394	320	714

[a]Listed according to particle morphology and type of nucleic acid. Data from Francki *et al.* (1991).

Table 16.2 Provisional Classification of Plant Viruses in Families and Genera[a]

Families	Genera
Rhabdoviridae	Two genera (subgroups A and B) and a possible genus (subgroup C)
Bunyaviridae	Tospovirus
Reoviridae	Phytoreovirus
	Fijivirus
	Oryzavirus
Cryptoviridae	Two genera (subgroups I and II)
Geminiviridae	Three genera (subgroups I, II and III)
Tombusviridae	Tombusvirus
	Carmovirus
Comoviridae	Comovirus
	Nepovirus
	Fabavirus
Bromoviridae	Bromovirus
	Cucumovirus
	Ilarvirus
	Alfamovirus
Potyviridae	Potyvirus (transmitted by aphids)
	Bymovirus (transmitted by fungi)
	Rymovirus (transmitted by mites)
	Ipomovirus (possible genus; transmitted by whiteflies)

UNGROUPED POSSIBLE GENERA
DNA Viruses
  Caulimovirus
  Badnavirus

RNA viruses

Isometric particles	*Rod-shaped particles*
Necrovirus	Tobamovirus
Dianthovirus	Furovirus
Tymovirus	Hordeivirus
Luteovirus	Tobravirus
Sobemovirus	*Filamentous particles*
Marafivirus	Closterovirus
Maize chlorotic dwarf v.	Capillovirus
Pea enation mosaic v.	Carlavirus
Parsnip yellow fleck v.	Potexvirus
	Tenuivirus

[a]From Martelli (1992) and Barnett (1991).

2.3 A BIONOMIAL SYSTEM

A species name indicates uniqueness, whereas a genus name can remind us of relatives and their properties. As a mechanism for information retrieval, I believe that a simple binomial system will turn out to be the most useful and versatile in the long run. A generic name with a virus (species) epithet, together with a strain or pathovar designation, should give a complete, unambiguous, and internationally understood designation of a particular virus.

In successive editions of the Reports of the ICTV, virus names in the index have been listed by the vernacular name (usually English) followed by the family, genus, or group name—for example, tobacco mosaic *Tobamovirus*, Fiji disease *Fijivirus*. This method for naming a plant virus is becoming increasingly used in the literature. It is, of course, a binomial system but with the virus (species) name coming before the genus name, or what will be the genus name following the rationalization indicated in Table 16.2. The fact that the two terms in the binomial are in the reverse order compared with the binomial names of all cellular organisms does not detract from its utility as an information retrieval system. Nevertheless, it becomes somewhat awkward if we wish to designate a strain or pathovar, e.g., tobacco mosaic *Tobamovirus* vulgare.

2.4 CRITERIA AVAILABLE FOR CLASSIFYING VIRUSES

Criteria for delineating viruses and virus strains were discussed in Chapter 12, Section 3. The existing classification of viruses built up by the ICTV relied almost entirely on phenotypic characters such as particle morphology, amount and kind of nucleic acid, size and nature of proteins present, serological relationships, presence or absence of lipid, and biological properties such as kind of host, host range, disease symptoms, and kind of vector.

The recent rapid accumulation of complete nucleotide sequence information for many viruses in a range of different families and groups is having a profound influence on virus taxonomy in at least three ways: (1) most of the virus families and groups previously delineated by the ICTV can now be seen to represent clusters of viruses with a relatively close evolutionary origin; (2) with the discovery of previously unsuspected genetic similarities between viruses infecting different host groups, the unity of virology is now quite apparent; (3) genotypic information is now more important for many aspects of virus taxonomy than phenotypic characters. However, there are several limitations to the use of genotype (sequence) data alone. One is the practical problem that with current technology, it is a relatively time-consuming and expensive task to determine the complete nucleotide sequence of a viral genome. In many kinds of study, the virus present in hundreds of samples must be identified over a short period. Another problem is that in the present state of knowledge, it is very difficult or impossible to predict most phenotypic properties of a virus on the basis of sequence data alone.

Any classification of viruses should be based not only on evolutionary history, as far as this can be determined from the genotype, but should also be useful in a practical sense. Most of the phenotypic characters used today in virus classification will remain important even when the nucleotide sequences of most viral genomes have been determined.

3 SPECULATION ON ORIGINS

There is no compelling reason to suppose that all viruses (or even parts of viruses) arose in the same way.

3.1 DESCENDANTS OF PRIMITIVE PRECELLULAR LIFE FORMS

The idea that viruses might be descendants from primitive precellular forms of life on earth is an old one. Until recently, however, the idea was given little credence, for two main reasons: (1) how could a highly specialized obligate parasite of cells exist before cells? and (2) although virus particles are relatively *simple* in structure, there was nothing known about their chemistry or mode of replication to suggest that they are more *primitive* in an evolutionary sense than cellular organisms.

Nevertheless, current views on the origin of life open up the possibility that some RNA viruses at least, or parts of them, may be descended from prebiotic polymers. Until fairly recently, precellular life was considered to involve both proteins and RNAs. However, following the discovery of introns in eukaryotic genes, it was found that the splicing out of introns from RNA transcripts, and ligation of the exons, could take place in the absence of protein. This would provide RNA with a vital evolutionary capacity it would otherwise lack—the ability to produce new combinations of genes. Thus, an evolutionary proposal was put forward in which RNAs were the only polymers in the prebiotic stage of evolution. On this general theory, RNA viruses or parts of them might represent greatly modified descendants of prebiotic RNAs, which later parasitized the earliest cells.

3.2 DEVELOPMENT FROM NORMAL CONSTITUENTS OF CELLS

Because viruses carry genetic information and because it is known that some viruses can exchange genetic material with their hosts, it has been proposed that they arose as host genes that escaped from the control mechanisms of the cell. These replicating agents are considered to have then developed the means of being transferred efficiently to other host cells. This view is now held by many virologists, and several lines of evidence support it.

3.2.1 Comparison of Bacterial Plasmids and Viruses

The strongest support for this kind of origin comes from the close parallels between certain extrachromosomal genetic elements in bacteria, and some bacterial viruses. *Plasmids* are autonomous extrachromosomal double-stranded DNA (dsDNA) genetic elements that are found naturally in many types of bacteria. Some can integrate into the bacterial chromosome and are replicated passively along with the chromosomal DNA. An integrated plasmid may become excised from the host DNA and replicate independently again. During this process, some host DNA, including complete genes, may be excised along with the plasmid DNA, and subsequently replicate with it. Several viruses infecting bacteria, and in particular the phages λ, Pl, and Mu, have properties like those of plasmids. Furthermore, it is possible experimentally to convert some viruses (e.g., phage λ) into plasmidlike entities by deleting genes other than those essential in replication. Conversely, by the addition of appropriate viral genes, such as those required for packaging, it may be possible to convert a plasmid into a viruslike entity.

3.2.2 Integration of Viral Genes into Eukaryotic Genomes

The fact that the DNA of several groups of viruses can integrate into the host cell genome makes it easy to believe that some viruses could have originated there.

3.2.3 Transposable Elements in Plants

Transposable genetic elements or transposons are now known to occur in many kinds of eukaryotic organisms, including many plant species. In the context of viral evolution, the class of transposable elements known as retrotransposons are of particular interest. These DNA elements are organized like retrovirus DNAs. There is good evidence that they carry out transposition in the cellular genome via an RNA intermediate and reverse transcription. Recently, retrotransposonlike mobile elements have been isolated from plants, including one from tobacco and other solanaceous species. This transposon, called Tnt1, is a mobile retrotransposonlike element 5334 nucleotides long. Its nucleotide sequences and open reading frames (ORFs) are organized like those of a similar element in Drosophila called *copia*, and there is a high degree of amino acid sequence similarity, indicating that they share a common origin. It is not impossible to imagine such elements evolving into an infectious retrovirus, with an extracellular phase.

3.2.4 Amino Acid and Nucleotide Sequence Similarities between Viruses and Cells

There are several reports of sequence similarities between proteins coded by viruses infecting vertebrates and host proteins. There are also reports of promoter sequences in plant viral genomes having remarkable similarities to control regions in certain eukaryotic genes.

3.3 ORIGIN BY DEGENERATION FROM CELLS

The idea that viruses are extremely degenerate parasitic forms that have evolved from cellular organisms has been put forward on a number of occasions, but has lost favor in recent years, mainly because of the similarities between the behavior of certain viruses and plasmids, as outlined in the previous section. However, the large size of the poxviruses, their complex structure, including many enzymes within the virus particle, and their ability to replicate in the cytoplasm independent of host nuclear functions, makes it plausible that these viruses arose from cells. However, no viruses with the properties of the animal poxviruses are known to infect plants.

3.4 SUMMARY FOR VIRUSES

On current evidence, it seems probable that different groups of present-day viruses originated in different ways. Some of the very large DNA viruses infecting animals probably descended by a degenerative process from very simple cellular parasites. Other, smaller viruses, especially those with RNA genomes, most probably evolved from host genes via transposons or by other means. Some present-day RNA viruses or parts of them might be direct descendants from a prebiotic RNA world.

3.5 ORIGIN OF VIROIDS

A variety of origins has been proposed for viroids, based mainly on nucleotide sequence similarities: (1) from the small nuclear RNAs that are believed to play a role in the processing of the primary transcription products of split genes; (2) from introns; there is some sequence similarity suggesting such an origin; (3) from transposable genetic elements. Sequence analysis of a group of viroids has shown some striking similarities with transposable genetic elements, including the proviruses of retroviruses; (4) from satellite RNAs. Although viroids and the viroidlike satellite RNAs show little sequence similarity, they have several properties in common: small size, circular single-stranded RNA (ssRNA), and lack of mRNA activity. In addition, some satellite RNAs probably have a rolling-circle model for replication involving greater-than-unit-length RNAs. On the other hand, viroids replicate in the nucleus, whereas viroidlike satellites do so in the cytoplasm. In principle, a viroid might have arisen from a viroidlike satellite by becoming independent of the helper virus for replication; (5) as living fossils of prebiotic evolution. It is conceivable that modern viroids, because of their small size, circular conformation, inherent stability, their RNA→RNA replication cycle, and lack of protein-coding capacity, may be descended from a type of RNA molecule that existed before the evolution of DNA.

3.6 ORIGIN OF SATELLITE VIRUSES

Apart from their dependence on a helper virus, and their small size, satellite viruses appear to belong in the same category of agents as other viruses. They most probably arose from an independent virus by degenerative loss of functions in a mixed infection that provided the helper virus.

3.7 ORIGIN OF SATELLITE RNAs

Most satellite RNAs fall into one of two groups. The first, typified by satellite RNAs of cucumber mosaic virus (CMV) and tomato blackring virus (TBRV), have terminal structures like that of the helper virus RNAs, replicate via a unit-length, negative-sense template, depend on the helper virus replicase, and appear to code for one to three proteins.

The second group, typified by satellite tomato ringspot virus (STRSV), appear to have no mRNA function and are replicated by a viroidlike, rolling-circle mechanism. The predicted secondary structure of the viroidlike satellite RNAs mimics the rod-shaped configuration of the viroids, but there is little base sequence homology between them, except for a conserved GAAAC occurring in some members of each group of agents. Thus, it is possible that satellite RNAs have arisen from two different lineages—one from satellite viruses and one from viroids.

The satellite RNA C of turnip crinkle virus (TCV) (Fig. 9.9) is of particular interest from the evolutionary point of view. Its structure appears to demonstrate that recombination can take place between a satellite RNA and the helper virus genome to generate a new satellite with different biological properties.

4 EVOLUTION

4.1 MECHANISMS FOR VIRAL EVOLUTION

The molecular mechanisms underlying genetic variation in viruses were summarized in Chapter 12, Section 2. We can envisage viral evolution proceeding in both a micro and a macro manner. In micro-evolution, existing viral genes accumulate small changes by such mechanisms as nucleotide substitutions, additions, or deletions. In macroevolution of a virus, a sudden major change may take place, by recombination with a related virus, by duplication of an existing gene, or by acquisition of a gene from the host or an unrelated virus, processes that have been termed *modular evolution*. Available evidence suggests that such a process may have occurred in viruses infecting plants. For example, in the genome of barley yellow dwarf (BYDV), the coat protein shows some distant similarities to those of some other icosahedral viruses, such as tomato bushy stunt (TBSV) and carnation mottle virus (CarMV). The 50-K ORF is similar to the readthrough protein of the beet necrotic yellow vein virus (BNYVV) coat protein. These various viruses have quite different general properties. Thus, BYDV could well be a mosaic of modules and gene-expression strategies arranged quite differently from those of other known viruses.

4.2 EVIDENCE FOR VIRUS EVOLUTION

There is a marked lack of intermediate types between most of the virus families and groups delineated by ICTV. The close similarities in particle morphology, genome strategy, and the three-dimensional structure of proteins, leave no room for doubt that the individual viruses within the families and groups delineated by the ICTV had a common ancestor at some time in the past. For example, among the definite members of the *Tobamovirus* group, all measures of their relatedness correlate with that indicated by the amino acid composition and amino acid sequence of their coat proteins; thus, they appear to form a distinct group of viruses with a common ancestor.

Knowledge of nucleotide sequences in viral genomes, and the corresponding sequences of amino acids in the encoded proteins, has provided powerful confirmation of the evolutionary basis for most of the families and groups of viruses delineated by the ICTV. For example, sonchus yellow net (SYNV) is similar but not identical to animal rhabdoviruses in the order of structural genes and in the nucleotide sequences at the gene junctions. The intergenic and flanking-gene sequences are conserved. They consist of a central core of 14 nucleotides (3'-UUCUUUUU-GGUUGU/A-5'). This sequence is similar to the sequence at the gene junctions in vesicular stomatitis and rabies viruses.

These conserved features argue strongly for a common origin for all members of the Rhabdoviridae family.

4.3 EVOLUTIONARY TREES DERIVED FROM SEQUENCE DATA

Sequence data can be used to estimate degrees of evolutionary relationship, to develop *trees* indicating possible lines of evolution for viruses within a family (e.g., Fig. 12.6), and to discover unexpected relationships (Section 5 following). Whereas this is a very valuable approach, it should be remembered that there are significant difficulties in deriving and interpreting trees constructed from sequence data.

4.3.1 Inherent Difficulties

There are inherent difficulties in constructing meaningful trees, especially where more than four taxa are involved. This is partly because of the large number of possible trees.

4.3.2 Nonconstant Rates of Evolution

The molecular clock hypothesis, which assumes a constant rate of change in sequence over evolutionary time, has often been used in the interpretation of trees. However, rates of change in different genes and lineages may vary widely. There is little doubt that different proteins, or parts of a protein coded by a viral genome, may evolve at different rates. In addition, some noncoding sequences in the genome may be highly conserved, particularly those recognition sequences essential for genome replication.

4.3.3 Sequence Convergence

Sequence similarity between two genes does not necessarily indicate evolutionary relationship (homology). Without other evidence, it may be impossible to establish whether sequence similarity between two genes is the result of a common evolutionary origin or convergence, especially where the degree of similarity is not great.

4.3.4 Proteins Approaching the Limits of Change

As two proteins with a common evolutionary origin diverge, they may approach a limit of change that retains common amino acid positions in excess of those expected for random sequences. This limit on change would be owing to functional requirements for the protein. When this limit is reached, convergence or back-mutations and parallel mutations become as common as divergent mutations. As two diverging proteins approach and remain in this steady-state condition, sequence differences no longer reflect evolutionary distance.

4.4 RATES OF EVOLUTION

The rate of point mutation for RNA viruses has been estimated to be approximately 10^{-3}–10^{-4} per nucleotide per round of replication, with some variation between different viruses. This contrasts with estimates of 10^{-7}–10^{-8} for DNA polymerases. However, the idea that the error rate in eukaryotic viral DNA synthesis is significantly lower than that for RNA viruses has been questioned. In theory, the measured mutation rates would allow for very rapid change in viral genomes of either RNA or DNA, and it is known that virus cultures contain large numbers of sequence variants. It is very difficult to relate mutation rates to the actual rates of change in viruses that might be going on in the field at present, or over past evolutionary time. The reasons for this include the following.

4.4.1 Selection Pressure by Host Plants

Conditions within a given host species or variety will exert pressure on an infecting virus against rapid and drastic change. Viral genomes and gene products must interact in highly specific ways with host macromolecules during virus replication and movement. These host molecules, changing at a rate that is slow compared with the potential for change in a virus, will act as a brake on virus evolution. The few field studies that are available support the idea that a given plant host species tends to stabilize a virus population.

4.4.2 Variation in Rates of Change in Different Parts of a Viral Genome

Noncoding regions of viral genomes, particularly at the 5′ and 3′ termini, that function as recognition sites in viral RNA translation and replication, may be highly conserved in the members of a virus family or group. On the other hand, in viruses with multipartite genomes, one coding genome segment may be conserved and the other highly variable. Coat-protein genes may be much more strongly conserved in some regions than in others by functional requirements in the proteins.

4.4.3 Uneven Rates of Changes over a Time Period

The environment that dictates the selection pressure on the replication and movement of a virus within a plant consists almost entirely of the internal milieu of the host. Other selection pressures on survival involve transmission from plant to plant, for example, by invertebrate or other vectors. It is possible that a switch to a new host-plant species may induce rapid evolution of a virus over quite a short period. Recombination between viral genomes with mutations in different parts of the genome may speed this process. This stage may be followed by a further period of stability for the new virus in the new host.

One way in which a virus may gain a foothold in a new host species is through coinfection with another virus that normally infects that species. For example, TMV does not normally infect wheat, but can do so if the plants are already infected with barley stripe mosaic virus (BSMV).

4.5 COEVOLUTION OF VIRUSES AND THEIR HOST PLANTS

Our knowledge of viruses infecting plants below the angiosperms is very limited, and gives us little information concerning possible coevolution of viruses and their host plants on a geological time scale.

Viruslike particles (VLPs) have been observed in thin sections of many eukaryotic algal species belonging to the Chlorophyceae, the Rhodophyceae, and the Phaeophyceae. The particles are polygonal in outline, varying in diameter from about 22 to 390 nm. Some have tails reminiscent of bacteriophages. The most studied viruses are those infecting a chlorellalike green algae. The virus known as PBCV-1 has a very large dsDNA genome (about 300,000 base pairs) with features like those of the poxviruses. The icosahedral particle morphology is complex, with at least 50 structural proteins, and is like that of the Iridoviridae family. Thus the largest and most complex virus infecting eukaryotic photosynthetic organisms is found in the simplest host— a *Chlorella*-like green alga.

On the basis of cytological and chemical similarities, land plants (embryophytes) are considered to have evolved from a charophycean green alga. Molecular evidence concerning the presence and arrangement of introns in members of the Charophyceae and *Marchantia* strongly support a Charophycean origin for land plants. The geological record suggests that the Charophyceae may have acquired introns 400–500 × 10⁶ years ago.

A virus infecting *Chara australis* (CAV), a morpholgically complex member of the Charophyceae, has been described from Australia. CAV is a rod-shaped virus that shares features of genome organization and sequences with several groups of rod-shaped viruses infecting angiosperms. It has no known angiosperm host, however, and thus it is most unlikely that CAV originated in a recent transfer of some rod-shaped virus from an angiosperm host to *Chara*.

In summary, the existence of an ss positive sense RNA virus infecting the genus *Chara* suggests an ancient origin for this type of virus. Other than this example, the meagre information about viruses infecting photosynthetic eukaryotes below the angiosperms can tell us very little about the age and course of evolution among the plant viruses. Chloroplast DNA sequence data suggest that the monocotyledons and dicotyledons diverged from a common stock about 200 (±40) million years ago. Thus, over this period of divergence, the present-day viruses infecting angiosperms presumably also evolved, at least with respect to their main host specificities.

There is evidence to suggest that coevolution of viruses and host plants continues at the present time. An area containing many species and varieties of a plant genus is taken to represent a site where evolution within that genus has occurred. For example, the *Solanum* species related to the potato probably evolved in the Andean region, with a center in the Lake Titicaca area of South Peru and Northern Bolivia, and were first domesticated there. Most of the viruses infecting potatoes are restricted in nature to the genus *Solanum*, for example, potato S *Carlavirus*, potato X *Potexvirus*, Andean potato mottle *Comovirus*, Andean potato latent *Tymovirus* and potato moptop *Furovirus*. These viruses have almost certainly been spread from the Andes to other parts of the world in potato tubers. To the extent that they have been examined in detail, these viruses show marked variation within the Andean region. This diversity of virus strain, climatic adaptation, and geographic restriction of some strains to parts of the region support the idea that the section of the genus *Solanum* composing the potatoes, and the major viruses of potatoes, have coevolved and are coevolving in the Andean region.

4.6 POSSIBLE ORIGIN OF PLANT VIRUSES IN INSECTS

The most recent major evolutionary explosion, as evidenced in the fossil record, took place about the beginning of the Cretaceous ($\approx$135 million years ago). Monocotyledons and dicotyledons emerged, and various orders of mammals and birds appeared. Some orders of insects had evolved by the Devonian (400 million years ago), but several important orders produced large numbers of new types as the higher plants emerged. Many of these coevolved with their angiosperm food plants. Thus, some viruses that were already present in insects may have adapted to replicate in the evolving mammals and angiosperms during Cretaceous times.

A virus known as leafhopper A virus (LAV) has particles similar to those of fijiviruses infecting plants. LAV infects and multiplies in the leafhopper *Cicadulina bimaculata*, and is transmitted through its eggs. It does not multiply in the maize host plants of the insect. However, when infected hoppers feed on a plant, virus is injected into, and circulates transiently in the plant. This virus can infect healthy hoppers that feed simultaneously on the same plant. Thus, the maize plant can be regarded as a circulative but nonpersistent vector of an insect virus. It is not unreasonable to suppose that over a long period, a virus such as LAV might occasionally acquire a gene or genes that would allow it to establish and replicate in the plant as well as the insect.

The Reoviridae family of viruses has members that infect both vertebrates and invertebrates, and others that infect both invertebrates and plants. This taxonomic distribution of hosts suggests that the more ancient invertebrates may have been the original source of this virus family. A number of features support the view that plant reoviruses originated in the leafhopper vectors or their ancestors. (1) Fijiviruses are morphologically similar to viruses such as LAV. (2) All known plant reoviruses replicate in their hopper vectors. (3) Plant reoviruses are not seedborne, nor are they transmitted by mechanical means, except in special circumstances. Thus, they are entirely dependent on the hopper vectors for survival. (4) The plant species infected by reoviruses are usually the prime food and breeding hosts of their hopper vectors. (5) The plant reoviruses appear more closely adapted to their hopper hosts because (a) they replicate to higher titer in the insects; (b) several plant reoviruses are transmitted through insect eggs, but none is transmitted through plant seed; (c) the percentage of virus-carrying insects in a given vector population is higher than the percentage of plants that can be infected by feeding single hoppers on them; and (d) some cause cytopathic effects in the hopper vectors, but in general these viruses have less severe pathogenic effects in the insects than in their plant hosts. In fact, most can be considered to cause latent infections.

Like the Reoviridae, members of the Rhabdoviridae family all have a very similar particle morphology and genome strategy, and infect either vertebrates or invertebrates, or invertebrates and plants. Thus, a common origin for this family among the insect vectors is indicated. The viruses are not seed transmitted, but are transmitted through the eggs of hopper vectors. They do not appear to cause disease in their insect vectors. The situation is somewhat more complex than that of the plant reoviruses since different plant rhabdovirus have hopper, aphid, piesmid, or mite vectors. Other plant viruses that may have originated as invertebrate viruses are TSWV and the tenuiviruses and marafiviruses.

Adaptation of viruses to their invertebrate vectors has probably involved several different processes on different time scales, and these may be very difficult to unravel. These processes include (1) evolutionary origins by descent on a geological time scale; (2) adaptations that may be of quite recent origin, to particular vector species, for example, through mutational changes in the viral coat protein or in a helper protein, in a group such as the potyviruses; and (3) evolutionary origins that involve both coevolution and descent, and direct colonization of new vector groups.

5 GENOME AND AMINO ACID SEQUENCE SIMILARITIES BETWEEN VIRUSES INFECTING PLANTS AND ANIMALS

Given the very similar particle structures among rhabdoviruses and their replication in both plants and animals, it is not surprising that some genome sequence similarities have been revealed. In fact, within this family, there is good reason to believe that the sequence similarities are in fact homologies, i.e., have a common evolutionary origin. It is much more surprising that amino acid sequence similarities in nonstructural proteins have been revealed between various RNA plant virus groups that have diverse particle morphology, and between these viruses and certain viruses infecting vertebrates. This has led to the idea that many plant plus-strand RNA virus groups may be classified into two major superfamilies, and that viruses within these superfamilies may have a common evolutionary origin. There are also some sequence similarities between members of the two superfamilies. A third superfamily centered on the *Luteovirus* group has recently been proposed. Sequence similarities have also been found between caulimoviruses, animal retroviruses, and hepadnaviruses.

5.1 THE PICORNALIKE PLANT VIRUSES

Figure 16.1 summarizes the similarities in genome organization and sequence that have been revealed between the following viruses: *Poliovirus* (a member of the animal virus family Picornaviridae), cowpea mosaic virus *Comovirus*, tomato black ring *Nepovirus*, and tobacco vein mottling virus *Potyvirus*. These viruses have the following features in common: (1) positive-sense ssRNA genomes; (2) a VPg at the 5′ terminus, and a polyA tract at the 3′ terminus; (3) single long ORFs coding for polyproteins, which are processed by viral-coded proteases to give the functional gene products; (4) several nonstructural proteins that have similar functions, and significant amino acid sequence similarity (>20%); and (5) the genes for these conserved proteins have a similar arrangement for all the genomes.

5.2 THE SINDBISLIKE PLANT VIRUSES

Several plant RNA viruses have been found to show similarities in amino acid sequence and gene arrangement with Sindbis virus, an *Alphavirus* with a lipoprotein envelope that infects vertebrates. This collection of plant virus groups is quite variable in genome structure and strategy. They are the tobamoviruses (TMV), tombusviruses (CuNV), carmoviruses (CarMV), tobraviruses (TRV), hordeiviruses (BSMV) and the three closely related groups—bromoviruses (BMV), cucumoviruses

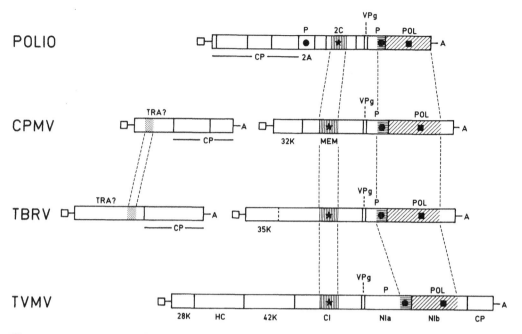

Figure 16.1 Comparison of the genomic RNAs of picornaviruses (POLIO) and the picornalike plant viruses. The superfamily of picornalike plant viruses includes comoviruses (CPMV), nepoviruses (TBRV), and potyviruses (TVMV). Open reading frames are represented as open bars; VPg, as open squares, and poly(A) tails, as A. Regions of significant (>20%) amino acid sequence similarity in the gene products are indicated by similar shading. TRA, transport function; CP, capsid protein(s); P, protease; MEM, membrane-binding; POL, core RNA-dep. RNA polymerase; HC, helper component; *, nucleotide-binding domain; ●, cysteine protease domain; ■, polymerase domain. From Goldbach and Wellink (1988).

(CMV) and AMV. Furoviruses and Potexviruses also have sequence similarities that place them in this superfamily. All these viruses have a 5′ cap, but the 3′ termini vary. Nevertheless most of them specify three proteins with significant sequence similarity to the three nonstructural proteins nsP1, nsP2, and nsP4 of Sindbis virus. These viruses are compared with Sindbis virus in Fig. 16.2.

5.3 SEQUENCE SIMILARITIES BETWEEN PICORNALIKE AND SINDBISLIKE VIRUSES, AND SOME CELLULAR PROTEINS

There is a low level of sequence similarity between certain sites in proteins of viruses belonging to the two RNA virus superfamilies and some eukaryote proteins. These similarities reside in

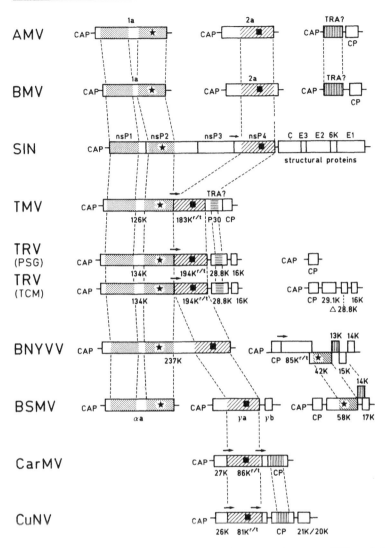

Figure 16.2   Comparison of the genomic RNAs of Sindbis virus (SIN) and the Sindbislike plant viruses. The superfamily of Sindbislike plant viruses includes the alfalfa mosaic virus group (AMV), bromoviruses (BMV), tobamoviruses (TMV), tobraviruses (TRV, strains PSG, and TCM), furoviruses (BNYVV), hordeiviruses (BSMV), carmoviruses (CarMV) and tombusviruses (CuNV). Open reading frames are represented as open bars, and regions of amino acid sequence similarity in the gene products are indicated by similar shading. For reasons of simplicity, closely adjoining or slightly overlapping genes in the TMV (p30 and CP) and CarMV (CP) genome are drawn contiguously. →, leaky termination codon; r/t, read-through. For other symbols, see Fig. 16.1. From Goldbach and Wellink (1988).

1. *A nucleotide triphosphate-binding site* A sequence of amino acids has been found in a protein of both virus superfamilies similar to that found in various guanosine triphosphate- or adenosine triphosphate-using enzymes, some of which possess helicase activity. They are assumed to have a role in DNA and RNA replication.

2. *RNA-dependent RNA polymerase domains* A conserved segment of about 14 amino acids has been identified in members of both Picorna and Sindbis superfamilies as well as other viruses. The sequence is probably an active site or a recognition site for polymerases in general.

3. *Serine proteases* A set of three conserved amino acid residues His, Asp/Glu, and Cys/Ser has been identified in the chymotrypsin-like serine proteases and in the cysteine proteases of some positive-strand RNA viruses.

5.4 A THIRD POSSIBLE SUPERFAMILY OF RNA VIRUSES

Based on sequence motifs of nucleic acid helicases and RNA polymerases, a third superfamily of RNA viruses has been proposed which includes the *Luteovirus, Carmovirus,* and *Tombusvirus* groups (Habili and Symons, 1989). Arranging the virus groups into three superfamilies based on the RNA polymerase sequence motif alone gave rise to exactly the same arrangement as that based on the helicase motif alone. These similarities lend support to the idea that there may be a real basis for the three superfamilies.

5.5 SIGNIFICANCE FOR VIRUS CLASSIFICATION
AND EVOLUTION

Structure of the virus particle, strategy of viral replication, nucleotide sequences, and amino acid sequences of viral-coded proteins will remain important for placing viruses into groups and families. Amino acid sequences of viral coat proteins often appear to be more characteristic of individual viruses than do nonstructural viral-coded proteins. These coat proteins must have evolved to give a satisfactory balance between three important functions: (1) the ability to self-assemble around the RNA; (2) stability of the intact particle inside the cell, and during transmission to a fresh host plant; and (3) ability to disassemble to the extent necessary to free the RNA for transcription and translation.

Thus, serological relationships based on coat proteins will remain important for characterizing individual viruses and many virus strains.

The genome and amino acid sequence similarities discussed above suggest relationships that cut across many of the accepted criteria for classifying viruses: (1) host groups—plant and vertebrate viruses; (2) morphology—viruses with rod-shaped particles, icosahedral particles, and particles with a lipoprotein envelope; (3) kind of nucleic acid— DNA and RNA; (4) numbers of genome segments—viruses with monopartite, bipartite, and tripartite RNA genomes.

The situation we observe among present-day viruses has probably arisen as a consequence of three major processes—divergent evolution of viruses from a common ancestor; convergent evolution of proteins with particular functions; and modular evolution involving the acquisition of genes from other viruses, or the host genome. In the present state of knowledge, we can only speculate as to the relative importance of these three processes. Nevertheless, some distinctions may be emerging. The amino acid sequence similarities between the enzymes discussed in Section 5.3 could well have arisen by convergent evolution, driven by a requirement for particular structures in their active sites. By contrast, it would be very difficult to imagine the similarities in amino acid sequence and gene order seen between, say, CPMV and poliovirus (Fig. 16.1) as arising by convergent evolution. Here divergence from a common ancestor seems most probable.

Among the viruses with RNA genomes, common evolutionary ancestry may exist on two different geological time scales: (1) Sindbis virus has an icosahedral core containing the RNA genome surrounded by a tightly applied lipoprotein envelope with two glycosylated proteins. TMV is a helical rod with a single coat protein. These two viruses could hardly be more different in morphology, and yet they show clear similarities in their genomes. Therefore, the ss positive sense RNA viruses may represent the most ancient viruses, evolving slowly to give rise to the existing superfamilies. Relationships between members within these superfamilies infecting plants or vertebrates may be real, but distant. There is no compelling evidence for evolution of these proposed ssRNA virus superfamilies in their insect vectors. The ancestral viruses may have predated the evolutionary separation of plants and animals some 10^9 years ago. (2) Members within the Reovirus and Rhabdovirus families infecting plants or vertebrates are much more closely related to each other than are the ss positive sense RNA viruses within each superfamily just discussed. Therefore, they are likely to have had a much more recent common origin. It is reasonable to suppose that the immediate ancestors of these virus families were present in the early insects and that as insects, angiosperms, and mammals coevolved during Cretaceous times, viruses within each family diverged sufficiently to replicate in two of these three major host groups.

It is unlikely that all this new sequence information will disturb the existing plant virus taxonomy in any major way. In fact, the new information is helping to improve existing taxonomy, for example, by providing stronger evidence that the bromoviruses, cucumoviruses, AMV, and ilarviruses should be in a single family, and by supporting the need for genera within the geminiviruses.

If modular evolution has been widespread in the evolutionary history of viruses, then it may not be possible to develop a taxonomy above the level of genus, based on evolutionary history that is represented by a single nested hierarchy. On this idea, superfamily, and perhaps family relationships may be properly represented only by two or more hierarchies, each for a particular part of the viral genome. In spite of this possibility, as the sequences of more viral genomes and host genes become known, it may be possible to develop a true systematics for viruses, that is, a taxonomy based on evolutionary relationship.

FURTHER READING

Francki, R. I. B., Fauquet, C., Knudson, D. L., and Brown, F. (1991). "Classification and Nomenclature of Viruses." Fifth Report of the International Committee for Taxonomy of Viruses. *Arch. Virol. Supplementum 2.* Springer-Verlag, Wien.

Goldbach, R. (1987). Sequence similarities between plant and animal RNA viruses. *Microbiol. Sci* 4, 197–202.

Matthews, R. E. F. (1991). "Plant Virology," 3rd Ed. Academic Press, New York.

Martelli, G. P. (1992). Classification and nomenclature of plant viruses: state of the art. *Plant Disease.* In press.

Zimmern, D. (1988). Evolution of RNA viruses. *In* "RNA Genetics, Vol. II" (E. Domingo, J. J. Holland, and P. Ahlquist, eds.), pp. 211–240. CRC Press, Boca Raton, Florida.

Ward, C. W., and Shukla, D. D. (1991). Taxonomy of potyviruses: Current problems and some solutions. *Intervirology* 32, 269–296.

FUTURE PROSPECTS FOR PLANT VIROLOGY

So far I have given a condensed account of the present state of knowledge in the field of plant virology. In this last chapter, I shall take a brief look at the past and then outline what I see as some important and exciting fields for research as we head toward the twenty-first century.

1 A BRIEF LOOK AT THE PAST

I have divided the past 100 years of research on plant virus diseases and plant viruses into five periods. These are arbitrary in that the various phases have overlapped, but they will give a brief overview of the way the subject has developed.

1. 1890–1935: The major activity in this period was the description of new virus diseases in plants. The fact that a disease agent could pass through a bacteria-proof filter categorized it as a filterable virus. The nature of these agents was unknown, and confusion regined in distinguishing between the viruses themselves and the diseases they caused.

2. 1935–1956: The first virus was isolated in 1935, and shown to be a ribonucleoprotein in 1936. In the following years, numerous plant viruses were isolated, and their physical and chemical properties were studied. They were all shown to consist of RNA and protein, and in 1956, the RNA alone was shown to be sufficient for infection.

3. 1956–1970: RNA viruses with multipartite genomes and a double-stranded DNA (dsDNA) plant virus were discovered, but the major developments in the period involved the use of electron microscopy. Greatly improved microscopes and specimen-preparation techniques were developed. Negative staining allowed some substructure in purified virus particles to be revealed. Thin sectioning of diseased tissue revealed some of the effects of virus infection on cells.

4. 1970–1980: The development of high-resolution X-ray crystallography allowed the three-dimensional structure of the protein shells of some viruses to be revealed at atomic resolution. The development of protoplast systems allowed significant advances to be made in the study of the way plant viruses replicate in cells. *In vitro* translation studies using viral RNAs began to provide information about the strategies of plant viral genomes.

5. 1980–1990: This period was dominated by developments in the nucleotide sequencing of viral RNA genomes. The ability to produce infectious RNA transcripts from cDNA copies of viral RNA genomes allowed the techniques of gene manipulation to be applied to their study. Nucleotide sequence data gave us an understanding of the genome organization of most groups of plant viruses. Improved diagnostic techniques based on nucleic acid hybridization were also developed. Finally, hopes were raised for new methods for the control of virus diseases, based on the incorporation of parts of viral genomes into transgenic crop plants.

2 TOWARD THE TWENTY-FIRST CENTURY

The scope for new developments over the next decade, covering both basic and practical aspects of plant virology, appear almost unlimited. This prospect arises almost entirely from the exploitation of techniques based on gene-manipulation technology.

2.1 STRUCTURAL STUDIES

X-ray crystallographic analysis has provided us with a detailed picture of the fine structure of a range of small viruses. So far, the technique has been essentially passive. Site-directed mutagenesis now makes it possible to replace one amino acid with another at any position in a viral protein. This could lead to a much more detailed and dynamic understanding of the relationship between structure and function in both structural and nonstructural viral-coded proteins.

2.2 VIRUS REPLICATION

Many aspects of virus replication *in vivo* are not yet adequately understood. These include (1) the sites, timing, and mechanisms by which infecting viral genomes become uncoated; (2) the site and mechanism whereby virus particles are assembled *in vivo* from pools of nucleic acid and protein; (3) the mechanisms by which the infecting nucleic acid establishes and organizes characteristic replication sites within the cell, some of which may involve specific modifications to host organelles; (4) the roles of host intracellular membranes in virus replication; (5) the way in which a virus preempts some or all of the protein-synthesizing apparatus of the host cell; (6) the way in which some viruses redirect the metabolic processes of the cell to provide building-blocks and energy for virus synthesis; (7) how the timing and location of synthesis of viral gene products in the cell are controlled; (8) the events determining the overall quantity of virus produced in a cell; and (9) the molecular mechanisms that control virus movement from cell to cell and in the conducting tissue.

Techniques and systems for study are now available that should allow answers to many of these and other related questions over the next few years. These include

1. *Transgenic organisms*   Expression of single viral genes in transgenic plants opens up many possibilities for studying various aspects of virus replication *in vivo.* In addition, a vector system has been developed that would also allow the expression of plant viral genes, singly or in combination, in a variety of mammalian systems.

2. In vitro *mutagenesis and recombinant viruses*   The ability to introduce, *in vitro,* base changes, deletions, or insertions at defined sites in a viral genome, and the construction of hybrid viruses *in vitro* opens up many possibilities for experiments that will illuminate the functions of viral genomes and the proteins for which they code.

3. In situ *transcription*   Under appropriate conditions, mRNA can serve as a template for reverse transcriptase *in situ,* within fixed tissue sections. An oligonucleotide complementary to a sequence in the mRNA of interest is prepared and hybridized *in situ.* This provides the primer for reverse transcriptase action. During this step radioactive nucleotides are included to allow for subsequent radioautographic localization of the messenger RNA (mRNA) of interest. The transcripts may then be eluted from the section and used in other procedures, for example, cloning.

4. *The polymerase chain reaction* is a powerful new procedure that opens up the possibility of studying nucleic acid molecules that may be present in only one or a few copies in each cell.

5. *Defective interfering RNAs*   DI RNAs have now been found in infections with several small RNA plant viruses. These are usually deletion mutants of the viral genome, which consist of the 5′ and 3′ terminal sequences of the genome with large internal deletions.

They will provide very useful tools for the study of various aspects of replication, especially regulatory controls and mechanisms of particle assembly, and may also provide agents for the modification or control of virus disease.

2.3 INDUCTION OF DISEASE

The role of viral genes and host genes in the induction of disease is currently one of the most interesting and practically important areas of plant virus research. The techniques noted in the previous section and others will allow new approaches to be made to many aspects of this problem. To give one example: The nature of dark green islands of tissue containing little or no virus that occur in mosaic diseases has not been established. Because these dark green tissues are often resistant to superinfection, it was suggested many years ago that some process like lysogeny might be taking place to give the tissue this resistance. Until recently, no technique has been available to test the idea. With the advent of the polymerase chain reaction, it should be possible to search for RNA viral nucleotide sequences in DNA form in dark green tissue. This technique should be sensitive enough to detect one copy of a viral genome (or part of it) per cell, if it exists.

It should also be possible to apply such techniques as immunogold cytochemistry and *in situ* nucleic acid hybridization to explore further the nature of dark green tissue over the period of expansion in a leaf showing mosaic disease.

2.4 HOST RANGE

The fact that a particular virus will infect some plant species and varieties and not others, even closely related ones, has fascinated virologists for many years. Some answers are emerging, but many more experiments are possible. For example, when a mixture of strains of potato virus X (PVX) is inoculated from tobacco into another solanaceous host, *Cyphomandra betacea*, only a certain kind of strain from the mixture (as judged by symptoms on tobacco) moves systemically in the new host. The mechanism for this strain selection is unknown, but may involve the compatibility of a viral-coded movement protein with the host in question. This idea could be tested by determining the nucleotide sequences of relevant isolates. If a difference were found only in a gene required for movement, the significance of the change could be established by exchanging the gene between strains.

2.5 ECOLOGY

Although a few studies have been made on viruses in their natural habitats, there is substantial scope for further investigations. Such studies might broaden our understanding of how viruses have maintained themselves and evolved under natural conditions, and how human activities have disturbed natural patterns of infection. Different kinds of natural ecosystems need to be studied, and a battery of diagnostic tools should be available—perhaps ideally, a set of nucleic acid hybridization probes that includes at least one from each family and group of plant viruses.

Another area of virus ecology needing much more detailed study is the role of waterborne viruses, especially with respect to their possible role in the decline of forests.

2.6 METHODS FOR DISEASE CONTROL

Disease control through the use of transgenic plants is one of the most exciting prospects for the coming decade. Virus-resistant plants offer the ideal method for the control of virus diseases in agricultural and horticultural crops. Since no chemicals are involved, there is no risk of environmental pollution. Substantial progress has been made, but various questions remain to be answered, as outlined in Chapter 15. Nevertheless, the stage has probably been reached from which further progress in the commercialization of some transgenic plants resistant to particular viruses will depend mainly on nontechnical aspects, such as regulatory requirements, the cost of the large-scale, multisite field tests that are required, and proprietary protection for genetically engineered cultivars (Gasser and Fraley, 1989). However, Gasser and Fraley predict that commercial introductions of genetically engineered soybean, cotton, rice, and alfalfa are likely by the mid-1990s, and that corn hybrids with resistance to viruses, insects, and herbicides should be available by the year 2000.

APPENDIX

LIST OF STANDARD ACRONYMS FOR SELECTED PLANT VIRUSES AND VIROIDS

Virus names, subgroups, and acronyms were taken from Hull *et al.* (1991). The group or genus names approved by the ICTV have been placed immediately following the virus names. This provides a binomial identification of the virus concerned. Tentative allocations of a virus to a group or genus are indicated by ?. Certain virus names, which do not appear in this list, are referred to in the text. These are usually from the older literature, and positive identification in relation to the present list may not be possible.

The italics for group or genus names indicate names approved by the ICTV. In the text of this book, I have used italics where a group is named (e.g., *Geminivirus*). Where viruses in the group are referred to in the plural, italics have not been used (e.g., geminiviruses). Rhabdovirus indicates that a virus has been allocated by the ICTV to the family Rhabdoviridae, but this is not an approved generic name for plant viruses and is therefore not italicized. The dash (–) after the virus name indicates that the virus has not yet been allocated to a group or genus by the ICTV. In the last column, a dash indicates that no AAB description has been published.

Subgroups for some viruses are indicated by a number or vector type after the family or group name. *Geminivirus:* subgroup I, II, or III as defined in Chapter 8, Section 2. Rhabdovirus subgroups: (A) mature and accumulate in the cytoplasm; (B) bud at the inner membrane of the nuclear envelope and accumulate in the perinuclear space. *Potyvirus:* (Aphid) (Mite) (Fungus) indicate the kind of vector. Those with mite or fungal vectors are listed by the ICTV as possible members of the group (Francki *et al.*, 1991).

Table 1 Viruses

Virus name	Acronym	AAB description no.
Abutilon mosaic *Geminivirus* (III)	AbMV	—
African cassava mosaic *Geminivirus* (III)	ACMV	297
Agropyron mosaic *Potyvirus* (mite)	AgMV	118
Alfalfa mosaic AMV group	AMV	229
American wheat striate mosaic rhabdovirus (A)	AWSMV	99
Andean potato latent *Tymovirus*	APLV	—
Andean potato mottle *Comovirus*	APMV	203
Arabis mosaic *Nepovirus*	ArMV	16
Asparagus virus 1 *Potyvirus* (aphid)	AV1	—
Barley stripe mosaic *Hordeivirus*	BSMV	68
Barley yellow dwarf *Luteovirus*	BYDV	32
Barley yellow mosaic *Potyvirus* (fungus)	BaYMV	143
Barley yellow striate mosaic rhabdovirus (A)	BYSMV	312
Bean common mosaic *Potyvirus* (aphid)	BCMV	337
Bean golden mosaic *Geminivirus* (III)	BGMV	192
Bean pod mottle *Comovirus*	BPMV	108
Bean yellow mosaic *Potyvirus* (aphid)	BYMV	40
Beet curly top *Geminivirus* (II)	BCTV	210
Beet mild yellowing *Luteovirus*	BYMV	—
Beet mosaic *Potyvirus* (aphid)	BtMV	53
Beet necrotic yellow vein *Furovirus*	BNYVV	144
Beet western yellows *Luteovirus*	BWYV	89
Beet yellows *Closterovirus*	BYV	13
Blueberry leaf mottle *Nepovirus*	BLMV	267
Broad bean mottle *Bromovirus*	BBMV	101
Brome mosaic *Bromovirus*	BMV	180
Cacao swollen shoot	CSSV	10
Carnation etched ring *Caulimovirus*	CERV	182
Carnation Italian ringspot *Tombusvirus*	CIRSV	—
Carnation mottle *Carmovirus*	CarMV	7

Table 1 (*continued*)

Virus name	Acronym	AAB description no.
Carnation ringspot *Dianthovirus*	CRSV	308
Carrot red leaf *Luteovirus*	CaRLV	249
Cassava latent virus = African cassava mosaic virus		
Cauliflower mosaic *Caulimovirus*	CaMV	243
Chara australis virus	CAV	—
Cherry leaf roll *Nepovirus*	CLRV	306
Chloris striate mosaic *Geminivirus* (I)	CSMV	221
Citrus tristeza *Closterovirus*	CTV	33
Clover yellow vein *Potyvirus* (aphid)	CLYVV	131
Cowpea chlorotic mottle *Bromovirus*	CCMV	49
Cowpea mosaic *Comovirus*	CPMV	197
Cucumber green mottle mosaic *Tobamovirus*	CGMMV	154
Cucumber mosaic *Cucumovirus*	CMV	213
Cucumber necrosis *Tombusvirus*	CuNV	82
Cymbidium mosaic *Potexvirus*	CyMV	27
Dahlia mosaic *Caulimovirus*	DMV	51
Digitaria striate *Geminivirus* (I)	DSV	—
Figwort mosaic *Caulimovirus*	FMV	—
Fiji disease *Fijivirus*	FDV	—
Grapevine fanleaf *Nepovirus*	GFLV	28
Groundnut rosette —	GRV	—
Groundnut rosette assistor *Luteovirus*	GRAV	—
Hop latent *Carlavirus*	HpLV	261
Lettuce big vein —	LBVV	—
Lettuce mosaic *Potyvirus* (aphid)	LMV	9
Lettuce necrotic yellows rhabdovirus (A)	LNYV	26
Lettuce speckles mottle (ungrouped)	LSMV	—
Lilac chlorotic leaf spot *Closterovirus*?	LCLV	202
Lilly symptomless *Carlavirus*	LSV	96
Lucern Australian latent *Nepovirus*	LALV	225
Lucern transient streak *Sobemovirus*	LTSV	224
Maize chlorotic dwarf MCDV group	MCDV	194
Maize dwarf mosaic *Potyvirus* (aphid)	MDMV	—
Maize rayado fino *Marafivirus*	MRFV	220
Maize rough dwarf *Fijivirus*	MRDV	72
Maize streak *Geminivirus* (I)	MSV	133

Table 1 (*continued*)

Virus name	Acronym	AAB description no.
Narcissus mosaic *Potexvirus*	NMV	45
Oat blue dwarf *Marafivirus*	OBDV	123
Oat mosaic *Potyvirus* (fungus)	OMV	145
Onion yellow dwarf *Potyvirus* (aphid)	OYDV	158
Papaya mosaic *Potexvirus*	PapMV	56
Papaya ringspot (=watermelon mosaic virus 1) *Potyvirus* (aphid)	PRSV	292
Parsnip yellow fleck	PYFV	129
Passionfruit woodiness *Potyvirus* (aphid)	PWV	122
Pea early-browning *Tobravirus*	PEBV	120
Pea enation mosaic PEMV group	PEMV	257
Pea seed-borne mosaic *Potyvirus* (aphid)	PSbMV	146
Peanut stunt *Cucumovirus*	PSV	92
Pepper mottle *Potyvirus* (aphid)	PepMoV	253
Pepper ringspot *Tobravirus*	PepRSV	—
Plum pox *Potyvirus* (aphid)	PPV	70
Potato A *Potyvirus* (aphid)	PVA	54
Potato leafroll *Luteovirus*	PLRV	291
Potato M *Carlavirus*	PVM	87
Potato mop-top *Furovirus*	PMTV	138
Potato S *Carlavirus*	PVS	60
Potato X *Potexvirus*	PVX	4
Potato Y *Potyvirus* (aphid)	PVY	242
Potato yellow dwarf rhabdovirus (B)	PYDV	35
Raspberry ringspot *Nepovirus*	RRSV	198
Red clover mottle *Comovirus*	RCMV	74
Red clover necrotic mosaic *Dianthovirus*	RCNMV	181
Ribgrass mosaic *Tobamovirus*	RMV	152
Rice dwarf *Phytoreovirus*	RDV	102
Rice grassy stunt *Tenuivirus*	RGSV	320
Rice ragged stunt plant reovirus (no genus name)	RRSV	248
Rice stripe *Tenuivirus*	RSV	269
Rice tungro bacilliform	RTBV	—
Rice tungro spherical ? MCDV group	RTSV	67
Satellite of TNV	STNV	15
Soil-borne wheat mosaic *Furovirus*	SBWMV	77

Table 1 (*continued*)

Virus name	Acronym	AAB description no.
Sonchus yellow net rhabdovirus (B)	SYNV	205
Southern bean mosaic *Sobemovirus*	SBMV	274
Sowthistle yellow vein rhabdovirus (B)	SYVV	62
Soybean mosaic *Potyvirus* (aphid)	SbMV	93
Sugarcane mosaic *Potyvirus* (aphid)	SCMV	88
Sun-hemp mosaic *Tobamovirus*	SHMV	152
Tobacco etch *Potyvirus* (aphid)	TEV	258
Tobacco mild green mosaic *Tobamovirus*	TMGMV	—
Tobacco mosaic *Tobamovirus*	TMV	151
Tobacco necrosis *Necrovirus*	TNV	14
Tobacco rattle *Tobravirus*	TRV	12
Tobacco ringspot *Nepovirus*	TRSV	309
Tobacco streak *Ilarvirus*	TSV	307
Tobacco stunt	TStV	313
Tobacco vein mottling *Potyvirus* (aphid)	TVMV	325
Tobacco yellow vein assistor ? *Luteovirus*	TYVAV	—
Tomato aspermy *Cucumovirus*	TAV	79
Tomato black ring *Nepovirus*	TBRV	38
Tomato bushy stunt *Tombusvirus*	TBSV	69
Tomato golden mosaic *Geminivirus* (III)	TGMV	303
Tomato mosaic *Tobamovirus*	ToMV	156
Tomato ringspot *Nepovirus*	ToRSV	290
Tomato spotted wilt *Tospovirus*	TSWV	39
Tulip breaking *Potyvirus* (aphid)	TBV	71
Turnip crinkle *Carmovirus*	TCV	109
Turnip mosaic *Potyvirus* (aphid)	TuMV	8
Turnip yellow mosaic *Tymovirus*	TYMV	230
Velvet tobacco mottle *Sobemovirus*	VTMoV	317
Watermelon mosaic virus 2 *Potyvirus* (aphid)	WMV2	291
Wheat streak mosaic *Potyvirus* (mite)	WSMV	48
White clover mosaic *Potexvirus*	WCLMV	41
Wound tumor *Phytoreovirus*	WTV	34
Zucchini yellow mosaic *Potyvirus* (aphid)	ZYMV	282

Table 2 Viroids

Viroid name	Acronym	AAB description no.
Apple scar skin	ASSVd	—
Avocado sun-blotch	ASBVd	254
Burdock stunt	BSVd	—
Carnation stunt-associated	CSAVd	—
Chrysanthemum chlorotic mottle	CCMVd	—
Chrysanthemum stunt	CSVd	—
Citrus cachexia	CCaVd	—
Citrus exocortis	CEVd	226
Coconut cadang-cadang	CCCVd	287
Coconut tinangaja	CTVd	—
Columnea latent	CLVd	—
Cucumber pale fruit	CPFVd	—
Grapevine 1B	GVd1B	—
Grapevine yellow speckle	GYSVd	—
Hop latent	HLVd	—
Hop stunt	HSVd	326
Potato spindle tuber	PSTVd	66
Tomato apical stunt	TASVd	—
Tomato bunchy top	TBTVd	—
Tomato planta macho	TPMVd	—

BIBLIOGRAPHY

Ahlquist, P., French, R., Janda, M., and Loesch-Fries, L. S. (1984). Multicomponent RNA plant virus infection derived from cloned viral cDNA. *Proc. Natl. Acad. Sci. U.S.A.* **81**, 7066–7070.

Ahlquist, P., Allison, R., Dejong, W., Janda, M., Kroner, P., Pacha, R., and Traynor, P. (1990). Molecular biology of Bromovirus replication and host specificity. *In* "Viral Genes and Plant Pathogenesis" (T. P. Pirone and J. G. Shaw, eds.), pp. 144–155. Springer Verlag, New York.

Angenent, G. C., Posthumus, E., Brederode, F. Th., and Bol, J. F. (1989). Genome structure of tobacco rattle virus strain PLB: Further evidence on the occurrence of RNA recombination among tobraviruses. *Virology* **171**, 271–274.

Angenent, G. C., Van den Ouweland, J. M. W., and Bol, J. F. (1990). Susceptibility to virus infection of transgenic tobacco plants expressing structural and non-structural genes of tobacco rattle virus. *Virology* **174**, 191–198.

Atabekov, J. G., and Taliansky, M. E. (1990). Expression of a plant virus-coded transport function by different viral genomes. *Adv. Virus Res.* **38**, 201–248.

Barnett, O. W. (1991). Potyviridae, a proposed family of plant viruses. *Arch. Virol.* **118**, 139–141.

Beachy, R. N. (1988). Virus cross-protection in transgenic plants. *In* "Temporal and Spatial Regulation of Plant Genes" (D.P.S. Verma and R. B. Goldberg, eds.), pp. 313–331. Springer Verlag, Wien, New York.

Beck, D. L., Guilford, P. J., Voot, D. M., Andersen, M. T., and Forster, R. L. S. (1991). Triple gene block proteins of white clover mosaic Potexvirus are required for transport. *Virology,* **183**, 695–702.

Bol, J. F., and van Kan, J. A. L. (1988). The synthesis and possible functions of virus-induced proteins in plants. *Microbiol. Sci.* **5**, 47–52.

Bos, L. (1978). "Symptoms of Virus Diseases in Plants," 3rd Ed. Pudoc, Wageningen, The Netherlands.

Bos, L. (1982). Crop losses caused by viruses. *Crop Protection* **1**, 263–282.

Boulton, M. I., and Markham, P. G. (1986). The use of squash-blotting to detect plant pathogens in insect vectors. *In* "Developments and Applications of Virus Testing"

(R. A. C. Jones and L. Torrance, eds.), pp. 55–69. Association of Applied Biologists, Wellsbourne, England.

Branch, A. D., Benenfeld, B. J., and Robertson, H. D. (1988). Evidence for a single rolling circle in the replication of potato spindle tuber viroid. *Proc. Natl. Acad. Sci. U.S.A.* **85,** 9128–9132.

Branch, A. D., Levine, B. J., and Robertson, H. D. (1990). The brotherhood of circular RNA pathogens: Viroids, circular satellites, and the delta agent. *Sem. Virol.* **1,** 143–152.

Briand, J.-P. (1978). Contribution à l'etude de la structure des tymovirus et a l'organisation de leur genome. Thèse D.Sc. l'Université Louis Pasteur de Strasbourg, France.

Butler, P. J. G. (1984). The current picture of the structure and assembly of tobacco mosaic virus. *J. Gen. Virol.* **65,** 253–279.

Caspar, D. L. D., and Klug, A. (1962). Physical principles in the construction of regular viruses. *Cold Spring Harbor Symp. Quant. Biol.* **27,** 1–24.

Chen, Z., Stauffacher, C., Li, Y., Schmidt, T., Bomu, W., Kamer, G., Shanks, M., Lomonossoff, G., and Johnson, J. E. (1989). Protein–RNA interactions in an icosahedral virus at 3.0 Å resolution. *Science* **245,** 154–159.

Chu, P. W. G., and Francki, R. I. B. (1982). Detection of lettuce necrotic yellows virus by an enzyme-linked immunosorbent assay in plant hosts and the insect vector. *Ann. Appl. Biol.* **100,** 149–156.

Citovsky, V., Knorr, D., and Zambryski, P. (1991). Gene I, a potential cell-to-cell movement locus of cauliflower mosaic virus, encodes an RNA-binding protein. *Proc. Natl. Acad. Sci. U.S.A.,* **88,** 2476–2480.

Clark, M. F., and Adams, A. N. (1977). Characteristics of the microplate method of enzyme-linked immunosorbent assay for the detection of plant viruses. *J. Gen. Virol.* **34,** 475–483.

Conti, M. (1984). Epidemiology and vectors of plant reo-like viruses. *Curr. Top. Vector Res.* **2,** 112–139.

Crick, F. H. C., and Watson, J. D. (1956). Structure of small viruses. *Nature (London)* **177,** 473–475.

Cusack, S., Miller, A., Krijgsman, P. C. J., and Mellema, J. E. (1981). An investigation of the structure of alfalfa mosaic virus by small-angle neutron scattering. *J. Mol. Biol.* **145,** 525–543.

Davies, J. W., and Stanley, J. (1989). Geminivirus genes and vectors. *Trends Genet.* **5,** 77–81.

Davies, J. W., Wilson, T. M. A., and Covey, S. N. "The Molecular Biology of Plant Viruses." Chapman and Hall, in press.

de Haan, P., Wagemakers, L., Peters, D., and Goldbach, R. (1990). The S RNA segment of tomato spotted wilt virus has an ambisense character. *J. Gen. Virol.* **71,** 1001–1007.

De Zoeten, G. A. (1968). Application of scanning microscopy in the study of virus transmission of aphids. *J. Virol.* **2,** 745–751.

Diener, T. O. (ed.) (1987). "The Viroids." Plenum Press, New York.

Ding, S.-W., Keese, P., and Gibbs, A. (1989). Nucleotide sequence of the ononis yellow mosaic tymovirus genome. *Virology* **172,** 555–563.

Ding, S., Howe, J., Keese, P., Mackenzie, A., Meek, D., Osorio-Keese, M., Skotnicki, M., Srifah, P., Torronen, M., and Gibbs, A. (1990). The tymobox, a sequence shared by most tymoviruses: Its use in molecular studies of tymoviruses. *Nucleic Acids Res.* **18,** 1181–1187.

Dollet, M., Accotto, G. P., Lisa, V., Menissier, J., and Boccardo, G. (1986). A geminivirus, serologically related to maize streak virus from *Digitaria sanguinalis* from Vanuatu. *J. Gen. Virol.* **67**, 933–937.

Dougherty, W. G., and Carrington, J. C. (1988). Expression and function of potyvirus gene products. *Annu. Rev. Phytopathol.* **26**, 123–143.

Duffus, J. E. (1971). Role of weeds in the incidence of virus diseases. *Annu. Rev. Phytopathol.* **9**, 319–340.

Duffus, J. E., Larsen, R. C., and Liu, H. Y. (1986). Lettuce infectious yellows virus—A new type of whitefly transmitted virus. *Phytopathology* **76**, 97–100.

Eglington, G., and Hamilton, R. J. (1967). Leaf epicuticular waxes. *Science* **156**, 1322–1335.

Esau, K., and Cronshaw, J. (1967). Relation of tobacco mosaic virus with host cells. *J. Cell Biol.* **33**, 665–678.

Esau, K., Cronshaw, J., and Hoefert, L. L. (1966). Organisation of beet-yellows inclusions in leaf cells of *Beta. Proc. Natl. Acad. Sci. U.S.A.* **55**, 486–493.

Etessami, P., Saunders, K., Watts, J., and Stanley, J. (1991). Mutational analysis of complementary-sense genes of African cassava mosaic virus DNA A. *J. Gen. Virol.* **72**, 1005–1012.

Fauquet, C., and Beachy, R. N. (1989). "International Cassava-trans Project. Cassava Viruses and Genetic Engineering." Occasional publication of ORSTOM, 213 Rue Lafayette, 75489 Paris Cedex 10, France.

Finch, J. T., and Klug, A. (1966). Arrangement of protein subunits and the distribution of nucleic acid in turnip yellow mosaic virus. II. Electron microscopic studies. *J. Mol. Biol.* **15**, 344–365.

Forster, A. C., Davies, C., Sheldon, C. C., Jeffries, A. C., and Symons, R. H. (1988). Self-cleaving viroid and newt RNAs may only be active as dimers. *Nature* **334**, 265–267.

Fraenkel-Conrat, H., and Singer, B. (1957). Virus reconstitution. II. Combination of protein and nucleic acid from different strains. *Biochim. Biophys. Acta* **24**, 540–548.

Fraenkel-Conrat, H., and Williams, R. C. (1955). Reconstitution of active tobacco mosaic virus from its inactive protein and nucleic acid components. *Proc. Natl. Acad. Sci. U.S.A.* **41**, 690–698.

Francki, R. I. B. (1985). Plant virus satellites. *Annu. Rev. Microbiol.* **39**, 151–174.

Francki, R. I. B., Fauquet, C., Knudson, D. L., and Brown, L. (1991). "Classification and Nomenclature of Viruses." Fifth Report of the International Committee for Taxonomy of Viruses, *Arch. Virol. Supplementum 2.* Springer-Verlag, Wien.

Francki, R. I. B., and Randles, J. W. (1980). Rhabdoviruses infecting plants. *In* "Rhabdoviruses," Vol. III (D. H. L. Bishop, ed.), pp. 135–165. CRC Press, Boca Raton, Florida.

Francki, R. I. B., Milne, R. G., and Hatta, T. (editors) (1985). Plant Rhabdoviridae. *In* "Atlas of Plant Viruses," Vol. I. pp. 73–100. CRC Press, Boca Raton, Florida.

Fraser, R. S. S. (1985). Genetics of host resistance to viruses and of virulence. *In* "Mechanisms of Resistance to Plant Disease" (R. S. S. Fraser, ed.), pp. 62–79. Martinus Nijhoff/W. Junk, Dordrecht, The Netherlands.

Fraser, R. S. S. (1987a). Resistance to plant viruses. *Oxf. Surv. Plant Mol. Cell Biol.* **4**, 1–45.

Fraser, R. S. S. (1987b). "Biochemistry of Virus-Infected Plants." Research Studies Press, Letchworth, England.

French, R., and Ahlquist, P. (1988). Characterisation and engineering of sequences controlling *in vivo* synthesis of brome mosaic virus subgenomic RNA. *J. Virol.* **62**, 2411–2420.

Fulton, J. P., Gergerich, R. C., and Scott, H. A. (1987). Beetle transmission of plant viruses. *Annu. Rev. Plant Pathol.* **25**, 111–123.

Fulton, R. W. (1975). The role of top particles in recombination of some characters of tobacco streak virus. *Virology* **67**, 188–196.

Gasser, C. S., and Fraley, R. T. (1989). Genetically engineering plants for crop improvement. *Science* **244**, 1293–1299.

Goldbach, R. (1987). Sequence similarities between plant and animal RNA viruses. *Microbiol. Sci.* **4**, 197–202.

Goldbach, R., and Wellink, J. (1988). Evolution of plus-strand RNA viruses. *Intervirology* **29**, 260–267.

Golemboski, D. B., Lomonossoff, G. P., and Zaitlin, M. (1990). Plants transformed with a tobacco mosaic virus non-structural gene sequence are resistant to the virus. *Proc. Natl. Acad. Sci. U.S.A.* **87**, 6311–6315.

Greber, R. S., Klose, M. J., and Teakle, D. S. (1991). High incidence of tobacco streak virus in tobacco and its transmission by *Microcephalothrips abdominalis* and pollen from *Ageratum houstonianum. Plant Disease* **75**, 450–452.

Gross, H. J., Domdey, H., Lossow, C., Jank, P., Raba, M., Alberty, H., and Sänger, H. L. (1978). Nucleotide sequence and secondary structure of potato spindle tuber viroid. *Nature (London)* **273**, 203–208.

Habili, N., and Symons, R. H. (1989). Evolutionary relationship between luteoviruses and other RNA plant viruses based on the sequence motifs in their putative RNA polymerases and nucleic acid helicases. *Nucleic Acids Res.* **17**, 9543–9555.

Harlow, E., and Lane, D. (1988). "Antibodies: A Laboratory Manual." Cold Spring Harbor Laboratory Press, New York.

Harris, K. F. (1983). Sternorrhynchous vectors of plant viruses: Virus–vector interactions and transmission mechanisms. *Adv. Virus Res.* **28**, 113–140.

Harris, K. F., and Maramorosch, K. (eds.) (1982). Pathogens, vectors, and plant diseases. Approaches to control. Academic Press, New York.

Harrison, S. C. (1983). Virus Structure: High-resolution perspectives. *Adv. Virus Res.* **28**, 175–240.

Harrison, S. C., Olson, A. J., Schutt, C. E., Winkler, F. K., and Bricogne, G. (1978). Tomato bushy stunt virus at 2.9Å resolution. *Nature (London)* **276**, 368–373.

Haseloff, J., and Gerlach, W. (1989). Simple RNA enzymes with new and highly specific endoribonuclease activities. *Nature (London)* **334**, 585–591.

Hatta, T., and Francki, R. I. B. (1977). Morphology of Fijidisease virus. *Virology* **76**, 797–807.

Hatta, T., Bullivant, S., and Matthews, R. E. F. (1973). Fine structure of vesicles induced in chloroplasts of Chinese cabbage leaves by infection with turnip yellow mosaic virus. *J. Gen. Virol.* **20**, 37–50.

Hinegardner, R. (1976). Evolution of genome size. *In* "Molecular Evolution" (F. J. Ayala, ed.), pp. 179–199. Sinauer, Sunderland, Massachussetts.

Hull, R. (1989). The movement of viruses in plants. *Annu. Rev. Phytopathol.* **27**, 213–240.

Hull, R., Milne, R. G., and Van Regenmortel, M. H. V. (1991). A list of proposed standard acronyms for plant viruses and viroids. *Intervirology,* in press.

Hutchins, C. J., Rathgen, P. D., Forster, A. C., and Symons, R. H. (1986). Self-cleavage of plus and minus transcripts of avocado sunblotch viroid. *Nucleic Acids Res.* **14,** 3627–3640.

Jackson, A. O., Francki, R. I. B., and Zuidema, D. (1987). Biology, structure and replication of plant rhabdoviruses. *In* "The Rhabdoviruses" (R. R. Wagner, ed.), pp. 427–508. Plenum, New York.

Jones, A. T. (1987). Control of virus infection in crop plants through vector resistance: A review of achievements, prospects, and problems. *Annu. Appl. Biol.* **111,** 745–772.

Kaper, J. M., and Collmer, C. W. (1988). Modulation of viral plant diseases by secondary RNA agents. *In* "RNA Genetics," Vol. III (E. Domingo, J. J. Holland, and P. Ahlquist, eds.), pp. 171–193. CRC Press, Boca Raton, Florida.

Kauffmann, S., Legrand, M., and Fritig, B. (1989). Isolation and characterisation of six pathogenesis-related (PR) proteins of Samsun NN tobacco. *Plant Mol. Biol.* **14,** 381–390.

Keese, P., and Symons, R. H. (1985). Domains in viroids: Evidence of intermolecular RNA rearrangements and their contribution to viroid evolution. *Proc. Natl. Acad. Sci. U.S.A.* **82,** 4582–4586.

Kleczkowski, A. (1950). Interpreting relationships between concentrations of plant viruses and numbers of local lesions. *J. Gen. Microbiol.* **4,** 53–69.

Klug, A., and Caspar, D. L. D. (1960). The structure of small viruses. *Adv. Virus Res.* **7,** 225–325.

Köhler, G., and Milstein, C. (1975). Continuous cultures of fused cells secreting antibody of predefined specificity. *Nature* **256,** 495–497.

Koltunow, A. M., and Rezaian, M. A. (1989). Grapevine viroid 1B, a new member of the apple scar skin viroid group, contains the left terminal region of tomato planta macho viroid. *Virology* **170,** 575–578.

Lowe, A. D. (1964). The ecology of the cereal aphid in Canterbury. *Proc. N.Z. Weed Pest Cont. Conf.* **17,** 175–186.

McLean, G. D., Garrett, R. G., and Ruesink, W. G. (1986). Plant virus epidemics. Monitoring, modelling and forecasting outbreaks. Academic Press, New York.

Maramorosch, K., and Harris, K. F. (eds.) (1981). Plant diseases and vectors: Ecology and epidemiology. Academic Press, New York.

Martelli, G. P. (1992). Classification and nomenclature of plant viruses: State of the art. *Plant Disease.* In press.

Martelli, G. P., and Castellano, M. A. (1971). Light and electron microscopy of the intracellular inclusions of cauliflower mosaic virus. *J. Gen. Virol.* **13,** 133–140.

Matthews, R. E. F. (1949). Studies on potato virus X. II. Criteria of relationships between strains. *Annu. Appl. Biol.* **36,** 460–474.

Matthews, R. E. F. (1981). "Plant Virology," 2nd Ed. Academic Press, New York.

Matthews, R. E. F. (1982). "Classification and nomenclature of viruses." Fourth Report of the International Committee on Taxonomy of Viruses. Karger, Basel Switzerland.

Matthews, R. E. F. (1991). "Plant Virology," 3rd Ed. Academic Press, New York.

Matthews, R. E. F. (editor) (1992). Diagnosis of Plant Virus Diseases. CRC Press. Boca Raton, Florida. In press.

Maule, A. J. (1991). Virus movement in infected plants. *Crit. Rev. Plant Sci.* **9,** 459–473.

Mellema, J. E., and Amos, L. A. (1972). Three-dimensional image reconstruction of turnip yellow mosaic virus. *J. Mol. Biol.* **72,** 819–822.

Milne, R. G. Immunoelectron-microscopy for virus identification. *In* "Electron Microscopy of Plant Pathogens" (K. Mendgen and D. E. Lesemann, eds.). Springer-Verlag, New York, in press.

Milstein, C. (1990). Antibodies, a paradigm for the biology of molecular recognition. *Proc. R. Soc. Lond. B* **239**, 1–16.

Morris, T. J., and Hillman, B. I. (1989). "Molecular Biology of Plant-Pathogen Interactions." A. R. Liss, New York.

Morris, T. J., and Knorr, D. A. (1990). Defective interfering viruses associated with plant virus infections. *In* "New Aspects of Positive-Strand RNA Viruses" (M. A. Brinton and F. X. Heinz, eds.), pp. 123–127. Amer. Soc. Microbiol., Washington, D.C.

Mundry, K., Watkins, P. A. C., Ashfield, T., Plaskitt, K. A., Eisele-Walter, S., and Wilson, T. A. M. (1991). Complete uncoating of the 5' leader sequence of tobacco mosaic virus RNA occurs rapidly and is required to initiate cotranslational disassembly *in vitro*. *J. Gen. Virol.* **72**, 769–777.

Murant, A. F., and Kumar, I. K. Different variants of the satellite RNA of groundnut rosette virus are responsible for the chlorotic and green forms of groundnut rosette disease. *Ann. Appl. Biol.* **117**, in press.

Namba, K., Pattanayek, R., and Stubbs, G. (1989). Visualization of protein–nucleic acid interactions in a virus. Refined structure of intact tobacco mosaic virus at 2.9 Å resolution by X-ray fibre diffraction. *J. Mol. Biol.* **208**, 307–325.

Nault, L. R., and Ammar, E. D. (1989). Leafhopper and planthopper transmission of plant viruses. *Annu. Rev. Entomol.* **34**, 503–529.

Nuss, D. L., and Dall, D. J. (1990). Structural and functional properties of plant reovirus genomes. *Adv. Virus Res.* **38**, 249–306.

Okada, Y. (1986). Molecular assembly of tobacco mosaic virus *in vitro*. *Adv. Biophys.* **22**, 95–149.

Olson, A. J., Bricogne, G., and Harrison, S. C. (1983). Structure of tomato bushy stunt virus. IV. The virus particle at 2.9Å resolution. *J. Mol. Biol.* **171**, 61–93.

Ogawa, M., and Sakai, F. (1984). A messenger RNA for tobacco mosaic virus coat protein in infected tobacco mesophyll protoplasts. *Phytopath. Z.* **109**, 193–203.

Pfeiffer, P., Gordon, K., Fütterer, J., and Hohn, T. (1987). The life cycle of cauliflower mosaic virus. *In* "Plant Molecular Biology" (D. von Wettstein and N. H. Chua, eds.), pp. 443–458. Plenum, New York.

Powell, Abel, P., Nelson, R. S., De, B., Hoffmann, N., Rogers, S. G., Fraley, R. T., and Beachy, R. N. (1986). Delay of disease development in transgenic plants that express the tobacco mosaic virus coat protein gene. *Science* **232**, 738–743.

Rezaian, M. A. (1990). Australian grapevine viroid. Evidence for extensive recombination between viroids. *Nucleic Acids Res.* **18**, 1813–1818.

Ross, A. F. (1961). Localized acquired resistance to plant virus infection in hypersensitive hosts. *Virology* **14**, 329–339.

Rossman, M. G., and Johnson, J. E. (1989). Icosahedral RNA virus structure. *Annu. Rev. Biochem.* **58**, 533–573.

Rossmann, M. G., Abad-Zapatero, C., Murthy, M. R. N., Liljas, L., Alwyn-Jones, T., and Strandberg, B. (1983). Structural comparisons of some small spherical plant viruses. *J. Mol. Biol.* **165**, 711–736.

Saito, T., Hosokawa, D., Meshi, T., and Okada, Y. (1987). Immunocytochemical localization of the 130K and 180K proteins (putative replicase components) of tobacco mosaic virus. *Virology* **160**, 477–481.

Saitou, N., and Nei, M. (1987). The neighbour-joining method: A new method for reconstructing phylogenetic trees. *Mol. Biol. Evol.* **4**, 406–425.

Sambrook, K. J., Fritsch, E. F., and Maniatis, T. (1989). "Molecular Cloning, A Laboratory Manual," 2nd Ed. Cold Spring Harbor Laboratory Press, New York.

Samuel, G. (1934). The movement of tobacco mosaic virus within the plant. *Ann. Appl. Biol.* **21**, 90–111.

Semancik, J. S. (ed.) (1987). "Viroids and Viroid-like Pathogens." CRC Press, Boca Raton, Florida.

Simon, A. E., Engel, H., Johnson, R. P., and Howell, S. H. (1988). Identification of regions affecting virulence, RNA processing and infectivity in the virulent satellite of turnip crinkle virus. *EMBO J.* **7**, 2645–2651.

Sit, T. L., AbouHaidar, M. G., and Holy, S. (1989). Nucleotide sequence of papaya mosaic virus RNA. *J. Gen. Virol.* **70**, 2325–2331.

Shukla, D. D., and Ward, C. W. (1988). Amino acid sequence homology of coat proteins as a basis for identification and classification of the potyvirus group. *J. Gen. Virol.* **69**, 2703–2710.

Shukla, D. D., Frenkel, M. J., McKern, N. M., and Ward, C. W. (1991). Immunological and molecular approaches to the diagnosis of viruses infecting horticultural crops. *In* "Horticulture—New Technologies and Applications" (T. Prakash and R. L. M. Pierick, eds.), pp. 311–319. Kluwer Academic Publishers, Dortrecht, The Netherlands.

Solis, I., and Garcia-Arenal, F. (1990). The complete nucleotide sequence of the genomic RNA of the tobamovirus tobacco mild green mosaic virus. *Virology* **177**, 553–558.

Symons, R. H. (1989). Self-cleavage of RNA in the replication of small pathogens of plants and animals. *Trends Biochem. Sci.* **14**, 445–450.

Thresh, J. M. (1982). Cropping practices and virus spread. *Annu. Rev. Phytopathol.* **20**, 193–218.

Thresh, J. M. (1983). Progress curves of plant virus disease. *Adv. Appl. Biol.* **8**, 1–85.

Thresh, J. M. (1989). Insect-borne viruses of rice and the green revolution. *Trop. Pest Manag.* **35**, 264–272.

Townsend, R., Stanley, J., Curson, S. J., and Short, M. N. (1985). Major polyadenylated transcripts of cassava latent virus and location of the gene encoding the coat protein. *EMBO J.* **4**, 33–37.

van Hoof, H. A. (1977). Determination of the infection pressure of potato virus Y[N]. *Neth. J. Plant Pathol.* **83**, 123–127.

van Kammen, A., and Eggen, H. I. L. (1986). The replication of cowpea mosaic virus. *BioEssays* **5**, 261–266.

van Loon, L. C. (1987). Disease induction by plant viruses. *Adv. Virus Res.* **33**, 205–255.

van Loon, L. C. (1989). Stress proteins in infected plants. *In* "Plant–Microbe Interactions: Molecular and Genetic Perspectives" (T. Kosuge and E. W. Nester, eds.), pp. 198–237. McGraw-Hill, New York.

van Regenmortel, M. H. V. (1982). "Serology and Immunochemistry of Plant Viruses." Academic Press, New York.

van Regenmortel, M. H. V. (1986). The potential for using monoclonal antibodies in the detection of plant viruses. *In* "Developments and Applications in Virus Testing" (R. A. C. Jones and L. Torrance, eds.), pp. 89–101. Developments in Applied Biology I. Association of Applied Biologists, Wellesbourne, England.

van Vloten Doting, L., and Bol. J. F. (1988). Variability, mutant selection and mutant stability in plant RNA viruses. *In* "RNA Genetics," Vol. III (E. Domingo, J. J. Holland, and P. Ahlquist, eds.), pp. 37–51. CRC Press, Boca Raton, Florida.

Ward, C. W., and Shukla, D. D. "Taxonomy of potyviruses: Current problems and some solutions." *Intervirology* **32**, 269–296.

Watson, J. D., Tooze, J., and Kurtz, D. T. (1983). "Recombinant DNA: A Short Course." Scientific American Books, New York.

Watson, J. D., Hopkins, N. H., Roberts, J. W., Steitz, J. A., and Weiner, A. M. (1987). "Molecular Biology of the Gene," 4th Ed., Vol. I. Benjamin/Cummings, Menlo Park, California.

Watson, M. A., Heathcote, G. D., Lauchner, F. B., and Sowray, P. A. (1975). The use of weather data and counts of aphids in the field to predict the incidence of yellowing viruses of sugar-beet crops in England in relation to the use of insecticides. *Ann. Appl. Biol.* **81**, 181–198.

Wilson, T. M. A. (1984). Cotranslational disassembly of tobacco mosaic virus *in vitro. Virology* **137**, 255–265.

Wilson, T. M. A. (1989). Plant viruses: A tool-box for genetic engineering and crop protection. *BioEssays* **10**, 179–186.

Yoshikawa, N., Poolpol, P., and Inouye, I. (1986). Use of a dot immunobinding assay for rapid detection of strawberry pseudo mild yellow edge virus. *Ann. Phytopath. Soc. Japan* **52**, 728–731.

Zimmern, D. (1988). "Evolution of RNA Viruses, in RNA Genetics," Vol. II (E. Domingo, J. J. Holland, and P. Ahlquist, eds.), pp. 211–240. CRC Press, Boca Raton, Florida.

INDEX

Printed and bound by CPI Group (UK) Ltd, Croydon, CR0 4YY

03/10/2024

01040329-0009